The Biology of Reproduction

Reproduction is fundamental to life; it is the way life persists across the ages. This book offers new, wider vistas on this vital biological phenomenon, exploring how it works through the whole tree of life. It explores facets such as asexual reproduction, parthenogenesis, sex determination and reproductive investment, with a taxonomic coverage extending over all the main groups – animals, plants (including algae), fungi, protists and bacteria. It collates into one volume perspectives from varied disciplines – including zoology, botany, microbiology, genetics, cell biology, developmental biology, evolutionary biology, animal and plant physiology, and ethology – using as far as possible a common terminology. The book aims to identify the commonalities among reproductive phenomena, while demonstrating the diversity that exists even among closely related taxa. Its integrated approach makes this a valuable reference source for students and researchers, as well as an effective entry point for a more in-depth study of specific topics.

Giuseppe Fusco is Associate Professor of Zoology in the Department of Biology at the University of Padova, Italy. He is a researcher in evolutionary and developmental biology and has edited three volumes in these fields. He previously collaborated with Alessandro Minelli on *Evolving Pathways* (Cambridge, 2008).

Alessandro Minelli is a former Full Professor of Zoology and, in retirement, an affiliated senior scientist at the University of Padova, Italy. He has served as vice-president of the European Society for Evolutionary Biology and as specialty editor-in-chief for evolutionary developmental biology of *Frontiers in Ecology and Evolution*. He is the author of several books on evolutionary biology, including *The Development of Animal Form* (Cambridge, 2003) and *Plant Evolutionary Developmental Biology* (Cambridge, 2018).

The Biology of Reproduction

GIUSEPPE FUSCO
University of Padova

ALESSANDRO MINELLI
University of Padova

CAMBRIDGE
UNIVERSITY PRESS

CAMBRIDGE
UNIVERSITY PRESS

University Printing House, Cambridge CB2 8BS, United Kingdom

One Liberty Plaza, 20th Floor, New York, NY 10006, USA

477 Williamstown Road, Port Melbourne, VIC 3207, Australia

314–321, 3rd Floor, Plot 3, Splendor Forum, Jasola District Centre,
New Delhi – 110025, India

79 Anson Road, #06–04/06, Singapore 079906

Cambridge University Press is part of the University of Cambridge.

It furthers the University's mission by disseminating knowledge in the pursuit of
education, learning and research at the highest international levels of excellence.

www.cambridge.org
Information on this title: www.cambridge.org/9781108499859
DOI: 10.1017/9781108758970

© Giuseppe Fusco and Alessandro Minelli 2019

First published 2019

Printed in the United Kingdom by TJ International Ltd. Padstow Cornwall

A catalogue record for this publication is available from the British Library.

Library of Congress Cataloging-in-Publication Data
Names: Fusco, Giuseppe, 1965– author. | Minelli, Alessandro, author.
Title: The biology of reproduction / Giuseppe Fusco, University of Padova, Alessandro
   Minelli, University of Padova.
Description: Cambridge, United Kingdom ; N.Y., NY : Cambridge University Press, 2019.
Identifiers: LCCN 2019014259 | ISBN 9781108499859 (hardback) | ISBN 9781108731713
   (paperback)
Subjects: LCSH: Reproduction.
Classification: LCC QH471 .F95 2019 | DDC 591.56–dc23
LC record available at https://lccn.loc.gov/2019014259

ISBN 978-1-108-49985-9 Hardback
ISBN 978-1-108-73171-3 Paperback

Contents

Preface

In this book on the biology of reproduction, all the main facets of this large chapter of the life sciences – e.g. sexual and asexual reproduction, parthenogenesis, sex determination, reproductive investment – are addressed, with a taxonomic coverage extending over all the main groups: animals, plants (including algae), fungi, protists and bacteria. Information about the many topics presented in the book is usually scattered in volumes on general and systematic zoology, general and systematic botany, microbiology, genetics, cellular biology, developmental biology, evolutionary biology, animal and plant physiology, and ethology – where the individual topics are treated with variable levels of detail. For the student or the general reader interested in the subject, the obvious difficulty represented by the fragmentation of the relevant information is magnified by the various and often idiosyncratic approaches to the subject adopted in these works. This makes it difficult to identify the commonalities among reproductive phenomena and processes across the tree of life while, conversely and paradoxically, the treatment of the subject in this dispersed literature is often generalized to the extent that it hides the diversity of reproductive phenomena frequently found even among closely related taxa.

While dealing with subjects as varied as the binary division of unicellular algae, the separation of a sequoia's stolon from the mother plant, the mating of squids, the production of spores by boletus mushrooms, or the paternal care of Darwin's frog, we try to present all these phenomena using a common language for all living beings, at least in the most general aspects of their reproductive biology, thus overcoming the diversity of the technical terminology in different traditions, e.g. in botany, zoology, microbiology, transmission genetics.

The text is aimed primarily at university students with a basic knowledge of general biology, but it will also be useful to professional biologists and philosophers of biology, as well as to non-expert readers interested in the more general aspects of reproductive phenomena.

This work is a revised and updated translation of a book we published last year in Italian as *Biologia della riproduzione*. We are grateful to our Italian publishers Pearson Italia for their support in producing this English edition. At Cambridge University Press, Dominic Lewis endorsed our proposal from the first moment we floated the idea that eventually materialized in these pages. Along the way to publication, we enjoyed the precious assistance of Aleksandra Serocka, Jenny van der Meijden and Hugh Brazier.

Several colleagues have contributed significantly to the realization of this work. We would like to thank Wallace Arthur, Loriano Ballarin, Ferdinando

Boero, James DiFrisco, Diego Fontaneto, Adriana Giangrande, Diego Hojs-gaard, Kostas Kampourakis, Marta Mariotti, Koen Martens, Pietro Omodeo, Valerio Scali and Emanuele Serrelli, who reviewed one or more chapters of the book and provided useful tips for improving it, not necessarily sharing our choices or our point of view, as well as Giorgio Bertorelle, Roberto Carrer, Maurizio Casiraghi, Andrea Di Nisio, Carlo Foresta, Francesco Nazzi, Marco Passamonti, Andrea Pilastro, Irene Stefanini and Antonio Todaro, who provided information or help on specific issues.

Perhaps nobody can claim to be a professional in a subject as broad as biological reproduction, in all its multiple aspects and across the whole tree of life. The authors certainly cannot. We will be grateful to any readers who flag errors, imprecisions and significant omissions.

Figure Credits

The authors acknowledge the following sources of copyright material and are grateful for the permissions granted.

Fig. 1.1 Public domain; **Fig. 1.2** Gross 2006, *PLOS Biol.* 4: e353; **Fig. 1.3** Ethan Daniels / Stocktrek Images / Getty Images; **Fig. 1.4a** Courtesy Yuuji Tsukii; **Fig. 1.4b** Science Photo Library – Steve Gschmeissner / Brand X Pictures / Getty Images; **Fig. 1.5** Rolf Muller / Getty Images; **Fig. 1.6** Used under licence from Shutterstock.com; **Fig. 1.7** Pearson Italia; **Fig. 1.8** Pearson Italia (modified from Tagami 2013, *J. Acarol. Soc. Jpn.* 22: 91–99); **Fig. 1.9** Vlmastra, Wikimedia Commons; **Fig. 1.10** Pearson Italia; **Fig. 1.11** Danita Delimont / Getty Images; **Fig. 1.12** blickwinkel / Alamy Stock Photo; **Fig. 1.13** Level12 / 500Px Plus / Getty Images; **Fig. 1.14** Pearson Italia (modified from Santelices 1999, *Trends Ecol. Evol.* 14: 152–155); **Fig. 1.15** Pearson Italia (modified from Jones *et al.* 2014, *Nature* 505: 169–173); **Fig. 1.16** Chiara Salvadori / Moment / Getty Images; **Fig. 1.17** Used under licence from Shutterstock.com; **Fig. 1.18** Pearson Italia; **Fig. 1.19** Pearson Italia (modified from Henderson & Gottschling 2008, *Curr. Opin. Cell Biol.* 20: 723–728); **Fig. 1.20** Courtesy Udo Schmidt; **Fig. 1.21** Michael Dietrich / Getty Images; **Fig. 1.22** J. Zapell, Wikimedia Commons; **Fig. 1.23** Courtesy Carolina Biological Supply Company; **Fig. 2.1** Pearson Italia (modified from Lecointre & Guyader 2001, *La classification phylogénétique du vivant*, Belin); **Fig. 2.2** Pearson Italia (modified from Mauseth 2009, *Botany: an Introduction to Plant Biology*, 4th edn, Jones & Bartlett); **Fig. 2.3** Pearson Italia; **Fig. 2.4** Pearson Italia (modified from Mauseth 2009, *Botany: an Introduction to Plant Biology*, 4th edn, Jones & Bartlett); **Fig. 2.5** Pearson Italia; **Fig. 2.6** Pearson Italia; **Fig. 2.7** Courtesy Stefano Piraino; **Fig. 2.8** Pearson Italia; **Fig. 2.9** Used under licence from Shutterstock.com; **Fig. 2.10** Pearson Italia; **Fig. 2.11** Used under licence from Shutterstock.com; **Fig. 2.12** Courtesy Fabio Gasparini; **Fig. 2.13** Algirdas, Wikimedia Commons; **Fig. 2.14** Pearson Italia (modified from Brooks 1946, *Ecol. Monogr.* 16: 409–447); **Fig. 2.15** Ger Bosma / Moment Open / Getty Images; **Fig. 2.16** Scott Portelli / Moment / Getty Images; **Fig. 2.17** Pearson Italia (modified from Heming 2003, *Insect Development and Evolution*, Comstock, Ithaca, NY); **Fig. 2.18** Dusty Pixel photography / Moment / Getty Images; **Fig. 2.19** Pearson Italia (modified from Dixon 1973, *The Biology of Aphids*, Edward Arnold); **Fig. 2.20** Juni, Wikimedia Commons; **Fig. 3.1** Pearson Italia; **Fig. 3.2** Courtesy Weon-Gyu Kho; **Fig. 3.3** Pearson Italia (modified from Dayal 1996, *Advances in Zoosporic Fungi*, MD Publications); **Fig. 3.4** Pearson Italia (modified from www.treccani.it/scuola/lezioni/scienze_naturali/riproduzione_viventi); **Fig. 3.5** Pearson Italia (modified from Zattara & Bely 2016, *Invertebr. Biol.* 135: 400–414); **Fig. 3.6** Pearson Italia (modified from P. Fauvel 1959, Classe des annélides polychètes, in P. P. Grassé (ed.) *Traité de Zoologie*, Tome V, Fasc. 1, Masson); **Fig. 3.7** Brian Guzzetti / Design Pics / Getty Images; **Fig. 3.8** Hassan Merheb / EyeEm / Getty Images; **Fig. 3.9** Pearson Italia (modified from Mauseth 2009, *Botany: an Introduction to Plant Biology*, 4th edn, Jones & Bartlett); **Fig. 3.10** Courtesy Laure Apothéloz-Perret-Gentil; **Fig. 3.11** Pearson Italia; **Fig. 3.12** Used under licence from Shutterstock.com; **Fig. 3.13** Warren *et al.* 2014, *PLOS One* 9: e88364; **Fig. 3.14** blickwinkel / Alamy Stock Photo; **Fig. 3.15** Used under licence from Shutterstock.com; **Fig. 3.16** Antagain / E+ / Getty Images; **Fig. 3.17** Parent Géry, Wikimedia Commons; **Fig. 3.18** Pearson Italia (modified from Brusca & Brusca 2003, *Invertebrate Zoology*, 2nd edn, Sinauer); **Fig. 3.19** Pearson Italia (modified from F. Baldanza 2010, www.littlescience .it/Meiosi%20e%20riproduzione.html); **Fig. 3.20** Natalia Lashmanova / Alamy Stock Photo; **Fig. 3.21** Courtesy Romano Dallai; **Fig. 3.22** Courtesy Martina Magris; **Fig. 3.23a** Pearson Italia (modified from Mauseth 2009, *Botany: an Introduction to Plant Biology*, 4th edn, Jones & Bartlett); **Fig. 3.23b** Pearson Italia (modified from R. E. Koning 1994, Pollen and embryo sac. *Plant Physiology Website*. www1.biologie.uni-hamburg.de/b-online/ibc99/koning/ pollenemb.html); **Fig. 3.24** Pearson Italia (modified from Mauseth 2009, *Botany: an Introduction to Plant Biology*, 4th edn, Jones & Bartlett); **Fig. 3.25** Pearson Italia (modified from Wikimedia Commons); **Fig. 3.26** Pearson Italia; **Fig. 3.27** T. Shimamura 1937, *Cytologia, Fujii Jubilee Volume*: 416–423); **Fig. 3.28** Pearson Italia (modified from Lampert & Schartl 2008, *Phil. Trans. R. Soc. B* 363 and Avise 2008, *Clonality*, OUP); **Fig. 3.29** Used under licence from Shutterstock.com; **Fig. 3.30** animatedfunk / E+ / Getty Images; **Fig. 3.31** Simon McGill / Moment / Getty Images; **Fig. 3.32** Pearson Italia (modified from E. Korschelt & K. Heider 1902–1920, *Lehrbuch der vergleichenden Entwicklungsgeschichte der wirbellosen Tiere: Allgemeiner Theil*, Fischer); **Fig. 3.33** Jasius / Moment / Getty Images; **Fig. 3.34** A. Martin UW Photography / Moment Open / Getty Images; **Fig. 3.35** Guenter Fischer / Getty Images; **Fig. 3.36** Pearson Italia (modified from Barrett & Hodgins 2006, in Harder & Barrett *Ecology and Evolution of Flowers*, OUP, pp. 239–255); **Fig. 3.37** Federica Grassi / Moment / Getty Images; **Fig. 3.38** Michael Nolan / Getty Images; **Fig. 3.39** Jedimentat44, Wikimedia Commons; **Fig. 3.40** Danita Delimont / Getty Images; **Fig. 3.41** Antagain / E+ / Getty Images; **Fig. 3.42** Pearson Italia (modified from Cardoso Neves *et al.* 2016, *Zool. Anz.* 265: 141–170); **Fig. 3.43** Pearson Italia (modified from O'Neill 2004, *PLOS Biol.* 2 (3): e76); **Fig. 3.44** Pearson Italia (modified from van Dijk 2009, in Schön *et al. Lost Sex*, Springer, pp. 47–62); **Fig. 3.45** Teresa Perez / Moment Open / Getty Images; **Fig. 3.46** Francesco Ruggeri / Photographer's Choice RF / Getty Images;

Fig. 3.47 Nnehring / E+ / Getty Images; **Fig. 3.48** a.collectionRF / Getty Images; **Fig. 3.49** H. Krisp, Wikimedia Commons; **Fig. 4.1** Courtesy James Lindsey; **Fig. 4.2** Courtesy Carlo Brena; **Fig. 4.3** Used under licence from Shutterstock.com; **Fig. 4.4** Lingbeek / E+ / Getty Images; **Fig. 4.5** Pearson Italia (modified from Sterrer 1986, *Marine Fauna and Flora of Bermuda*, John Wiley & Sons); **Fig. 4.6** Hien Nguyen / 500px / Getty Images; **Fig. 4.7** Pearson Italia; **Fig. 4.8a** Used under licence from Shutterstock.com; **Fig. 4.8b** Corey Ford / Stocktrek Images / Getty Images; **Fig. 4.8c** Libor Vaicenbacher / 500Px Plus / Getty Images; **Fig. 4.8d** John Lund / Photodisc / Getty Images; **Fig. 4.9** Used under licence from Shutterstock.com; **Fig. 4.10** Courtesy Thomas Palmer; **Fig. 4.11** Steve Gschmeissner / Science Photo Library / Getty Images; **Fig. 4.12** Chaithanya Krishna Photography / Moment / Getty Images; **Fig. 4.13** Simon Murrell / Cultura / Getty Images; **Fig. 4.14** Wilkinson *et al.* 2013, *PLOS One* 8: e57756; **Fig. 4.15** Courtesy Jorge Almeida; **Fig. 4.16** Used under licence from Shutterstock.com; **Fig. 4.17** Pearson Italia (modified from G. O. Sars 1891, Pycnogonidea. Norwegian North Atlantic Expedition, 1876–1878. 6 (Zool. 20): 1–163); **Fig. 4.18** credit: colourofrice / 500px / Getty Images; **Fig. 4.19a** Classen Rafael / EyeEm / Getty Images; **Fig. 4.19b** Emrah Turudu / Stockbyte / Getty Images; **Fig. 4.19c** RedHelga / E+ / Getty Images; **Fig. 4.19d** YinYang / E+ / Getty Images; **Fig. 5.1** Science History Images / Alamy Stock Photo; **Fig. 5.2** Forest & Kim Starr, Wikimedia Commons; **Fig. 5.3** Pearson Italia; **Fig. 5.4** Pearson Italia; **Fig. 5.5** Richard Robinson 2006, *PLOS Biol.* 4 (9): e304; **Fig. 5.6** Used under licence from Shutterstock.com; **Fig. 5.7** Pearson Italia (modified from Raikov 1994, *Eur. J. Protistol.* 30: 253–269); **Fig. 5.8** Pearson Italia (modified from Barton & Charlesworth 1998, *Science* 281: 1986–1990); **Fig. 5.9** Pearson Italia (modified from Furuya & Lowy 2006, *Nat. Rev. Microbiol.* 4: 36–45); **Fig. 5.10** Pearson Italia; **Fig. 5.11** Pearson Italia; **Fig. 5.12** Pearson Italia; **Fig. 5.13** Pearson Italia; **Fig. 5.14** Matteo Colombo / Moment / Getty Images; **Fig. 5.15** Pearson Italia (modified from Sano *et al.* 2011, *Dev. Growth Differ.* 53: 816–821); **Fig. 5.16** Courtesy Yuuji Tsukii; **Fig. 5.17** Pearson Italia; **Fig. 5.18** Pearson Italia (modified from Stenberg & Saura 2009, in Schön *et al.*, *Lost Sex*, Springer, pp. 63–74); **Fig. 5.19** MedioTuerto / Moment / Getty Images; **Fig. 5.20** Pearson Italia (modified from Stenberg & Saura 2009, in Schön *et al.*, *Lost Sex*, Springer, pp. 63–74); **Fig. 5.21** Pearson Italia (modified from Stenberg & Saura 2009, in Schön *et al.*, *Lost Sex*, Springer, pp. 63–74); **Fig. 5.22** Pearson Italia (modified from Stenberg & Saura 2009, in Schön *et al.*, *Lost Sex*, Springer, pp. 63–74); **Fig. 5.23** Used under licence from Shutterstock.com; **Fig. 5.24** Pearson Italia (modified from Stenberg & Saura 2009, in Schön *et al.*, *Lost Sex*, Springer, pp. 63–74); **Fig. 5.25** Pearson Italia (modified from Simon *et al.* 2003, *Biol. J. Linn. Soc.* 79: 151–163); **Fig. 5.26** Pearson Italia (modified from Simon *et al.* 2003, *Biol. J. Linn. Soc.* 79:

151–163); **Fig. 5.27** Pearson Italia (modified from Simon *et al.* 2003, *Biol. J. Linn. Soc.* 79: 151–163); **Fig. 5.28** Used under licence from Shutterstock.com; **Fig. 5.29** Pearson Italia; **Fig. 5.30** Pearson Italia; **Fig. 5.31** Courtesy Fatiha Abdoun; **Fig. 5.32** Pearson Italia (modified from Eisen *et al.* 2006, *PLOS Biol.* 4 (9): e286); **Fig. 5.33** Pearson Italia; **Fig. 5.34** Sergio Viana / Moment / Getty Images; **Fig. 5.35** Pearson Italia; **Fig. 6.1ab** Pearson Italia (modified from Kraak & Pen 2002, in Hardy (ed.), *Sex Ratios: Concepts and Research Methods*, CUP); **Fig. 6.1cd** Pearson Italia (modified from Ming *et al.* 2011, *Annu. Rev. Plant Biol.* 62: 485–514); **Fig. 6.2** Pearson Italia (modified from Silver 1995, *Mouse Genetics*, OUP, and Virkki *et al.* 1991, *Psyche*, 98: 373–390); **Fig. 6.3** Pearson Italia (modified from Bachtrog *et al.* 2011, *Trends Genet.* 27: 350–357); **Fig. 6.4** Pearson Italia; **Fig. 6.5a** Creativeye99 / E+ / Getty Images; **Fig. 6.5b** Life on White / Stockbyte / Getty Images; **Fig. 6.6a** Life on White / Photodisc / Getty Images; **Fig. 6.6b** Used under licence from Shutterstock.com; **Fig. 6.7** Pierre-Louis Crouan, Wikimedia Commons; **Fig. 6.8** Tier Und Naturfotografie J und C Sohns / Photographer's Choice RF / Getty Images; **Fig. 6.9** Used under licence from Shutterstock.com; **Fig. 6.10** Used under licence from Shutterstock.com; **Fig. 6.11** Pearson Italia; **Fig. 6.12** Pearson Italia (modified from Beukeboom & van de Zande 2010, *J. Genet.* 89: 333–339); **Fig. 6.13** Pearson Italia (modified from Sabelis & Nagelkerke 2002, in Hardy (ed.), *Sex Ratios: Concepts and Research Methods*, CUP, pp. 235–253); **Fig. 6.14** Pearson Italia (modified from Kraak & Looze 1993, *Neth. J. Zool.* 43: 260–273) and Universal Images Group North America LLC / Alamy Stock Photo; **Fig. 6.15** Pearson Italia; **Fig. 6.16** Used under licence from Shutterstock.com; **Fig. 6.17** Used under licence from Shutterstock.com; **Fig. 6.18** Used under licence from Shutterstock.com; **Fig. 6.19** James Gerholdt / Stockbyte / Getty Images; **Fig. 6.20** isoft / E+ / Getty Images; **Fig. 6.21** Courtesy James K. Adams; **Fig. 6.22** Dartmouth Electron Microscope Facility; **Fig. 7.1** Pearson Italia; **Fig. 7.2** Pearson Italia; **Fig. 7.3** Pearson Italia (modified from Purves *et al.* 1998, *Life: the Science of Biology*, 5th edn, Freeman); **Fig. 7.4** Pearson Italia; **Fig. 7.5** Pearson Italia; **Fig. 7.6** Pearson Italia; **Fig. 7.7** Pearson Italia; **Fig. 7.8** Pearson Italia; **Fig. 7.9** Pearson Italia (modified from Mauseth 2009, *Botany: an Introduction to Plant Biology*, 4th edn, Jones & Bartlett); **Fig. 7.10** Pearson Italia; **Fig. 7.11** Pearson Italia; **Fig. 7.12** Pearson Italia; **Fig. 7.13** Pearson Italia; **Fig. 7.14** Pearson Italia; **Fig. 7.15** Pearson Italia; **Fig. 7.16** Pearson Italia; **Fig. 7.17** Pearson Italia (modified from Bayer & Owre 1968, *The Free-Living Lower Invertebrates*, Macmillan); **Fig. 7.18** Pearson Italia (modified from Hickman *et al.* 1993, *Integrated Principles of Zoology*, 9th edn, McGraw-Hill); **Fig. 7.19** Pearson Italia (modified from Ebert 2005, *Ecology, Epidemiology, and Evolution of Parasitism in Daphnia*, National Center for Biotechnology Information); **Fig. 7.20** Pearson Italia (modified from Flint 1998, *Pests of the Garden and Small Farm*, University of California Press).

Introduction

A cat gives birth to her kittens, two earthworms mate and exchange sperm, a bee brings the pollen collected from the stamens of a buttercup to the pistils of another buttercup – diverse scenes from the same biological phenomenon, reproduction.

Reproduction is the subject of this book. At first sight, it seems to be a well-defined subject of study, which includes everything concerning the ways, times and mechanisms by which living beings produce their descendants. We all have an intuitive idea of what reproduction is, but this is likely to be based on the behaviour of a few organisms (especially animals) familiar to us. In fact, by widening our perspective on reproduction to less well-known organisms, up to and including the whole of the living world, we reach a point where the boundaries of reproductive phenomena become less and less distinct and finally blend into other aspects of biology. If we think we have a clear idea of the boundary between reproductive and growth processes based on what we know about humans or giraffes, the distinction between the two processes becomes more difficult to define for a strawberry plant or a marine annelid. If it seems easy to establish with reasonable clarity the identity of the individual that reproduces in the case of an eagle or a mosquito, in the case of a coral it is far from obvious where the boundaries between two individuals lie. Not to mention an ant colony – not a good choice of system to reassure us in our belief that we can always easily distinguish between the reproduction of an individual and the reproduction of a society.

These difficulties are unavoidable, and the iterated introduction of *ad hoc* definitions is an exercise in taxonomic arbitrariness.

Of course, we cannot do without definitions if we want to communicate, but the boundaries we establish from time to time between the objects of our study will not always and necessarily correspond to 'natural' boundaries emerging unequivocally from the biology of the organisms we study. There is no other way out, then, than to take a pragmatic approach. Definitions help, but they only work within delimited areas, beyond which they can be more of a hindrance than a help. That's life.

1

Along our journey we will meet many problems of boundaries and definitions precisely because this book deals with reproductive phenomena in all kinds of living beings: the binary fission of a unicellular alga, the stolon of a sequoia that separates it from the mother plant, the mating between two kangaroos, the production of spores by a mushroom, the encounter between the pollen of a thistle and the ovule of a conspecific plant, and more. We try to present all these phenomena using a common language for all living beings, at least in what are the most general aspects of their reproductive biology. This is not always easy, because in the specialized literature concerning the different groups the reproductive phenomena are often described from non-overlapping perspectives, using only marginally congruent terminology.

Against this background we have had to make choices in establishing the structure of this book and the boundaries of its subject matter.

A first and most fundamental choice was to limit the treatment to a 'phenomenology of reproduction'. We fully acknowledge the interest of many other topics – e.g. the adaptive value of different reproductive modes or strategies and the possible scenarios of their evolution, or the vast subject of sexual selection – but we had to leave them out, because they would easily form the subject of another book. Reading tips will be provided for these topics.

A second, unavoidable basic question for a book on the reproduction of living beings is to decide who the living are. Trying to answer this question would lead to a very different book, and we therefore made a choice in line with the pragmatic approach outlined above. Although reproduction can unite material systems of different kinds, here we deal only with living beings in the strict sense, that is, with biological systems made up of one or more cells. Therefore, viruses, prion proteins and transposable genetic elements are excluded. We mention their reproduction only occasionally, when it is relevant to the reproductive phenomena of living beings in the strict sense.

There are, then, many possible ways of classifying different reproductive modes. And there are many different criteria, all equally justifiable, that could be applied to give a relative weight to each of the many topics into which the subject is divided. We give some prominence to those aspects of reproduction that have appreciable effects on evolutionary processes. For example, the association between the genetic system and the reproductive system of an organism determines the quantity and structure of individual variation that is produced in each generation – and this, in turn, constitutes the raw material on which natural selection and other mechanisms of evolutionary change operate.

Without a doubt, the theme of reproduction opens up a huge number of other biological themes, from ecology to applications in medicine, agriculture and animal husbandry, but we have to leave these subjects to other books.

Other aspects of the delimitation of the subject matter to be discussed depend on the taxonomic group considered. The whole life cycle of a living organism can be seen as a set of processes that contribute to a more or less faithful production of copies of itself. There is no aspect of the life of an organism that is not related to reproduction, directly or indirectly. Traditionally, however, under *reproductive strategies* (or *reproductive modes*) we refer only to certain aspects of genetics (e.g. sex chromosomes), anatomy (e.g. reproductive organs), physiology (e.g. sex hormones), life cycle (e.g. reproductive phases), or behaviour (e.g. courtship) that seem more directly involved in reproduction. These aspects vary greatly from species to species. It is not surprising, therefore, that textbooks on systematic biology offer very different contents, in the sections dedicated to the reproduction of each group of organisms. Examples are the structure of the flower in the case of plants, the origin of the gonoducts in annelids, the molecules involved in chemical communication between partners in insects, the strategies of courtship in birds, the duration of gestation in mammals.

Even here we could not escape space constraints, and owing to the specificity of some aspects we had to leave these out, despite their interest. For example, we refer only in passing to many topics in vertebrate ethology such as courtship and parental care, or to the physiology of the production of seeds and fruit in plants. Suggested readings will also be provided for these topics.

One last note concerns taxonomy. For our broad-spectrum taxonomic treatment it was necessary to adopt a classification scheme that is up to date as far as possible, but at the same time reasonably consolidated. For convenience of exposition, not all taxonomic groups we discuss are strictly monophyletic. Among the most common paraphyletic or polyphyletic groups that we mention are prokaryotes (eubacteria plus archaea), protists (unicellular eukaryotes), polychaetes (a paraphyletic grouping of annelids), crustaceans (in the traditional sense that excludes insects), reptiles (in the traditional sense that excludes birds), algae (photoautotrophic protists, rodophytes, phaeophytes, chlorophytes and other minor groups), bryophytes (non-tracheophyte embryophytes), pteridophytes (non-spermatophyte tracheophytes), gymnosperms (non-angiosperm spermatophytes) and plants in the widest sense (algae of various groups plus embryophytes). The appendix at the end of the book provides a phylogenetic classification of the taxa mentioned in the text.

Here, finally, is a preview of the contents of the chapters that follow.

In Chapter 1 we introduce some fundamental concepts, starting with a tentative definition of reproduction that will be revised and enriched in the following chapters, and with the traditional distinction between sexual and asexual reproduction. We address the delicate issues relating to the notions of biological individual, generation and life cycle (but we refrain from discussing these topics from a philosophical perspective). We also deal with the not always clearly defined relationship between reproductive and developmental processes, in particular those related to regeneration.

Chapter 2 is devoted to the relationships between reproduction and life cycle. We start by illustrating the classical division of life cycles on the basis of the alternation of nuclear phases (haplontic, diplontic, haplodiplontic). Then we move on to the alternation between sexual and asexual generations (metagenetic cycles), amphigony and parthenogenesis (heterogonic cycles), gonochoric and hermaphrodite (heterogenic cycles), solitary and colonial, unicellular and multicellular. We conclude with short sections on the alternation of generations dependent on seasonal polyphenism and the different ways in which different reproductive phases can be distributed within one generation.

Chapter 3 is dedicated to the natural history of reproduction. A first section on asexual reproduction deals with the different forms of cell division in unicellular prokaryotes and eukaryotes. A short interlude introduces the notion of sex (itself in some respects problematic) and describes sexual phenomena uncoupled from reproductive processes, both in prokaryotes and in unicellular eukaryotes. The main types of sexual reproduction (gametogamy, gamontogamy, autogamy) are described, as are the distinction between sexes and mating types and the different ways in which the individual sexual condition (e.g. unisexual or hermaphrodite) can be distributed within the population. Short paragraphs are devoted to secondary sexual characters and to conditions such as aneuploidy, gynandromorphism and intersexuality. Attention is then shifted to the reproductive organs of the metazoans and the morphology of eggs and spermatozoa, and also to the reproductive organs of plants and the morphology of their gametes. Finally, the fate of gametes is described, both in typical biparental reproduction, with particular regard to its ecological context, and in uniparental sexual reproduction (self-fertilization, parthenogenesis, gynogenesis, androgenesis and hybridogenesis).

The short Chapter 4 deals with investment in reproduction by organisms, first considering the destiny and care of the products of reproduction and the alternative between oviparity and viviparity, followed by a brief mention of the forms of parental care given to the offspring by one or other parent, or both. Some energetic and metabolic aspects of reproduction, in particular vitellogenesis in animals and the formation of endosperm in flowering plants, are also considered, concluding with a short section on the various strategies of parental investment, including different fecundity/fertility levels, considered against different environmental contexts.

Chapter 5 deals with the genetics and cytogenetics of reproduction, starting again from asexual reproduction. Genetic variation due to new mutations, recombination, stochastic segregation or epigenetic causes is discussed. Moving to sexual reproduction, we discuss mechanisms of genetic exchange in the prokaryotes, but eventually focus more closely on sexual reproduction in the eukaryotes. First, the various sources of genetic variation (independent assortment of chromosomes and chromatids, crossing over and gene conversion at meiosis, syngamy) are discussed, thus addressing the genetics of hereditary transmission through different modes of sexual reproduction (amphigony, self-fertilization, meiotic and ameiotic parthenogenesis, gynogenesis, hybridogenesis, androgenesis). The last paragraphs of this chapter are devoted to sexual leakage and some special cases of sex in eukaryotes (conjugation in ciliates, parasexual cycle in fungi, chimerism).

In Chapter 6 we discuss the determination of sex and mating type, considering both genetic systems of sex determination and those dependent on environmental factors such as temperature or interactions with conspecific individuals, to end with cases of maternal determination of sex and so-called mixed sex-determination systems. We devote only brief notes to sexual differentiation, to conclude with the mating types of fungi and protists.

In Chapter 7 we present an overview of the reproductive phenomena that occur in the different phyla.

Chapter 1: Introductory Concepts

Ever since living beings first arose from simple non-living organic compounds on a primordial planet, more than three and a half billion years ago, an unceasing multitude of organisms has flourished by means of the reproduction of pre-existing organisms. The generational change through reproduction is an essential element of the continuity of life. Not surprisingly, the ability to reproduce is considered one of the most important properties to characterize living systems, which in this way generate other material systems that to some extent resemble themselves. But first things first.

Compared to material systems belonging to the domains of rocks and minerals, living beings are material systems with a relatively modest degree of physical persistence, and their existence depends on some capacity for 'renewal' through time. The plants that cover the slopes of Monte Antelao, in the Italian Dolomites, are not the same individual plants that covered those slopes even just a thousand years ago. And even those individuals growing there this year (2019) that were already there last year (2018) are 'the same' only to a limited extent, having been subject to a considerable flow of matter through growth and metabolism, which has profoundly modified their constitution at the molecular level. On the other hand, Monte Antelao has maintained its identity for millions of years, still composed of the same atoms (mainly calcium, carbon, hydrogen, oxygen and magnesium), which mostly have remained in the same spatial relationships, affected only to a negligible degree by movements of blocks of rock along fault lines and the phenomena of surface erosion and transport.

Living systems compensate for their limited capacity for persistence with a high capacity for renewal – but how is this renewal process achieved? The answer seems obvious: through reproduction! A pine tree, before disappearing as a material system, generates other pines, and the pine forest endures, at least for a while, certainly for much longer than the time that the single pine tree persists. However, although apparently there is nothing controversial or problematic in the concept of reproduction, the diversity of living beings makes it difficult to devise a universal definition of this process, as well as to delineate it with respect to other biological processes that are only apparently completely distinct. The philosopher of biology Peter Godfrey-Smith has argued that 'the idea of reproduction is surrounded by uncertainties and puzzle cases' (2009, p. 69).

A rational exploration of the reproductive processes and their interactions with other biological processes is the subject of the entire book, while the objective of this chapter is to equip ourselves with some conceptual tools that will be needed along our journey. We start from a few definitions, in order to avoid possible ambiguities and introduce a shared terminology that will be useful for the comparative approach, with broad taxonomic coverage, developed in this book. Therefore, the aim of these definitions is not to adopt arbitrary resolutions in a controversial subject, but rather to create the necessary conditions for starting our exploration of the biological phenomena. Later in the book, these definitions will hopefully find justification, or at least will be further clarified.

1.1 A First Definition of Reproduction

As a starting point we can try to sketch an informal or intuitive concept of reproduction, i.e. a concept close to common sense. This concept has deep roots in human history, as it emerged through acquiring knowledge of the life cycles of the plants and animals most familiar to us, ourselves included. The roots of our notion of reproduction are thus clearly pre-scientific.

In biology, **reproduction** is often defined as *the process by which new individuals are produced from pre-existing individuals*. This very concise definition is based on a couple of assumptions that, in the common-sense view again, are nearly always taken for granted. Namely, that these 'new individuals' (i) are materially generated by portions of the body of pre-existing individuals, which thus take the role of parents, and (ii) somehow qualify as entities of the same kind as their parents. This simple definition, with the accompanying specifications, allows us to delimit this process with respect to other types of production of biological material (or material of biological origin) that we would not count as reproduction, such as (i) the individual's body growth, (ii) the production of metabolic waste products, (iii) the secretion of organic matter such as the silk used in the construction of cocoons and spider webs, and (iv) the emergence of new individuals directly from the abiotic world, so-called *spontaneous generation* (Figure 1.1; Box 1.1).

Two main ideas appear to contribute to the concept of reproduction. One is that 'new individuals' are added to the set of existing ones. In this **demographic** concept of reproduction, 'new' should be understood as a *quantitative addition to the number of entities that already exist*. The second is the idea of producing 'individuals that are new', compared to the existing ones. In this

Figure 1.1 Title page of Francesco Redi's book of 1688, in which spontaneous generation in insects is refuted.

Box 1.1 Spontaneous Generation

In his *Exercitationes de generatione animalium* of 1651, William Harvey (1578–1657) recognized a fundamental divide between the world of living beings and the rest of nature, affirming a fundamental principle based precisely on the phenomenon of reproduction: *omnia ex ovo*, all living beings are born of an egg. With this expression, Harvey distanced himself neatly from a long tradition, still alive in his day, that admitted the possibility of *spontaneous generation*, that is, the direct formation of living organisms (at least the simplest ones) from inanimate matter, in particular from mud or rotting organic material. A few years earlier, his Flemish contemporary Jan Baptist van Helmont (1577–1644) had claimed that from a dirty shirt enclosed in a box together with some grains of wheat, mice could be born, and that, in more natural conditions, frogs are born from the mud of ponds. Until then, it was commonly thought that the boundary between the inanimate and the animate worlds could be easily crossed. In the second half of the sixteenth century, the Italian polymath Ulisse Aldrovandi (1522–1605) had observed how blowflies and flesh flies gave birth to 'worms' on meat, but this did not lead him to rule out the possibility of spontaneous generation. Similarly, the philosopher Pierre Gassendi (1592–1655) was also ambiguous in this regard. While recognizing that the worms found in a fruit's pulp derive from eggs laid on the flowers by an insect, he did not abandon the old fable, also to be found

Box 1.1 (cont)

in Virgil's (70–19 BC) *Georgics*, according to which bees are generated from the rotten blood of dead bulls.

Having rejected spontaneous generation, William Harvey distinguished three possible reproductive modes: *oviparous*, *viviparous* and *vermiparous*. The last of these would indicate the possibility of an animal giving birth to 'worms' (in Latin, *vermes*), indicating animal forms that differed from the reproducing animal itself. There would therefore exist a kind of 'ambiguous generation' (*generatio œquivoca*) that gives rise to offspring unlike the parents. By means of their galls, for instance, some plants are able to generate small insects. Even Francesco Redi (1626–1697), the scholar who with his experiments inflicted a mortal blow on the doctrine of spontaneous generation, shared the same mistaken opinion about galls.

On the other hand, Redi set up and carried out a real experimental programme designed to reveal whether 'any exudate from a rotten corpse, or any filth of something putrefied, generates worms'. Redi put pieces of meat or small dead animals in tightly closed containers. He observed that the 'worms' (insect larvae) developed on the rotting material and saw that these, when grown to maximum size, were transformed into a sort of large 'eggs' (pupae) that, depending on the case, took a red or black colour. From these 'eggs', picked up and isolated in glass containers, green blowflies or black flesh flies emerged within 1–2 weeks. Redi at this point began to suspect that the 'worms' derived not from the meat, but from flies similar to those into which they themselves were transformed. Thus he set up two groups of flasks: those in the first group were plugged with paper and string immediately after inserting a piece of meat or a dead animal, while those in the second group, in which were also placed small animals or pieces of meat, were kept open. In accordance with his expectations, 'worms' developed in the second set of containers, accessible to the flies, while none was found in the plugged vessels. But there was still a possibility to consider: perhaps, in the closed containers, the 'worms' had failed to develop because of a lack of air rather than because flesh flies and blowflies did not visit them. Redi therefore set up a further series of tests, in which open vessels were compared to vessels closed only by a very light fabric that let in air, but not insects. Once again, these experiments confirmed his hypothesis: in the containers closed with a veil no 'worms' appeared, although flesh flies and blowflies, attracted by the smell of rotting flesh, were often observed on the veil, laying eggs that could not penetrate the vessel. Hence Redi's conclusion: 'Therefore, from what I have shown, dead animals do not generate worms.'

continues

Box 1.1 (cont)

On the question of spontaneous generation, however, the last word had not yet been pronounced. There was still uncertainty – maybe it was possible for the smallest organisms that the microscope was revealing at the time. The problem had to be confronted again by the Italian biologist Lazzaro Spallanzani, and later by Louis Pasteur, before it was fully confirmed that all existing life forms are generated always and only by other organisms similar to themselves.

Around the middle of the eighteenth century, John Turberville Needham (1713–1781) had shown that in boiled infusions, kept in hermetically sealed containers, tiny organisms (*animalcula*) of various kinds could still develop, a result that seemed to suggest some residual validity for the doctrine of spontaneous generation. For Lazzaro Spallanzani (1729–1799), however, these results were more than doubtful. He therefore repeated his adversary's experiment, proving that simple heating of an infusion kept in a well-closed container is sufficient for no more *animalcula* to form in it. These, therefore, do not originate from organic particles present everywhere, but only from 'eggs' or 'seeds', i.e. from specific airborne 'animated primordia'.

In 1810 the brewer Nicolas Appert (1752–1841) introduced the practice of sterilizing containers used for the storage of foodstuffs in hot water, using glass jars that had been previously closed with as much air as possible excluded. For this, a special commission convened by the Ministry of the Interior of the French Empire awarded Appert a prize of 12,000 francs. However, Louis-Joseph Gay-Lussac (1778–1850), a member of the ministerial commission, while recognizing the validity of Appert's method, believed that it was not heat that killed the germs of putrefaction, but the absence of oxygen. It was necessary to wait until 1836, when Franz Schulze (1815–1873) published the results of his experiments, to see that presence or absence of air did not influence the growth of the small organisms in the infusions.

In 1859 Felix-Archimède Pouchet (1800–1872) published a book entitled *Hétérogénie ou traité de la génération spontanée*, which stimulated the Académie des Sciences to launch, on 30 January 1860, a competition on the subject. The prize was awarded in 1862 to Louis Pasteur (1822–1895), for his demonstration that the air is full of 'organized corpuscles', and that the previously sterilized liquids do not become contaminated if they come into contact at room temperature with air that, in turn, has been sterilized.

innovative concept of reproduction, 'new' should be understood as *qualitatively different from the pre-existing kinds of entities*. These two concepts are based on two different meanings of the idea of renewal in a population. According to the *demographic* concept, the focus is on replacing the individuals that inevitably perish, thus maintaining or possibly increasing the size of the population. In contrast, according to the *innovative* concept, the focus is on the appearance of 'something new under the sun'. These two distinct aspects of reproduction are found together in some forms of reproduction, for instance in the most common forms of sexual reproduction, but not in all forms. As we will see in the next section, some forms of reproduction consist only of a process of renewal in the demographic sense, while innovation, although occurring in all organisms, is not necessarily associated with reproduction.

This first, very general definition of reproduction is not intended to establish what reproduction in general is, but rather to provide a definition that is adequate to describe and delineate this phenomenon in terms that are applicable to all living beings. There are non-biological systems (e.g. digital information stored in a memory chip) and biological systems at levels of organization other than organisms (e.g. at the level of the organism's genome, or a society of individual organisms) which 'reproduce' in some sense but to which our definition does not apply, or is somehow inappropriate (Figure 1.2; Box 1.2).

This first definition of reproduction, along with further clarification provided later in this chapter, will accompany us in our exploration of the phenomena of generation in the living world. More informed, we will return to the concept of reproduction, in the closing pages of the book.

Figure 1.2 Some DNA sequences, known as transposons, are able to reproduce like living organisms, albeit using different mechanisms. The photo shows the effects of the propagation of these sequences on the colour of kernels in an ear of maize.

Box 1.2 Other Reproductions

There are material systems that are not considered living organisms, or do not qualify as living cellular systems – for instance viruses, prion proteins (*prions*) and transposable DNA sequences (*transposons*) – which can reproduce, although differently from living organisms. An important difference with respect to organismal reproduction is that these systems can generate 'copies of themselves' without contributing constitutive matter to their descendants. In other words, the causal link between 'parents' and 'offspring' does not pass through the transformation of a part of the parent into the offspring (Godfrey-Smith 2009).

For instance, in retroviruses (such as HIV), the genetic material consists of an RNA sequence. When infecting a cell, the genome of a viral 'parent unit' is retro-transcribed into the cell's DNA, while at the same time it induces the cell to produce viral capsid proteins. The nucleotide sequence inserted into the host genome is then transcribed into RNA molecules that will constitute the genomes of the viral 'offspring units'. Neither the genetic material nor the capsid of the 'offspring generation' are formed from the parent's molecules. The parent virus is causally responsible for the production of a new viral unit, but without contributing to it with parts of the material system of which it is made.

Similarly, some types of transposable genetic elements, called *retrotransposons*, are first transcribed from DNA into RNA, then the RNA thus produced is retrotranscribed into a DNA copy that inserts in a new position in the genome. In this case also, the DNA sequence of the new inserted element does not materially originate from the parent sequence.

Prions are particular conformational isomers of some glycoproteins capable of converting other molecules of the same protein into their isoform. In this way, prions can induce a chain reaction that leads to their multiplication in an infectious way. Prions induce other molecules (their isomers) to assume their own tertiary and quaternary structure (they 'make copies of themselves') without any material relocation. Some prions are associated with pathological conditions of the mammal nervous system – for instance bovine spongiform encephalopathy (BSE), also known as 'mad cow disease'.

The reproduction of a material system *a* can therefore consist in inducing another system *b* to produce a copy of *a* (as in the case of retroviruses and retrotransposons), or to transform into *a* (as in the case of prion proteins).

1.2 Asexual and Sexual Reproduction

The different ways in which living beings reproduce can be classified on the basis of different features of the process, and as a consequence different classifications are produced. Traditionally, the most fundamental distinction is between **asexual reproduction** and **sexual reproduction**. This division is largely based on the association of the production of new individuals with the production of new genetic variants in a population. This topic is developed in detail in Chapter 5, but it is better to clarify in advance some concepts that we will encounter repeatedly throughout the book. Unfortunately, as for other biological subjects, a discordant terminology is in use in the literaure, which differs in the assignment of some modes of reproduction to either sexual or asexual reproduction. See Box 1.3 for classifications different from that adopted in this book.

Box 1.3 Sexual or Asexual?

Different authors have different opinions on where to draw the line between sexual and asexual reproduction. In particular, what is contentious is whether to consider parthenogenesis and some other forms of uniparental reproduction as sexual or asexual reproduction, and the species or the individuals that practise it as sexuals or asexuals.

Following Boyden (1950), in this book we have adopted a terminology that takes into account both the role of sex and the formation of sex cells. Cases of uniparental reproduction involving sexual processes (such as recombination in self-fertilization), or deriving from processes typical of sexual reproduction (such as egg formation in parthenogenesis), are treated in the context of sexual reproduction, even if they may have a clonal outcome (Section 5.2.3).

However, in the non-specialist scientific literature, but also in a considerable part of the technical literature, a different terminology is in use, which differs in the assignment of some modes of reproduction to either sexual or asexual reproduction. One school of thought makes asexual reproduction coincide with uniparental reproduction, so that sexual reproduction would coincide with amphigony. For instance, organisms reproducing by parthenogenesis or by self-fertilization are classified among the asexuals (e.g. Avise 2008). Another school labels as asexual all modes of reproduction with a clonal outcome, such as ameiotic parthenogenesis or self-fertilization in 100% homozygous individuals (Section 5.2.3). In this case there would be sexual (meiotic) and asexual (ameiotic) forms of parthenogenesis, and sexual and asexual eggs (e.g. Bell 1982).

---

continues

> **Box 1.3** (cont)
>
> All these different schools, included the one followed here, have however to cope with the fact that many forms of reproduction represent difficult or intermediate cases, hard to fit into any too rigid classification.
>
> In any case, the reader is alerted to be careful in exploring the vast literature on the subject: *sexual* and *asexual* do mean different things to different people.

The mode of reproduction, asexual or sexual, is of enormous importance for the biology of an organism, both in its relationship to its environment and for its evolution (although evolution can also, in turn, affect reproduction). The distinction between asexual and sexual reproduction is fundamentally based on the characterization of the *sexual processes*.

At the genetic level (the one that is relevant here) **sex** can be defined as the set of biological processes (**sexual processes**) through which new combinations of genetic material are created from different sources. These phenomena of genetic reassortment may or may not be associated with reproduction (Ghiselin 1974a). They include the union of genomes of different origins (e.g. from two different individuals, through the fusion of two gametes, or the horizontal transfer of genes) and genetic recombination in the broad sense (e.g. as occurring during meiosis in eukaryotes) (Section 5.2.2.1). The phenomenon of *horizontal gene transfer* (HGT), also known as *lateral gene transfer* (LGT), i.e. the non-genealogical (non-vertical) transmission of genetic material from one organism to another, is extremely widespread among prokaryotes, but it has also been described for many eukaryotes (e.g. in insects; Peccoud *et al.* 2017), often in relation to endosymbiosis (Figure 1.3). Sexual processes are thus distinguished, at least in principle, from the processes of change of genetic information that occur in a single individual without the contribution of exogenous DNA, such as gene and chromosomal mutations and gene transposition. Sexual processes, associated or not with reproduction, are found in virtually all major taxa of living beings.

Asexual reproduction (in plants, also called **vegetative reproduction**) is a mode of *uniparental reproduction* (i.e. from a single parent) that does not involve sexual processes or the production of gametes, not even in derived or residual form. Therefore, in the generation of a new individual, the mixing of different genomes and the processes of genetic recombination through meiosis and syngamy, typical of sexual reproduction, are absent. As a first approximation, asexual reproduction generates individuals genetically identical to each other and identical to the parent, thus forming a *clone*. For this reason

Figure 1.3 The genome of tunicates (here, the solitary sea squirt *Rhopalaea crassa*) contains a gene for cellulose synthesis derived from symbiotic bacteria by horizontal gene transfer. This event of genetic reassortment not associated with reproduction would have occurred in a common ancestor of all tunicates, more than 500 million years ago.

it is also called *clonal reproduction*. However, asexual reproduction may in some cases not produce perfectly clonal descendants (Section 5.1), while sexual reproduction occasionally has a clonal outcome (see below and Section 5.2.3).

Asexual reproduction is the most common form of reproduction among unicellular organisms, prokaryotes and eukaryotes alike, but it is also very common among multicellular organisms. In some groups, for instance in bacteria and some protists, e.g. the euglenozoans, this is the only mode of reproduction (*obligate asexual reproduction*). However, more often, asexual reproduction coexists with sexual reproduction in the same species, either as the only form of reproduction in a specific phase of its life cycle (e.g. in the polyp phase of cnidarians with a typical metagenetic cycle; see Section 2.2), or as a reproductive option (*facultative asexual reproduction*) co-occurring with the sexual one (e.g. in many plants; Section 2.8). In any case, the exclusively asexual reproduction of some organisms does not rule out the possibility of having sex that is not associated with reproduction, such as conjugation in bacteria (Section 5.2.1).

A widely accepted distinction contrasts *symmetric asexual reproduction*, such as binary fission, where the parent's body is divided equally between the two offspring individuals, and *asymmetric asexual reproduction*, such as budding, where the parent persists as a distinct individual across the reproductive act while a minor portion of its body becomes its offspring. In the symmetric binary fission of many protists (Figure 1.4a), the two cells that are thus

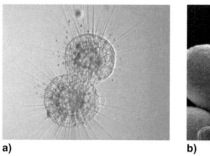
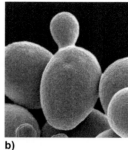

a)   b)

Figure 1.4 (a) Symmetric asexual reproduction (binary fission) in the protist *Actinosphaerium*, and (b) asymmetric asexual reproduction (budding) in the yeast *Saccharomyces cerevisiae*.

obtained are considered sisters, descendants of an individual which, by dividing, has ceased to exist. But in the budding of a cell of the common bread yeast (*Saccharomyces*), the larger cell is called the mother cell, while the smaller cell which detaches from it is called the daughter cell (Figure 1.4b). This unequal treatment might seem rationally unsound. Firstly, it is possible to imagine a complete range of intermediate cases of allocation of resources between the individuals resulting from a division process, and it is not obvious where the boundary between considering them as siblings or as parent and offspring should be placed. Secondly, there is no unequivocal criterion for assigning the products of reproduction to the offspring or the parental generation (Section 1.3.1). However, this distinction might be justified, at least in certain cases, by the different behaviour of the products of reproduction with respect to senescence. The two *Euglena* daughter cells, like two sisters, have the same life expectancy, but the yeast mother cell generates an individual with a longer life expectancy than her own current value, exactly as it should be for a mother's offspring. We elaborate on this aspect of asexual reproduction in Section 1.5. In asymmetric asexual reproduction, the generic term **propagule** can be used to indicate the part of the parent's body that differentiates into what will become the offspring's body, regardless of both its own unicellular or multicellular constitution and the unicellular or multicellular constitution of the parent.

Sexual reproduction is a form of reproduction that generates new individuals with a genetic make-up resulting from the association and/or the reassortment of genetic material of different origins. In its more derived forms, some distinctive features of sexual reproduction are retained, e.g. the production of gametes, despite limited or no genetic reassortment.

In the most canonical form of sexual reproduction, the new genome is formed by the union of (partial) copies of the genomes of two parents through fertilization (Figure 1.5). This is called *amphigony* or *biparental sexual*

Figure 1.5 Mating common terns (*Sterna hirundo*): an example of the initial phase of the most canonical form of biparental sexual reproduction.

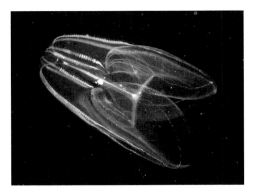

Figure 1.6 Most ctenophores are simultaneous hermaphrodites capable of self-fertilization, a form of uniparental sexual reproduction.

reproduction. However, there may also be forms of *uniparental sexual reproduction*, in which the genome of the single parent is modified and reorganized during the process that leads to offspring production, as for instance in parthenogenesis (Section 5.2.3.3) or in self-fertilization (Figure 1.6, Section 5.2.3.2). Sexual reproduction is found in all multicellular eukaryotes and in most protists (excluding, for instance, ciliates and euglenozoans), but it does not necessarily preclude other forms of sex or other forms of reproduction. The literature on the subject is vast; Bell (1982) remains a valuable starting point.

With reference to the dual (demographic and innovative) concepts of reproduction discussed in the previous section, the demarcation between sexual and asexual reproduction we have just established entails both the possibility of innovation without reproduction and without demographic growth (through sex alone), and the possibility of reproduction with or without sex, and with or without the introduction of genetic novelty.

1.3 Generation, Life Cycle, Development

The need to introduce some definitions is not limited to the problem of reproduction. Throughout this book we frequently use other terms for which multiple discordant definitions have been given in the literature. The 'correct' meaning to be attributed to each of these terms could open an endless debate. However, following the pragmatic approach we have adopted here, we provide at this point some operational definitions. These definitions are not intended to impose our personal view, but rather to serve the practical purpose of helping the reader by reducing ambiguities to a minimum, wherever possible, while highlighting the slippery use in the literature of the terminology employed to describe these phenomena. Among the concepts to be clarified through operational definitions are those of *generation, life cycle* and *development*.

1.3.1 Generation

In common language, the verb 'to generate' can be used as a synonym for 'to produce', but here we use it in a more restrictive sense, referring only to *the production of offspring through reproduction*. By restricting the meaning of the verb in this way, the noun 'generation' can correspondingly indicate both the *act of producing something through reproduction*, i.e. the process of reproduction itself, and *a set of individuals that come into being through reproduction*, i.e. the result of a reproductive process.

While 'generation' in the sense of reproductive process does not seem to require further clarification, 'generation' in the second sense is somewhat ambiguous and its use in the scientific literature is rather inconsistent. Let's define here the **nth generation** as *the set of individuals generated through* n *reproductive events starting from an individual or a parent pair* (Figure 1.7). All the descendants of an individual that reproduces several times in the course of its life (an iteroparous individual; see Section 2.9) count as one generation only. For instance, all the offspring of a female elephant that has been reproducing over multiple reproductive seasons form a single generation (*offspring generation*) which follows the generation to which she herself belongs (*parental generation*). Similarly, two seeds of sequoia produced by the same mother plant 2000 years apart belong to the same generation. By contrast, we regard the following situation as involving three distinct generations: a mother plant (parental generation), a daughter plant developed from a vegetative propagule

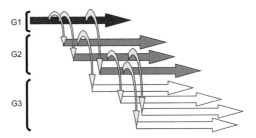

Figure 1.7 Schematic representation of the concept of generation adopted in the text. Horizontal arrows: development of individuals belonging to three generations (G1–G3). Curved arrows: events of reproduction. Individuals produced at different times by different individuals (or even by the same individual) of the same (parental) generation belong to a single (offspring) generation. It should be noted that individuals of a given generation may come into being before some individuals of a previous generation. For simplicity, a form of uniparental reproduction is assumed.

detached from it (first offspring generation) and a plant developed in turn from a vegetative propagule detached from the latter (second offspring generation). Belonging to a given generation is a relative characteristic, which depends on the choice of a reference individual ancestor.

This *genealogical* definition of generation is centred on the individual and will be useful in describing life cycles and reproductive processes in different organisms.

However, there are other ways in which generations are defined. In studies of population dynamics, all the individuals born during the same breeding season are considered to belong to the same generation. This *populational* concept of generation groups together individuals that should be more properly described as members of the same *cohort*.

In species with *non-overlapping generations* (also *discrete*, or *separate generations*), where for instance all adults of a given year die soon after reproduction, to be replaced by the individuals of the next generation, it is possible to establish a correspondence between the genealogical and the populational concepts of generation (i.e. between generations and cohorts). On the contrary, in species with *overlapping generations*, where mating between individuals of different ages is possible, the assignment of an individual to a specific generation in a genealogical sense can be undermined by sexual reproduction. To which generation does an aphid belong, which has been generated by the mating of a female and a male that descend from the same founder female through 11 and 8 generations of parthenogenetic females respectively (Section 2.3)? Or, to which generation do the offspring born from the mating of a female with her own son (*oedipal mating*) belong? This is something that happens regularly in some mites (e.g. *Histiostoma murchei*, Figure 1.8) and in some nematodes (e.g. *Gyrinicola batrachiensis*) (Adamson and Ludwig 1993).

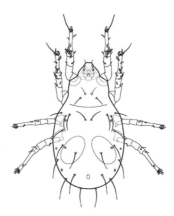

Figure 1.8 A female *Histiostoma* mite. *H. murchei* practises only a form of oedipal mating. Virgin females parasitize earthworm cocoons, in each of which they deposit 2–9 eggs. These will hatch within two or three weeks, producing only males. These males mature in about two days, mate with the mother, and die. The mother then lays about 500 fertilized eggs from which only females are born, which once developed will seek new earthworm cocoons to parasitize.

These individuals are at the same time children and half-siblings (having the same mother) of the father; and children and grandchildren of the mother!

Other authors have suggested still different concepts of generation. For Gorelick (2012), a change of generation occurs both at meiosis and at syngamy. Thus, all the gametes produced by a diploid individual would belong to a subsequent haploid generation, rather than representing a part of itself, as a product of its development, or a transitory intergenerational phase. The fusion of those gametes would then form a further generation of diploid individuals. By contrast, Minelli (2014) distinguished between a 'demographic generation', a group of individuals produced by sexual or asexual reproduction, and a 'genetic generation', a group of individuals produced by sexual reproduction or pure sexuality. An individual that has undergone genetic transformation through a sexual exchange would switch to a subsequent genetic generation, while remaining in the same demographic generation. We leave it to the interested reader to evaluate the advantages and limitations of these proposals.

The most important aspect of the concept of generation adopted here, which differs significantly from other points of view, is that it considers sexual and asexual reproduction as equivalent in their potential to produce new individuals of the next generation. For some authors (e.g. Janzen 1977), what is referred to here as asexual reproduction is nothing but a form of growth (or propagation) of the individual (Section 1.6.2), and what is considered here as a sequence of asexual generations would be qualified as a process of expansion and transformation of the soma of a single individual.

1.3.2 Life Cycle

Like the concept of reproduction, the concept of a life cycle seems well founded in common sense. As an example, let's see how the life cycle of an earthworm is commonly described. By pure convention, we begin the description starting from a fertilized egg, a zygote. The zygote is in the ground, within a protective case (cocoon or *ootheca*). Here it proliferates by mitosis, starting what is referred to as embryonic development and building up, through complex morphogenetic processes, the soma of an individual that at some point will be ready to interact with the external world. At hatching, this is a young worm, very similar to the future adult, which will live free in the soil, feeding and growing and continuing its development. At a certain point in this process of growth and maturation it will become able to reproduce, i.e. 'reproductively mature'. After finding a partner, it will pass its sperm to the partner and will simultaneously receive those of the latter (earthworms are insufficient simultaneous hermaphrodites; see Section 3.3.2.2). From the fertilization of its eggs by the sperm of its companion, and from the fertilization of the latter's eggs by its own sperm, the zygotes of a new generation of earthworms will form. A cycle described in this way traces the series of transformations and events that, starting from a given biological stage of a given organism, lead to the same stage in a successive (genealogical) generation: from egg to egg, but also from adult to adult, or from embryo to embryo. In a cyclical process, the choice of which stage to consider as the initial stage can only be arbitrary or conventional.

However, as an example of a relatively more complex life cycle, let's concisely describe that of a fern such as *Polypodium* (Figure 7.11), starting from the better-known phase represented by a macroscopic plant with roots and fronds. A mature leafy fern plant (the diploid phase called a *sporophyte*) reproduces sexually (by means of recombination) and uniparentally (i.e. without the need for a partner) by producing haploid spores by meiosis. Spores disperse and germinate on the ground, developing into tiny multicellular haploid plants called *prothalli* (the *gametophyte* phase). Prothalli, which bear both male and female reproductive organs, reproduce sexually and biparentally (i.e. through cross-breeding), producing gametes that will fuse to form diploid zygotes, the founding cells of the sporophytes of the next cycle. The embryonic sporophyte is retained by the parental gametophyte, which nourishes it during early development, until it produces the first leaves and the first roots, thus becoming independent (Figure 1.9). In the cycle of a fern there are at least two generations (a sporophyte and a gametophyte), which constitute two distinct **organizational forms**, i.e. two distinct *kinds of individuals*, or *kinds of generations* of the same species, each with its own ontogeny. In the case of

Figure 1.9 The two generations of the multigenerational life cycle of a fern (here, *Onoclea sensibilis*): the haploid gametophyte represented by the prothallus (the basal structure in the figure), and the diploid sporophyte, represented by the upright frond.

the fern, there is one that from a zygote develops into a macroscopic diploid leafy plant, and another that from a spore develops into a tiny haploid thallus. The two generations are separated by two reproductive phases: the production of spores by the sporophyte and the production of gametes by the gametophyte, followed by gamete fusion. The offspring of the sporophyte do not resemble their parents, but their grandparents. The same applies to the gametophyte's offspring.

Without going into an analysis that will be the subject of Chapter 2, we anticipate here an important distinction that serves as a basis for a more general concept of life cycles. The cycle of the earthworm described above is an example of a **monogenerational life cycle**, that is a cycle in which the same *developmental phase* (e.g. the young at the start of the free-living phase) of the single *organizational form* of the organism (here, the vermiform animal) is repeated after one generation. The same applies to the life cycle of a sea urchin or a beetle, irrespective of the complexity of their development, which includes metamorphosis: only one generation separates a given developmental phase of the single organizational form from the next. In contrast, the cycle of the fern is an example of a **multigenerational life cycle**, because it passes through a given *developmental stage* (e.g. the full-formed prothallus) of a given *organizational form* (in this case, the gametophyte) through more than one generation, in this case two. In multigenerational life cycles there are reproductive phases where offspring are generated that are not of the same kind (of the same organizational form) as the parent(s), so that

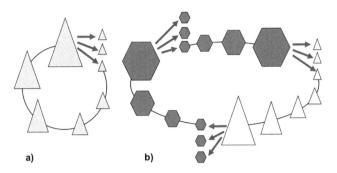

a) b)

Figure 1.10 Schematic representation of the life-cycle concept. (a) A monogenerational life cycle, in which there is only one generation, one organizational form (triangle), one developmental process, and one reproductive phase (arrows). The reproductive phase is counted as one even if the individual reproduces several times. (b) A multigenerational life cycle, in which multiple generations (in this case, three) occur, as well as multiple organizational forms (in this case two, triangle and hexagon), multiple developmental processes (one per generation), and multiple reproductive phases (one per generation, arrows). For simplicity, a form of uniparental reproduction is assumed.

more than one generation is needed to return to the starting form. Multigenerational cycles are called, in a broad sense, *cycles with alternation of generations*. In several organisms, the life cycle passes through an even larger number of generations that may differ in genetic make-up, morphology and living environment. The way in which a cycle is closed, returning to a conventional starting form, can be very tortuous (e.g. in digeneans, as described in Section 2.2).

We define here the **life cycle** (or **biological cycle**) of an organism as *the series of developmental transformations and reproductive phases that lead from a given developmental stage of a given organizational form, to the same developmental stage of the same organizational form in a following generation, through all the organizational forms of the organism.* The life cycle summarizes the processes of physical transformation of an organism and the processes of propagation that allow it to have descendants. It can include one or more developmental processes and one or more reproductive phases (Figure 1.10). In multigenerational life cycles, the number of generations is greater than or equal to the number of organizational forms. For instance, it is greater when an organizational form can reproduce asexually several times before generating the next form. This also entails the possibility of intraspecific variation in the number of generations to close the cycle.

The characterization of a life cycle rests on the possibility of distinguishing the reproductive events, which imply the transition to a new generation, from the processes of development, which are instead transformations of the same individual. We will see that this is not always easy (Section 1.6.3), but first we must say a couple of words about development.

1.3.3 Development

A widespread but rather simplistic, not to say wrong, definition of development sees it as the 'series of transformations of an individual from the egg to
the adult stage'. This definition has obvious limitations: (i) it only applies to
development processes that start from an egg, and (ii) it suggests an idea of
development that applies only to multicellular organisms. A minimal knowledge of the life cycles of plants, fungi and unicellular organisms reveals that
an egg is not necessarily the starting point of a development process, because
the latter can also be initiated from a spore or from a fragment of the parent's
body (Section 3.1). Moreover, the life cycles of unicellular organisms, like
those of multicellular organisms, include phases of growth and maturation
that are fundamental for the progression of the cycle. Only a form of 'multicellular chauvinism' would refuse to describe these unicellular life phases as
developmental stages, given also that they are similar to the developmental
processes at the cellular level in multicellular organisms.

Such a restrictive concept of development does not allow a broad comparative analysis of life cycles and reproductive processes. We would like to be able
to associate the idea of development with the embryonic, larval and adult
phases of a beetle's life, but also with the germination of a spore, with the
morphogenetic transformations of the single cell that constitutes the soma of
a trypanosome, and with the structural changes at the molecular level that
precede the division of a bacterium.

A possibility that we suggest here is to start from the definition of a life
cycle, and to consider as development all those processes of transformation
that are complementary to reproduction. In a life cycle we can have one or
more reproduction events, carried out by as many generations, within which
are intercalated an equal number of developmental sequences (Figure 1.10).
From this idea, a definition of **development** follows, as *the set of transformations of an individual from its individuation (however defined) until its disappearance (however defined)*. Notice the use of the terms 'individuation' and
'disappearance', rather than 'birth' and 'death'. Birth often refers to a particular
moment in the middle of the development of an organism, such as childbirth
in mammals, egg hatching in birds, and seed germination in seed plants.
Death generally indicates the termination of an individual due to trauma
(e.g. by being eaten by another organism) or ageing, and does not include
the end of a unicellular individual owing to its division into two daughter cells,
where, to adopt legal terminology, 'the body of evidence' is actually not found
(Section 1.2). This definition clearly focuses on development in the sense of a
sequence of transformations of the individual, disregarding the common
meaning that refers to the specific developmental process (or set of processes)

of a given developmental phase (e.g. embryonic cleavage) or a given body structure (e.g. limb development) (Minelli 2018).

Such a general definition of development obviously leads to many 'difficult boundaries', only some of which are of interest to us here. For instance, in a textbook of developmental biology it will be necessary to specify in detail which of the transformations an individual undergoes should be treated in the context of developmental processes (e.g. growth, differentiation, morphogenesis, ageing and regeneration), and which should be left to other biological disciplines such as physiology and the behavioural sciences, which are generally concerned with reversible transformations occurring on a relatively short timescale.

On the other hand, it is more important to discuss here the difficult boundary between the transformations of an individual that do not alter its identity, and those that, at some point, will result instead in a new distinct individual, somehow emerging from it. In the development of a bud of a hydra, at what point should we place the transition from 'lateral outgrowth of the mother's body' to 'individual offspring'? At the time of detachment or before? At what point does a *Drosophila* egg, formed during the development of a mature female, become the first cell, the founding cell, of an individual other than herself? At fertilization? Or later, when the zygotic genome starts to be expressed? This qualifies as a classical 'problem of origins'. Many examples of this and other 'difficult boundaries' (Section 1.6) are illustrated in Chapter 3.

The reader will certainly have noted that, as in the case of reproduction, a definition of development cannot be given without an explicit concept of the individual – the topic of the next section.

1.4 Reproduction and Individuality

However reproduction is defined, the concept of an *individual* occupies a central position. Both in the *demographic* meaning of reproduction, where emphasis is placed on the generation of 'new individuals', and in the *innovative* one, where emphasis is placed on the production of 'novel individuals', some formalization of the concept of the individual is obviously in order. But what is an individual in biology? Rivers of ink have flowed in the attempt to answer this question, through articles and monographs in biology and the philosophy of biology alike (e.g. Buss 1987; Wilson 1999; Godfrey-Smith 2009; Dupré 2010; Gilbert *et al*. 2012; Pradeu 2012, 2016; Fields and Levin 2018) – but the answer is that there is no unequivocal answer. In our analysis of the problem we follow the schematization proposed by Santelices (1999).

For a biological system at the level of the organism, we usually describe as an **individual** a well-integrated entity, reasonably well defined in space and time, characterized by genetic homogeneity and genetic uniqueness, as well as by physiological unity and autonomy. However, if you imagine that all living organisms possess all these attributes of individuality, or that only a handful of exceptions lack one or more of them – well, that is simply not the case. Let's examine these attributes one by one.

1.4.1 Genetic Uniqueness of the Individual

An individual can be characterized by the possession of a unique genome. Individuals lacking this feature are, nonetheless, commonplace – including all those that originated through one of those forms of reproduction with a clonal outcome mentioned in Section 1.2 (see also Section 5.1). Two amoebae that have just originated by fission from a parent amoeba share an almost identical genome, as do all the strawberry shoots derived from the same runner, and a pair of human identical twins. In all these cases, however, and particularly in the last, we would be inclined to recognize these as distinct individuals. All forms of asexual reproduction and all forms of sexual reproduction with a clonal outcome therefore represent a problem for a definition of individuality based on genetic uniqueness.

1.4.2 Genetic Uniformity of the Individual

An individual can be characterized by the possession of one or more copies of a single genome, but we should not interpret this too strictly. In fact, it is highly unlikely that the different nuclei of a multinucleated unicellular organism, or the nuclei of the many cells of a multicellular organism have 100% identical genomes. When referring to the genetic identity among the members of a clone (such as the cells of a multicellular organism), it is implied that the mutations accumulated in the subsequent divisions starting from the founder cell are overlooked. Similarly, we are discounting the differences due to the random distribution of the genomes of cytoplasmic organelles in a multicellular eukaryote (Section 5.1.3.2). However, beyond this obvious degree of *intraorganismal genetic heterogeneity* (IGH) shared by all multicellular organisms, individuals lacking genetic uniformity in a more substantial way include all those for whom genetic mosaicism or chimerism are characteristic traits of the normal process of development.

Genetic mosaicism is the condition of an individual carrying different genomes that originated from the genome of a single founder cell (Section 5.1.1.3). For instance, mitochondria of maternal origin and mitochondria of paternal origin (with their genomes) are found in the zygote of the common mussel (*Mytilus edulis*) and the Mediterranean mussel (*M. galloprovincialis*). However, in adult males, mitochondrial DNA of paternal origin is predominant

Figure 1.11 An old oak can exhibit high intraorganismal genetic heterogeneity, simply because of the long time and the large number of mitotic cycles that can separate two cell lines in this tree.

in the gonadal tissues, while mitochondrial DNA of maternal origin predominates in the somatic tissues (Section 5.2.3.1). But even the simple intraorganismal genetic heterogeneity that normally tends to be considered negligible may become relevant in very old or very large organisms (i.e. those with many cells), in which the last common ancestor of two cells in the same individual's body may be traced many mitotic cycles back. In an old oak tree, numerous mutations may have accumulated in the meristems that give rise to new branches and new flowers every year (Figure 1.11). Because of the modular organization of these plants, each mutation in a meristem is inherited by the whole branch that develops from it, and each terminal branch can be an independent site of sexual reproduction (and perhaps, from an evolutionary perspective, selection), and thus the genetic make-up of the cells that form pollen and ovules in a branch can be significantly different from that of similar cells in another branch of the same tree. The most recent common ancestor of two reproductive cells of the same oak might have lived centuries ago, rather like the progenitor of a whole population of individuals in a species with an annual life cycle. In organisms such as plants, where during development there is no early segregation of the cells destined to produce gametes, we recognize another difficult border between development and evolution. But we cannot further expand on this issue here (see, for instance, Godfrey-Smith 2009).

A **chimera** (in biology, not in mythology) is instead a multicellular individual made of cell populations originating from more than one founder cell (Section 5.2.5.3). Chimeric gametophytes originating from the fusion of several spores have been described for the red alga *Gracilaria chilensis* (Santelices *et al.* 1996). Among cnidarians, conspecific coral larvae (planulae) often merge into one individual before differentiating into a polyp (Harrison 2011).

Figure 1.12 In the freshwater sponge *Spongilla* chimeric individuals may result from the fusion of multiple larvae.

The hydrozoan *Ectopleura larynx*, rather than developing colonies through asexual budding, as is typical of the group, after an initial phase of limited budding, develops larger colony sizes through the aggregation and fusion of sexually (non-clonally) produced polyps, sometimes genetically unrelated (Chang *et al.* 2018). The fusion between larvae of the same species has also been observed in freshwater sponges such as *Spongilla* (Figure 1.12; Brien 1973), whereas in the holothuroid *Cucumaria frondosa* fusion of different individuals occurs among hatched blastulae, never before hatching or at larval stages. The fully fused chimeric embryos are 2–5 times larger than non-chimeric embryos (Gianasi *et al.* 2018).

Among fungi, a chimeric individual can easily originate from cytoplasmic fusion between the hyphae of distinct individuals. But chimerism is also known among mammals. Monkeys of the genus *Callithrix* (Section 5.2.5.3) generally give birth to two dizygotic (non-identical) twins. But these are not 'normal' twins. During pregnancy, connections between the two placentas are established so that cell exchanges occur between the two embryos. When the two little monkeys are born, each of them is a mixture of cells derived from the independent fertilizations of two distinct eggs (Ross *et al.* 2007). The case is even more remarkable when the two embryos are of different sexes. Thus, through this form of reproduction, two 'genetic individuals' and two 'physiological individuals' are obtained, but the two genetic individuals are distributed between the two physiological individuals.

Possibly little known is the fact that a form of chimerism also frequently affects the adult females of our own species (together with those of other placental mammals). Following pregnancy, blood cells from the fetus may remain in circulation and multiply in the mother for decades after birth (*fetal microchimerism*; Evans *et al.* 1999).

1.4.3 Autonomy and Physiological Unity of the Individual

A biological individual could be characterized as an undivided morpho-functional living unit, able to relate to the environment independently, including the ability to properly respond to environmental stimuli and the faculty to reproduce. Individuals lacking these characteristics are the members of highly integrated colonies, such as those of some marine invertebrates. Colonial hydrozoans known as the Portuguese man o' war (*Physalia*) behave as an integrated unit to the point that they are often mistaken for individual jellyfish. In each colony different types of individuals (*zooids*) coexist and cooperate, only some of which (*gonozooids*) are able to reproduce. In *Volvox*, a colonial green alga, only the special reproductive cells called *gonidia*, which are located within the sphere formed by flagellate somatic cells, can reproduce asexually to form new daughter colonies.

Individuals lacking reproductive autonomy also include members of the sterile castes in some animal societies. Among these animals, which are called *eusocial*, there are many species of bees, wasps, ants, termites and, unique among mammals, the naked mole-rat (*Heterocephalus glaber*). Individuals not participating in reproduction also include the young *soldier nymphs* of some aphids (Stern 1994), the *soldier rediae* of some digeneans (Hechinger *et al.* 2011) and the *soldier larvae* of the tiny parasitoid wasps of the genus *Copidosoma* (Grbic *et al.* 1992). In other animals, some eggs, embryos or larvae are destined to serve as food for their siblings. This is the case of the *trophic eggs* of some ants and other social insects (Crespi 1992), the *trophic embryos* of the freshwater planarian *Schmidtea mediterranea* (Harrath *et al.* 2009) and the *trophic larvae* of some *Salamandra* populations (Buckley *et al.* 2007) (Section 4.4.4).

Colonies and advanced societies pose a problem for the interpretation of the reproductive process, when the unit of reproduction could be identified at more than one level of biological organization. For instance, it could be recognized either in the single zooid of a colony or in the colony as a whole, which in the latter case can be seen as a 'superorganism' (Figure 1.13). We will return to this topic in Section 2.5.

But autonomy is not a problem only for colonial species. In mosses and liverworts, the sporophyte lives at the expense of the female gametophyte that generated it, while in the seed plants both male and female gametophytes live at the expense of the sporophyte that generated them. In some abyssal fishes (Ceratioidei), the male, after a short independent life, attaches to a female, enters

Figure 1.13 The highly differentiated colonies of the hydrozoan known as the Portuguese man o' war (*Physalia physalis*) include different types of individuals (pneumatophore, gastrozooids, dactylozooids, gonozooids). Because of the high specialization and functional integration of individuals and their non-autonomy, the whole colony can be considered a 'superorganism'.

into intimate contact with her tissues and becomes an appendage of her body, fed by her through their conjoined circulatory systems, but able to produce sperm (Miya *et al.* 2010). The male loses his autonomy (*sexual parasitism*); while from this union a chimeric hermaphrodite is produced (Section 3.5.2.1).

1.4.4 How Many Kinds of Individual?

The link between reproduction and individuality is somehow inevitable. Reproduction is the production of new entities, and these must correspond to some kind of individuals that can be counted. But, as Godfrey-Smith (2009) suggests, on the question of what to recognize as an individual we should develop a more relaxed, or pluralistic, approach. The choice depends on the biological problem we intend to investigate or represent and/or on the biology of the taxonomic group under examination.

Santelices (1999) goes beyond listing the problems that make it difficult to define an individual, and tries to define different 'kinds of individual', each characterized by the presence or absence of the aforementioned features (Figure 1.14).

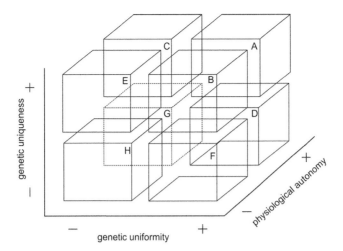

Figure 1.14 Eight types of individual characterized by presence (+) or absence (−) of each of three attributes: genetic uniqueness, genetic uniformity, physiological autonomy. (A) Individuals that possess all three attributes (e.g. *Drosophila melanogaster*). (B) Individuals lacking some form of autonomy (for example, a worker of *Apis mellifera*). (C) Individuals lacking genetic uniformity, developing as chimeric individuals (e.g. Wied's marmoset, *Callithrix kuhlii*) or becoming genetic mosaics in their lifetime (e.g. *Quercus robur*). (D) Individuals lacking genetic uniqueness, reproducing clonally (e.g. bdelloid rotifers). (E) Individuals characterized exclusively by genetic uniqueness, as chimeric individuals with sexual reproduction without autonomy (this category is possibly just hypothetical). (F) Individuals characterized exclusively by genetic uniformity (e.g. the individual zooids of *Physalia physalis*). (G) Individuals characterized exclusively by physiological autonomy, as chimeric individuals that can reproduce clonally (e.g. the freshwater sponge *Spongilla lacustris*). (H) Chimeric individuals with clonal propagation, genetic instability and lack of autonomy (e.g. the red alga *Gracilaria chilensis*).

Without going into the merits of this proposal, which nonetheless depends on the choice of the characteristics to be combined, it is clear that the units among the living beings that we call 'individuals' can have very different characteristics. In other words, there are different *kinds of individual*.

The difficulty we found in unequivocally defining the individual in reproductive biology echoes similar difficulties in developmental biology and evolutionary biology, which we consider here very briefly.

In the ecological approach to the processes of development (*ecological developmental biology*; Gilbert and Epel 2015), the boundaries of the individual emerge as imprecise because of their relationships with the environment in which they live. The most striking example is the large community of symbiotic microorganisms that 'inhabit' the body of most multicellular organisms, which is necessary for the normal development and the regular functioning of the organism. Our digestive system hosts a number of bacterial cells of the same order of magnitude as the cells (the 'human cells' in the strict sense) of our entire body ($\approx 3.5 \cdot 10^{13}$; Sender *et al.* 2016). These bacterial cells belong to at least 10^5 different 'species', collectively carrying an amount of genetic information much larger than ours (Locey and Lennon 2016).

Symbiosis is also a problem when you want to use the immune response to define an individual, as suggested for example by Pradeu (2010). For Pradeu, if an organism does not have an immune reaction to the presence of extraneous (*non-self*) cells, often necessary for its survival, these should be considered as part of the individual. Examples of these 'extensions' of the individual would be the photoluminescent bacteria in the light organs of some cephalopods, the intestinal bacteria of the animals mentioned above and the nitrogen-fixing bacteria or the fungi that facilitate the absorption of phosphorus in plants. On the contrary, non-self cells that are present inside the body and trigger an immune reaction should not be considered part of the individual. Examples of these 'invaders' of the individual would be all pathogenic microorganisms and parasites. However, if because of some deficiency in the immune system a parasite is not recognized as such by the host, should host and parasite be considered a single individual? Or, if an organism is affected by an auto-immune disease, does it constitute more than one individual? As seen above, the intimate, and often complex, relationships between the organism and its environment make it difficult to delimit the individual based on physiological criteria (Gorelick 2012).

The modern theoretical approach to the problem of levels of selection is part of an advancing multilevel selection theory (Okasha 2006). In evolutionary biology, the problem of defining the individual is associated with the question of identifying a putative fundamental unit that is the target of natural selection. At what level(s) does natural selection operate, and at what level(s) are manifested the adaptations that it produces? Does natural selection operate primarily among individual organisms, groups of individuals, genes or species? All these entities can fulfil an ontological concept of 'individual' (Ghiselin 1974b).

1.5 Reproduction and Senescence

Reproduction allows the persistence of a species even if its individual members are continually lost. The mortality rate per unit of time in a population is never zero. Individual living beings do not last forever, for two main reasons. Firstly, an individual may die by an accident related to its interactions with factors of its living environment, either biotic (e.g. predators and parasites) or abiotic (e.g. temperature and radiation beyond the tolerance threshold of the organism). Secondly, even in environmental conditions ideal for its survival, and in case of unlimited availability of resources for its maintenance, the individual cannot save itself from 'certain death'. This is the inescapable result of the developmental process called **senescence**

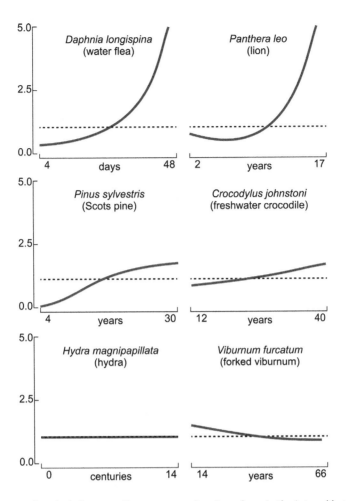

Figure 1.15 Examples of relative mortality curves as a function of age in the interval between reproductive maturity and the age at which only 5% of a cohort of individuals survive (individuals born the same year or, for those with a life cycle of less than 1 year, born in the same period of the same breeding season). Relative mortality is scaled relative to the mean mortality over the time interval considered. Species with very different lifespans may show the same relative course of senescence. *Hydra* and *Viburnum* do not show signs of senescence.

(or **ageing**), which appears as an increase in the probability of death with age (Figure 1.15; Shefferson *et al.* 2017).

For most living beings, the life expectancy of a newly generated individual is greater than that of an individual in an advanced phase of its ontogeny. What happens is that at a certain point in the life of an individual a process of deterioration of the functionality of the organism begins (progressively or abruptly, depending on the species), which eventually leads to death. In some cases, as in many annual plants, the octopus and the Pacific salmon, senescence is triggered quickly the first time the organism reproduces sexually.

Figure 1.16 The giant redwood *Sequoiadendron giganteum*, which can live more than 3500 years, is one of the organisms that do not show signs of senescence.

It is customary to say that the part of an individual that can 'survive' the inexorable degenerative process of senescence is at most the set of cells that, in the form of gametes, spores or other propagules, can contribute to generating the individuals of the next generation. However, whether this can count as a form of 'survival' or 'escape from death' for the individual is a matter of philosophical attitude, rather than a scientific question.

Whereas the problem of population depletion due to accidental causes concerns all living beings, which may all die, the phenomenon of senescence does not affect all organisms in the same way. Most prokaryotes and many protists, including several species of amoebozoans, cryptophytes, chlorophyceans, apicomplexans, euglenozoans and radiolarians, do not seem to experience senescence (Finch 1990). Also, there is no certain evidence of senescence in some plants that live for a very long time, over 4000 years, such as some conifers (*Pinus longaeva* and *Sequoiadendron giganteum*; Figure 1.16), some sponges and sea anemones, hydras (Martinez 1997), the queens of different species of social insects (bees, ants, wasps and termites), some tube-dwelling polychaetes (*Lamellibrachia*) and certain bivalves (*Arctica*) (Fahy 2010). The same may apply to the black coral *Leiopathes* sp., for which a maximum age of 4265 years has been estimated (Roark *et al.* 2009). In the life cycle of the hydrozoan *Turritopsis* the medusa can return to the previous stage of polyp, thus apparently escaping senescence (Piraino *et al.* 1996; but see Box 2.2). All these organisms are considered potentially immortal, but see Box 1.4 for remarks on the concept of immortality.

Box 1.4 Sex and Death

The very 'pulp' title of this box is not a catchy invention of the authors of this book: this expression, indeed, is not infrequent in articles and monographs on the relationship between sex and senescence (e.g. Bell 1988; Clark 1996; Biddle *et al.* 1997; Sterelny and Griffiths 1999). It alludes to the fact that sex, either associated with reproduction or not, is considered to be able to counteract senescence. This is a complex question that cannot be examined in detail here, but it invites us to at least approach the concepts of senescence and immortality more closely.

Without going into the details of a topic pertaining to developmental biology, *senescence* (or *biological ageing*) is a cumulative process of change, at different levels of body organization (from molecules to cells, tissues and organs), which progressively corrupts metabolism and body structures, producing a deterioration of the qualities of the organism that eventually leads to its death.

Actually, the problem of senescence is related more to the concept of immortality than to the fact of death, understood as the end of an individual's existence. In biology, to be immortal does not mean that an organism cannot die, a quality reserved for certain mythological figures and comic-book superheroes. All living things can die from trauma, disease, or simply being eaten by another living being. In biology, *immortality*, with reference to either a cell or an individual, is rather a potentiality. It is defined as the absence or the arrest of senescence, or alternatively, at a population level, as a non-increase in the mortality rate with age. An individual or cell that does not age, or ceases to age at some point in its existence, is said to be *biologically immortal*.

But let's try to understand more precisely what it means to be immortal. Most prokaryotes and many protists do not show any sign of senescence and are therefore rightfully listed among the immortals. However, even here, reproduction has some role in rejuvenation. Consider an amoeba, which reproduces by binary fission. The parent individual ceases to exist when, by dividing, it generates the two daughter cells. This disappearance, which could be considered a mere by-product of our arbitrary definitions of individual and generation, is in fact of great importance. The single amoeba could not live indefinitely without dividing, and therefore as an individual it is not immortal, not even potentially. During the life of an organism, especially if this is long, irreparable damage inevitably accumulates at the molecular level (not only in DNA) for purely accidental reasons. This unceasing degradation is inexorably ruled by the second law of thermodynamics, which would lead to the organism's death anyway. Only reproduction, by diluting the damaged

continues

Box 1.4 (cont)

molecules in the descendants, and by means of purifying selection acting on those descendants, is able to maintain the health of the clone. In a sense, reproduction has a rejuvenating effect even on organisms that do not age!

Multicellular eukaryotes are for the most part 'common mortals', but not all the cells in their body are subject to senescence in the same way. In particular, for species that show early separation between somatic and germline cells, the contrast between the mortal fate of the first and the potential immortality of the latter stands out. Gametes are seen as 'intergenerational lifeboats' capable of saving the genes of a ship (the soma) inevitably destined to sink (Avise 2008). But how do these cells have such different destinies? To some extent, the answer lies in the fact that germline cells are apparently not subject to a form of senescence that afflicts the cells of the somatic line, linked to the DNA replication that precedes cell division. Because in the replication of linear DNA molecules, such as those of the chromosomes of eukaryotes, DNA polymerase cannot complete the 5′ end of the new filaments, repeated replication cycles determine the formation of ever shorter DNA molecules. Cells solve this technical problem by confining this unavoidable erosion to the *telomeres*, sequences of non-coding DNA that are found at the ends of the chromosomes. However, at each cell division, the length of the telomeric DNA segment is reduced. Eventually, when the telomeric sequences have been consumed, the cell is no longer able to divide. In somatic cells, the maximum number of divisions (30–50) is known as the *Hayflick limit*. Shortening of telomeres due to DNA replication also affects germline cells, but here a particular enzyme (*telomerase*) is also expressed, which is capable of elongating the telomeric sequence. This allows germline cells to evade this form of senescence and to replicate virtually without limits. High levels of telomerase expression are also recorded at meiosis. Unfortunately, studies on the relationship between telomere erosion and senescence are still relatively few and limited to a small number of model organisms (Monaghan and Haussmann 2006).

These facts on the whole justify the particular significance of sex in the perpetuation of the life cycle of many eukaryotes in relation to senescence. However, even here, as in the case of clonal reproduction, the role of reproduction as such should not be overlooked. In fact, gametes are generally produced in a number much in excess of what 'survives' in the individuals of the next generation, and even here purifying natural selection plays an important role in counteracting the progressive deterioration of the molecular constitution of organisms, through the elimination of gametes with degenerate traits (Avise 2008).

On the other hand, for most living things, those affected by senescence, it is not enough for reproduction to generate new individuals, in addition to those already present, or to replace those that have died. Reproduction must also ensure that newborns are actually 'young', i.e. that they have, so to speak, 'turned back the clock of senescence', so that the population actually 'rejuvenates' through reproduction. Generating young individuals from old individuals is an imperative for the continuity of life (Turke 2013). This is accomplished in many different ways.

Sexual reproduction has this ability to rejuvenate. Through the cytogenetic processes that lead to the formation of gametes in multicellular organisms, or give a haploid unicellular individual competence to fuse with another, the senescence timer is effectively reset to zero. The life expectancy of a fertilized egg (zygote) is definitely higher than that of the two parents from which the two gametes were produced. Within certain limits, varying from species to species, it is also independent of the age of the parents. In multicellular organisms that reproduce only sexually, it is said that germline cells (those that give rise to gametes; Section 3.4) enjoy some hope of immortality through the successive generations, while the cells of the somatic line (all the other cells of the body) are destined to die with the death of the individual (but see above in this section). However, although in many organisms, for example in many animals, germinal cells irreversibly differentiate during early development, in other organisms, e.g. in plants, there is no clear and early separation between the cells of the two lines (Section 3.4).

This property of sexual reproduction seems to be an attribute also of sex in the broad sense. Ciliates reproduce only asexually, in many species by binary fission, but commonly practise a form of sex called *conjugation* (Figure 1.17, Sections 3.2.2 and 5.2.5). Here, two individuals (*conjugants*) unite temporarily,

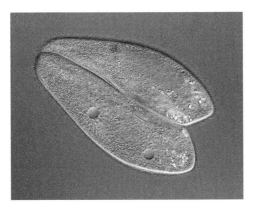

Figure 1.17 Two ciliates (*Paramecium caudatum*) in conjugation. The sexual exchange has the effect of 'rejuvenating' the two conjugants, erasing the effects of clonal senescence.

exchange genetic material, and then separate again. The result of this exchange is a pair of independent individuals (*ex-conjugants*) genetically identical to each other, but genetically different from both conjugants. In most ciliates, the clone that originates from an ex-conjugant after separating from its partner shows a form of senescence (**clonal senescence**), consisting of a limit to the number of cell divisions in the propagation of the clone. This number varies from species to species, but also between strains of the same species. In *Tetrahymena* this limit varies between 40 and 1500 divisions (Finch 1990). Moreover, the clone goes through different maturation stages that in a multicellular organism we would not hesitate to describe as developmental phases. During an initial period of 'sexual immaturity' of the clone (measured in number of divisions since the last conjugation) individuals can only multiply asexually, without being able to conjugate. Then follows a period of 'sexual maturity' during which they will be able to conjugate. Ex-conjugants will emerge from this event genetically modified, but also in some way rejuvenated, with an expected number of cell divisions equal to the maximum possible for the species or strain. The same effect can be obtained through another form of sex, *autogamy*, where a single individual recombines its own genome through meiosis and fusion of the products of the same meiosis, in a sort of self-fertilization. Individuals that do not conjugate, however, may continue to multiply, but will enter a period of gradual senescence, which will gradually slow down the rate of cell divisions and eventually lead to the extinction of the clone. If they conjugate during this phase of senescence, the ex-conjugants will have an expectancy of clonal propagation below the maximum value for the species (Figure 1.18).

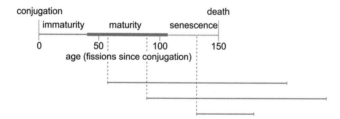

Figure 1.18 Clonal senescence in ciliates. During an initial period of 'sexual immaturity' of the clone (measured as number of cell divisions since the last conjugation) individuals can only multiply asexually, without being able to conjugate. Those individuals that undergo conjugation during the next phase of 'sexual maturity' of the clone (vertical dashed lines) will create new clones with a life expectancy (measured as number of cell divisions) equal to the maximum value for the species (in this example, 150). Those that do not undergo conjugation will enter a phase of senescence in which cell divisions slow down and eventually stop. If they undertake conjugation during this third, late phase of the propagation of the clone, the new resulting clones will have a life expectancy negatively affected by the age of the founding conjugants.

This ability of the sexual processes, whether or not they are associated with reproduction, to bring the process of senescence back to a starting point is often considered a prerogative of the rearrangement of the genetic material that is carried out through meiosis. However, even in some cases of asexual reproduction or in mechanisms of sexual reproduction that do not involve meiosis, a capacity for rejuvenation is observed. Examples are provided by many plants with vegetative reproduction. A branch broken away from an old willow has a good chance of taking root and developing into a new young individual with a life expectancy that does not depend on the age of the parent. There are clones of quaking aspen (*Populus tremuloides*) estimated to be more than 80,000 years old (Section 1.6.2), of the creosote bush (*Larrea tridentata*) of almost 12,000 years, and of bracken (*Pteridium aquilinum*) of almost 1500 years (Gardner and Mangel 1997).

This asymmetry in the effects of senescence between the 'generating' and the 'generated' individual in asexual reproduction can also emerge in the cell division of unicellular organisms. As seen in the previous section, the individual cells of the yeast *Saccharomyces cerevisiae* divide asymmetrically, so that the larger cell is called the mother cell, and the smaller the daughter cell (Figure 1.4b). The mother cell is subject to a form of senescence (*replicative senescence*), since there is a limit to the number of daughter cells (about 50) that it is able to produce before finally ceasing to divide. Until a certain 'age' of the mother cell (number of divisions undergone till then), the daughter cells will not be affected by the mother cell's previous history, and the replicative potential (the number of possible divisions) of a daughter cell, which will soon act as a mother cell, will be the maximum value for the species. However, as the age of the mother cell increases, the daughter cells will start to be progressively more and more affected by their mother's age, detaching from it with a replicative potential already reduced compared to the possible maximum (Figure 1.19; Henderson and Gottschling 2008). A form of senescence associated with asymmetric cell division has also been recently discovered in bacteria, where asymmetry may occur at the morphological level (*Caulobacter crescentus*; Ackermann *et al.* 2003), or even at the biochemical level (*Escherichia coli*; Stewart *et al.* 2005).

In other organisms, however, asexual reproduction seems to produce a form of clonal senescence. This is the case in the natural strains of the ascomycete *Podospora anserina*, some strains of the bread mould *Neurospora*, the oligochaete annelid *Paranais litoralis* and the free-living flatworm *Stenostomum incaudatum*, as well as the clones of several monogonont rotifer species (Gardner and Mangel 1997). A notable case is offered by different species of bamboo. In these long-lived plants, all the individuals of the same

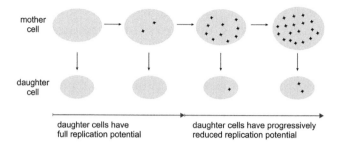

mother
cell

daughter
cell

daughter cells have
full replication potential

daughter cells have progressively
reduced replication potential

Figure 1.19 Asymmetric clonal senescence in the yeast *Saccharomyces cerevisiae*. At the start of the series of divisions, at each division the mother cell accumulates senescence factors (crosses), of which none is found in the daughter cell. Thus the mother cell will have a progressively declining replication potential (number of divisions still possible in the future), while the daughter cells retain full replication potential. However, in the second half of the sequence of divisions of the mother cell, an increasing amount of senescence factors will begin to be distributed to the daughter cells as well. As a consequence, these will emerge from the division from the mother cell with a reduced replication potential.

species bloom, produce seeds and then die in a synchronized way during the same season. There are species with flowering cycles of 30, 60 and even 120 years (Section 4.3). The plants born from the seeds produced in a certain year will be able to live until they flower. However, any new plants subsequently generated asexually from them, by rhizomes or cuttings, will also bloom and die together with those that have grown from seed, as if they were born with the age of their seed-grown progenitor.

The rejuvenating power of reproduction is also revealed by cases of sexual reproduction in the absence of meiosis and with a clonal outcome, as in ameiotic parthenogenesis (Section 5.2.3.3). Such forms of reproduction could not be perpetuated through generations, were it not for some form of rejuvenation, at least every few generations, as in the bdelloid rotifers (Gardner and Mangel 1997). These tiny invertebrates have been reproducing by ameiotic parthenogenesis for more than 60 million years, although a recent genomic study in *Adineta vaga* showed that they exchange DNA horizontally both within and between species (Debortoli *et al.* 2016). In Europe, some clones of parthenogenetic weevils (members of the beetle family Curculionidae) have persisted since at least the retreat of the Pleistocene glaciers (Figure 1.20; Hughes 1989). The parthenogenetic generations in the heterogonic life cycles (Section 2.3) of cladocerans and aphids (Figures 7.19 and 7.20) do not show any signs of clonal senescence. In all these cases of clonal reproduction, individuals certainly have a limited life span, but reproduction generates young offspring, as in the case of sexual reproduction.

Figure 1.20 Parthenogenetic populations of the weevil *Otiorhynchus scaber* have existed since at least the end of the Pleistocene.

1.6 Difficult Boundaries

Delimiting the biological process we call reproduction with respect to other biological processes is not as easy as may first appear. An exploration of these 'difficult boundaries' – or, if you prefer, of these 'grey areas' – is not merely an academic exercise. Close observation of the limitations of our descriptions of natural phenomena, in particular biological ones, serves to develop an awareness that each descriptive system, with its associated terminology, reflects, more or less explicitly, the arbitrariness of the classification of the natural phenomena under investigation. Each descriptive system can exhibit advantages and disadvantages, and be adequate to address some questions while at the same time being inadequate to provide answers to others. In some cases it could even prove to be an obstacle to the investigation itself, concealing the substantial unity of phenomena that for some reason are classified under separate labels. An in-depth discussion of these difficult boundaries would require anticipating in this chapter issues that are based on phenomena and concepts that we discuss in the rest of the book. We have already mentioned in Section 1.3.3 the critical boundary between reproduction and development. Here we outline the problem with respect to three other key concepts, referring to the following chapters for their more in-depth illustration through examples.

1.6.1 Reproduction vs. Transformation

Reproduction is a process of renewal of living matter. However, this cyclical 'restart' can take place through modes that fade into processes that, at first sight, seem to us completely different from reproduction, like the transformations an individual undergoes in the course of its life. The developmental process alone can involve very profound transformations of an individual's body constitution, at all levels of its organization, from the molecular to the morphological and functional. Are there transformations that should count as a passage from one generation to the next?

In accepting asexual reproduction among the possible forms of reproduction, we have established that reproduction can be obtained without introducing genetic novelty, thus considering the demographic aspect of the process as a condition sufficient to qualify it as a reproductive phenomenon. Should we equally consider genetic innovation to be sufficient for reproduction, such that it would be possible to have reproduction without the addition of new individuals? And if so, what transformations should count as a passage from one generation to the next? Once again, the crux of the question lies in the definition of 'individual'.

Few would maintain that the changes in the genetic make-up of a single individual that occur without the contribution of exogenous DNA (e.g. a gene mutation) should count as the production of a new individual. To question the identity of the individual after such a transformation is equivalent to denying the possibility that an individual can remain itself through a change in its qualities, and ultimately means denying the temporal extension of an individual and its historical continuity. However, there are different opinions regarding sexual processes, i.e. those changes in the genetic make-up of an individual that occur because of the recombination of DNA from different sources (Section 1.3.1).

The most widely shared view on this question, and the one adopted here, is that, on their own, sexual processes do not count as reproduction, which is equivalent to saying that there is a category of sexual processes separate from the category of sexual reproduction. The conjugation of bacteria and the conjugation of ciliates are examples of sex without reproduction. Apart from the greater consensus that exists on this point of view, it is also justified from a quantitative perspective. Many sex events, for instance horizontal DNA transfer, may be so inconspicuous as to affect the genome far less than ordinary, and much more frequent, point mutations. Should we say that a bacterium, after having acquired some DNA from the surrounding environment (a phenomenon of bacterial sex known as *transformation*; see Section 5.2.1) has generated a new individual, regardless of how much DNA it has acquired and how or to what extent this has been incorporated into its genome?

These are the problems at the boundary between reproduction and the genetic transformations of an individual. However, the same problem surfaces at the boundary between reproduction and the transformations of the body organization of an organism, which do not touch the genes at all. This is the case in some of the forms of metamorphosis examined in Section 2.2, and in particular at the boundary between metagenesis and metamorphosis discussed in Box 2.3.

1.6.2 Reproduction vs. Growth

According to some authors, genetic identity is the fundamental criterion for defining a biological individual. In this view, any form of clonal propagation, which produces multiple copies of an individual's genotype without modifying it (at least to any great extent), is not seen as reproduction, but rather as the growth of one individual. A lawn of dandelions (*Taraxacum*; Figure 1.21), a plant of the Asteraceae that propagates by apomixis (Section 3.6.2.9), would be seen as a 'large diffuse tree' that has not invested energy and material resources in building a woody trunk, branches and a persistent root system (Janzen 1977). The entire lawn would be a huge, although scattered, 'genetic individual', consisting of many physically separate 'physiological individuals', each having the capacity to grow further and possibly to reproduce. The difference between this and the growth of an oak tree would lie only in the fact that the oak grows by adding modules with reproductive capacity that remain physically connected.

Many plants, many invertebrates and many fungi, as well as most unicellular organisms, are able to propagate without introducing genetic novelty, by generating new individuals from portions of the parent's body or, in the alternative interpretation, by growing new, more or less strongly connected, body modules. This phenomenon is so widespread among plants that botanists have found it useful to introduce two distinct terms to indicate two different kinds of 'plant individual' (Harper and White 1974). A **genet** is a

Figure 1.21 A lawn of dandelions (*Taraxacum*). Because of their reproduction by apomixis, an entire population could be seen as a large 'distributed genetic individual' made up of numerous separate 'physiological individuals' (the individual plants born from as many seeds).

Figure 1.22 This forest of quaking aspen (*Populus tremuloides*) that grows in Utah is a single clone that shares a common root system. Considered as a single individual, it would be the heaviest and oldest living organism known on Earth.

set of genetically identical entities (which can be considered individuals or modules of an individual) derived by clonal multiplication from a single genetically unique individual. All the apple trees of the 'Red Delicious' variety, which are derived by cuttings from a single tree that lived in Iowa (USA) in the late 1800s, are part of a single genet. Similarly, all the polyps of a coral colony form a single genet. In contrast, a **ramet** is an anatomically and physiologically bounded biological entity, independent of its genetic constitution. As such, it may well be a member of a genet. Each 'Red Delicious' apple tree is a ramet; in a similar way, each polyp of a coral colony is a ramet.

An emblematic case of vegetative reproduction (or, from a different perspective, of individual growth) is provided by a 'vegetal entity' known as Pando (or the Trembling Giant; Figure 1.22). What looks like a forest of quaking aspen (*Populus tremuloides*) covering about 45 hectares in Utah, is nothing but a clone of a single (genetic) male individual that would constitute, according to some, a single living organism, which weighs about 6600 tonnes and includes ca. 47,000 trunks that continually decay and are regenerated by a single, gigantic root system. This organism would be about 80,000 years old and would thus be the heaviest and oldest known living organism. However, Pando poses a problem for the *Guinness Book of Records*, stemming precisely from the difficult boundary discussed in this section. Since Pando's root system has probably fragmented over time into a set of contiguous but disconnected subsystems, should it still be considered a single individual? Moreover, the somatic mutations accumulated over such a long period of time make it genetically very heterogeneous with respect to the level of genetic variation generally associated with a clone. Should we still regard as a single individual an entity with a level of genetic heterogeneity that is typical of a population of individuals?

1.6.3 Reproduction vs. Regeneration

Regeneration is a developmental process that allows an individual to replace a lost part of its body. Apparently, there is no danger of mistaking regeneration for reproduction, or vice versa. However, the distinction is not always so clear-cut, particularly for those organisms capable of complete regeneration (*whole-body regeneration*), as seen, among animals, in many cnidarians, annelids and flatworms (Hinman and Cary 2017). In the most common forms of asexual reproduction of multicellular organisms, a part of the parent's body differentiates into what will become the propagule (a mitospore, a bud, a lily bulbil, a piece of a catenulid flatworm), the founder of an individual of the offspring generation that eventually detaches from the parent to further develop and lead an independent life. However, in asexual reproduction by *architomy* (Section 3.1.2.3), at first a small piece of the individual parent, with the tissue organization of that part of the parent's body, detaches, and only after detachment does this piece (re)generate all that is missing to form another complete independent individual. This is what happens for example in the freshwater oligochaete *Lumbriculus* (Figure 1.23). It seems difficult to establish here a clear boundary between asexual reproduction by fragmentation (Sections 3.1.2.2 and 3.1.2.3) and regeneration.

This difficult boundary is not independent of the 'problem of origins' touched on in Section 1.3.3, of which we will see several examples in Chapter 3. In the development of a part of an individual's body that will become the founding cell, or the group of founding cells of a new individual through asexual reproduction, at which point do we place the transition from 'part of the parent's body' to 'new individual offspring'? The problem of regeneration

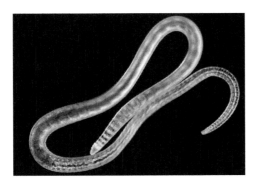

Figure 1.23 The freshwater oligochaete *Lumbriculus* can reproduce asexually by architomy.

adds to this problem a further reason for uncertainty. Should the answer depend on whether those cells are already committed to become another individual? In fact, while the formation of a bud in hydra anticipates its future detachment as an autonomous individual, there is nothing in the development of a portion of the body of certain annelids that qualifies it as a part destined for reproduction, although it may contribute to reproduction following accidental separation. However, on closer examination, it is observed that both regeneration and asexual reproduction depend on the availability of undifferentiated cells (or cells that can return to an undifferentiated state after differentiation) ready to multiply to rebuild a complete body through appropriate morphogenetic processes. There seem to be no factors that limit the activity of these cells to the exclusive service of one or the other process (Sections 3.1.2.2 and 3.1.2.3).

Chapter 2: Reproduction and Life Cycle

In Chapter 1 we defined the life cycle of an organism as the sequence of developmental transformations and reproductive phases that lead from a given stage of development of a given organizational form of that organism, to the same stage of development of the same organizational form in a following generation. Thus modes and times of reproduction contribute in a fundamental way to the characterization of a life cycle. Accordingly, this chapter is not an examination of how reproduction occurs in otherwise defined life cycles, but rather it aims to show how diversity of reproductive processes contributes to diversity of life cycles.

As we will see, life-cycle features vary greatly among living things, sometimes exhibiting a high degree of complexity. Although it may be difficult to produce a rigorous definition of life-cycle complexity, intuitively we might suggest that this is related to the number and scope of the changes the organism undergoes during its cycle. These may be changes in the organism's form, such as between the pluteus larva and the adult sea urchin, or changes in the environment, such as between the pelagic larva and the benthic adult of many aquatic molluscs, or changes in the reproductive mode, when multiple reproductive phases occur in a multigenerational cycle, as in many cnidarians. Furthermore, in some cycles, several reproductive alternatives or developmental options may be available at certain life stages, so that a following stage can be reached through different paths.

In descriptions and classifications of life cycles it is common to distinguish between simple cycles and cycles with 'alternation of generations'. However, in the latter category different forms of life-cycle complexity are generally included, such as changes in the nuclear phases (ploidy level), or the possibility to opt for alternative paths within the same cycle. In order to explore the relationships between reproductive processes and the structure of life cycles, without subverting a well-established nomenclature, we try here to distinguish in a more analytical way the different forms of alternation that can characterize a life cycle, either monogenerational or multigenerational. Most of these categories apply only to eukaryotes.

2.1 Alternation of Nuclear Phases

Traditionally, for eukaryotes that reproduce sexually, even non-exclusively, three main types of life cycle are distinguished, based on the relative predominance of nuclear phases with different numbers of sets of homologous chromosomes (**ploidy level**). These are typically one (**haploid phase**) or two sets (**diploid phase**) (Figure 2.1).

- **Haplontic cycle** (or **haplobiontic cycle**, or **haplo-homophasic cycle**). The organism is haploid for most of its life cycle, except for the zygote phase. Meiosis, which restores the haploid phase, is said to be *initial*, because it immediately follows syngamy, i.e. it occurs at the beginning of the dominant, haploid phase of the cycle. Reproduction is carried out in the haploid phase and, in sexual reproduction, syngamy precedes meiotic recombination.

- **Diplontic cycle** (or **diplobiontic cycle**, or **diplo-homophasic cycle**). The organism is diploid for most of its life cycle, with the exception of the gamete phase. Meiosis, which gives rise to the gametes, is said to be *terminal*, because it immediately precedes gamete production, i.e. it occurs at the end of the dominant, diploid phase of the cycle. Reproduction is carried out in the diploid phase and, in sexual reproduction, meiotic recombination precedes syngamy.

- **Haplodiplontic cycle** (or **haplodiplobiontic cycle**, or **heterophasic cycle**). In this life cycle there are two phases, one haploid, the other diploid, neither of which is transitory. In multicellular organisms both phases are multicellular. Meiosis, which occurs at the transition from the diploid to the haploid phase, is said to be *intermediate*. Reproduction can occur in both phases and the processes of sexual reproduction are distributed between the two phases: meiotic recombination occurs in the diploid phase, with production of spores (*meiospores*), while syngamy involves two gametes produced in the haploid phase. This alternation of nuclear phases necessarily results in a multigenerational cycle, so that a haplodiplontic cycle is in effect a cycle with alternation of generations in register with an alternation of phases.

The label assigned to the life cycle of an organism is commonly extended to the organism itself. Haplontic eukaryotes include many zygomycetes (e.g. *Rhizopus*, the black mould of bread), many green algae (e.g. *Spirogyra*) and some protists (e.g. gregarines and coccidia). All animals, some brown algae (e.g. *Fucus*), most protists and some fungi are diplontic. Most plants, including all embryophytes, are haplodiplontic, as are many brown algae, some fungi and some protists.

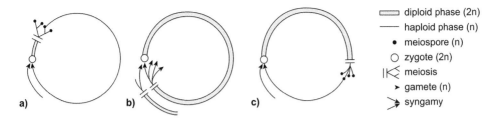

Figure 2.1 Classification of eukaryote life cycles based on the predominance of haploid or diploid nuclear phases. (a) Haplontic cycle: the organism is haploid for most of its life cycle, except for the zygote phase. (b) Diplontic cycle: the organism is diploid for most of its life cycle, except for the gamete phase. (c) Haplodiplontic cycle: haploid and diploid phases are represented by distinct generations. The haploid phase of the haplontic cycle, the diploid phase of the diplontic cycle and both phases of the haplodiplontic cycle can comprise several generations.

This tripartite system of classification applies equally to unicellular and multicellular organisms (Figure 2.2). Unicellular haplonts (Figure 2.2a) are found among green algae (e.g. *Chlamydomonas*), fungi (e.g. the ascomycete *Dipodascus*) and protists (e.g. the haemosporidian *Plasmodium*). Unicellular diplonts (Figure 2.2c) include many protists, e.g. *Amoeba* and diatoms, and some fungi (e.g. *Saccharomyces*). Unicellular haplodiplonts (Figure 2.2e) are found only among forams such as *Mixotheca*. In multicellular eukaryotes, a growth phase (mitotic cell proliferation) characterizes, as part of the development process, the haploid phase of the haplontic cycles (e.g. in the green algae *Ulothrix* and *Chara*, Figure 2.2b), the diploid phase of the diplontic cycles (e.g. the brown algae *Fucus*, Figure 2.2d), or both phases of the haplodiplontic cycles (e.g. the red algae *Rhodochorton*, Figure 2.2f).

All multicellular haplodiplonts are plants, and this is reflected in the nomenclature used for the different generations corresponding to the two nuclear phases. Haploid and diploid generations are therefore commonly referred to as **gametophyte** and **sporophyte**, respectively, to indicate the generation that produces gametes and the generation that, by meiosis, produces spores (for a recent review on the genetic basis of plant alternation of generations see Bowman *et al.* 2016).

In haplontic cycles, the haploid phase may include only one generation, but in general it includes multiple asexual generations. By contrast, in diplontic cycles, the diploid phase can comprise only one generation (e.g. *Homo sapiens*) or several generations (e.g. the scyphozoan *Aurelia*, or the aphid *Myzus*). In haplodiplontic cycles, the generations are always at least two (gametophyte and sporophyte, as we have seen), but either the haploid phase (e.g. in the peat moss *Sphagnum*) or the diploid phase (e.g. in the bamboo *Bambusea*) or both (e.g. in the ferns of the genera *Grammitis* and *Hymenophyllum*; Farrar 1990) can comprise more than one asexual generation.

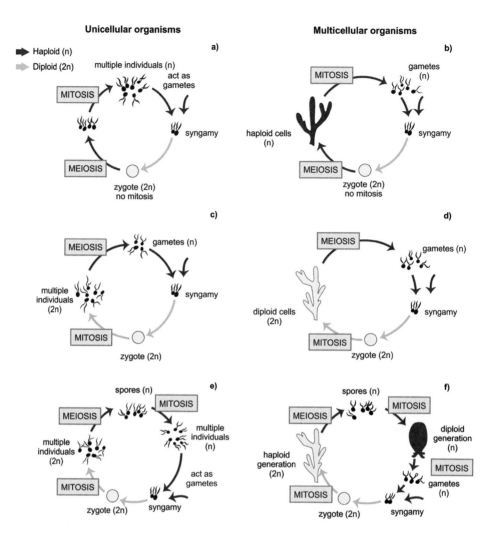

Figure 2.2 Schematic comparison between (a, b) haplontic, (c, d) diplontic and (e, f) haplodiplontic cycles in unicellular (a, c, e) and multicellular (b, d, f) organisms. In this scheme, the haplodiplontic cycles (e, f) are of heteromorphic type.

This classification into haplontic, diplontic and haplodiplontic cycles is essentially based on the position of the transition from the diploid to the haploid phase, which is accomplished through meiosis. Such a schematic interpretation inevitably suffers some limitations. For instance, in the haplontic cycle, when we say that meiosis 'immediately' follows syngamy, this is not to be understood in a strict temporal sense (minutes or seconds), but in the sense that between meiosis and syngamy no other equally significant events occur. In fact, there are many organisms for which the zygote represents a phase of resistance, called the *zygospore*, that allows the organism to get through an adverse season or to overcome a period of insufficient resources; or it may represent a phase of passive dispersal. This is the case, for instance, in the unicellular green alga *Chlamydomonas* (Figure 7.8). Similarly, in the diplontic cycle, when we say that meiosis occurs 'at the end' of the diploid phase, this does not necessarily imply the termination of the diploid phase. In many species gametogenesis lasts for a considerable part of the life of the organism, and the latter can even continue to live after the capacity to produce gametes has ceased, leading to a sterile post-reproductive period, as in the females of our species. Finally, the zygote phase of a life cycle does not necessarily take the form of a single cell with a diploid nucleus, but, for instance, it can consist of a structure that comprises many zygotic nuclei produced by as many fertilization events. Among others, this is the case of the haplontic cycle of many common moulds (zygomycetes; Figure 7.14). When the haploid hyphae of two distinct individuals come into contact, each of them forms a multinucleated *gametangium* with haploid nuclei. The two gametangia undergo plasmogamy (the fusion of their cytoplasms), forming a *zygosporangium* with haploid nuclei deriving from both parental hyphae. This (heterokaryotic) plasmodium develops a thick wall, thus becoming the resistant stage of the organism. Under favourable conditions, karyogamy (the fusion of the nuclei) occurs in the zygosporangium, with the formation of several zygotic nuclei, followed by meiosis, from which the spores of the next generation are produced to be dispersed into the environment. Similarly, in the haplodiplontic cycle of the foraminifera that reproduce through *gamontogamy* (Section 3.2.2), a plasmodium containing many gametic nuclei is formed by the fusion of two *gamonts*. Through multiple fertilization events between the nuclei deriving from the two gamonts, a plasmodium with many zygotic nuclei will form.

The tripartite system of classification of life cycles fits most unicellular and multicellular eukaryotes well, but not all of them. A notable exception is represented by ascomycetes and basidiomycetes. In these fungi, whenever the life cycle includes a transition from a diploid to a haploid phase, there is also a passage through a third nuclear phase. This is the **dikaryotic phase**, which occurs when two cells merge, sharing their cytoplasm (plasmogamy) but retaining their separate nuclei by delaying karyogamy. In most eukaryotes, this nuclear phase is short and transient, but in ascomycetes and basidiomycetes, karyogamy is regularly deferred with respect to plasmogamy (Figures 7.15 and 7.16). Two haploid cells (n) of two distinct hyphae join (plasmogamy), thus founding a new hypha with two haploid nuclei per cell (*dikaryotic phase*, n+n), one nucleus for each of the two types of hyphae that fused. This *heterokaryotic* hypha can grow for a very long time, producing a (dikaryotic) mycelium and developing the fruiting bodies of the fungus. The diploid phase

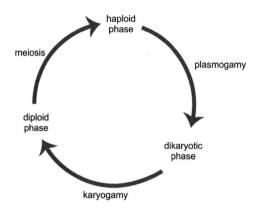

Figure 2.3 General scheme of the succession of nuclear phases in a biological cycle, connected by the most significant cytogenetic events that mark the transitions. In different organisms, different phases can be predominant over the others – for instance, the haploid phase in mosses, the dikaryotic in basidiomycetes, the diploid in vertebrates.

(2n) occurs exclusively in the sporangia, where, after karyogamy, the spores are formed by meiosis (*meiospores*).

One could construct a general life-cycle scheme, to replace the diagrams in Figure 2.1, which explicitly includes all the three phases of a cycle, haploid, dikaryotic and diploid (Figure 2.3). These may have different relative importance, depending on the taxon. However, although this schematization can claim greater generality, it is not universal either, and there will still be cases that can hardly be shoehorned into it, or for which further clarification is necessary. A simple example is offered by polyploid species, or by those species that have populations or individuals with different ploidy levels.

In species that do not reproduce sexually, for which sex is not a constitutive part of the life cycle, the alternation of nuclear phases can take place within the same generation, or even within the same cell. In the haplontic protist *Euglena*, the life cycle coincides with the cell cycle (Box 2.1). The diploid phase is reduced to the G2 growth phase of the interphase that follows DNA replication (phase S), during which the doubling of chromosomes occurs. A transitory dikaryotic phase occurs between the end of the mitotic telophase and the subsequent completion of cytokinesis, which occurs by binary fission and coincides with the completion of the reproductive process.

In obligate parthenogenetic species with ameiotic parthenogenesis, where meiosis is suppressed and the eggs are produced by a cell division basically indistinguishable from mitosis (Section 5.2.3.3), the haploid (gametic) phase is completely suppressed. However, given the progressive process of genetic and structural divergence that affects the homologous chromosomes in clonal

Box 2.1 Cell Cycle

In the proliferation by mitosis of eukaryotic cells, whether whole unicellular individuals or parts of a multicellular individual, a cyclic sequence of phases is typically recognized, which is, however, a generalized description of a process that can actually follow a number of different courses.

The cycle consists of an alternation between a *mitotic phase* (M) of nuclear division, possibly followed by the division of the cell, and an *interphase* (I), which generally extends over most of the duration of the entire cycle. In turn, the interphase can be subdivided into three sub-phases: (i) the *G1 phase* (*first gap phase*), which is a growth phase, followed by (ii) the *S phase* (*DNA synthesis phase*), during which the chromosomes are duplicated, followed in turn by (iii) the *G2 phase* (*second gap phase*), which is once again a growth phase (but with a double nuclear DNA content) and paves the way for the subsequent mitotic phase. Synthesis of RNA and proteins and multiplication of cytoplasmic organelles (including their genetic material; see Box 5.3) occur throughout the interphase, while the synthesis of nuclear DNA is restricted to the S phase.

In fact, many cells are in a phase other than those just listed, called *G0 phase* (*quiescence phase*, although this term is not always appropriate), which represents the condition of a cell that has exited the cell cycle and can no longer divide. The G0 condition can be irreversible or reversible. In the latter case, reintegration into the cell cycle depends on the reception of specific external inductive signals.

All the different phases of the cell cycle are strictly regulated by specific molecules present in the cytoplasm. Their concentration and state of activation depend, in turn, on appropriately transduced chemical and physical signals, either internal or external to the cell. Located at key points in the cell cycle are so-called *checkpoints*, where the cell cycle stops by default until a consensus signal is received, in the form of specific signal molecules in the cytoplasm, which allows the cell to progress from a mitotic phase to the next.

For a more in-depth treatment of the cell cycle and its regulation, refer to cell biology texts (e.g. Alberts *et al.* 2015).

reproduction, it could also be argued that the phase with a higher ploidy level is suppressed. Bdelloid rotifers are considered to be degenerate tetraploids (Gladyshev and Arkhipova 2010), and some species even have an odd number of chromosomes, for instance *Philodina roseola*, with 13.

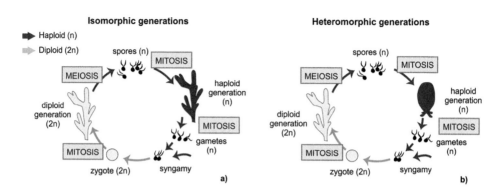

Figure 2.4 Schematic comparison between (a) haplodiplontic cycles with *isomorphic generations,* where sporophyte and gametophyte are morphologically similar, and (b) haplodiplontic cycles with *heteromorphic generations,* where sporophyte and gametophyte are morphologically distinct.

2.1.1 Isomorphic and Heteromorphic Haplodiplontic Cycles

In multicellular haplodiplontic eukaryotes, both the unicellular products of meiosis at the beginning of the haploid phase (the spores) and the unicellular product of syngamy at the beginning of the diploid phase (the zygote) undergo mitotic cell proliferation, as part of the developmental processes characteristic of each of the two generations. Besides the difference in ploidy level between haploid and diploid phases, the generations of the two phases can show different degrees of morphological divergence. **Generations** are said to be **isomorphic** when there is a substantial structural identity between the generations of the two phases, or **heteromorphic** when these are easily distinguishable morphologically (Figure 2.4). Among the chlorophytes, a classic example of an alga with isomorphic generations is *Ulva* (the common sea lettuce, Figure 7.9), whereas *Derbesia* or *Dictyota* have heteromorphic generations. All embryophytes are characterized by markedly heteromorphic generations, with a dominance of the gametophyte in bryophytes (Figure 7.10), in contrast to dominance of the sporophyte in tracheophytes (Figures 7.11–7.13).

2.1.2 Haplodiplontic Cycles with Homospory and Heterospory

In haplodiplonts, classifying a species as *gonochoric* (or *dioecious*, i.e. with *separate sexes*), *hermaphrodite* (or *monoecious*, i.e. with *combined sexes*), or with *indeterminate sex conditions* (i.e. with isogamety) (see Sections 3.2.1 and 3.3.1) is complicated by the presence of haploid (gametophyte) and diploid (sporophyte) generations that do not necessarily reproduce by the same mode. The characteristics of the products of meiosis in the diploid generation (spores) contribute to defining the reproductive modes of a haplodiplontic organism no less than those of the gametes produced by the haploid generation.

The sporophyte may produce a single type of spore (**homospores** or **isospores**), which gives rise to a single type of gametophyte, or two types of spore (**heterospores** or **anisospores**), generally identified as *microspores* and *megaspores*. These give rise to two types of gametophyte, *microgametophyte* and *megagametophyte*, respectively. The former condition is called **homospory** or **isospory** (characterizing a **homosporous/isosporous** organism or life cycle), the latter as **heterospory** or **anisospory** (characterizing a **heterosporous/anisosporous** organism/cycle).

It should be noted that the categories 'male' and 'female', strictly reserved for individuals producing anisogametes and the gametes themselves, is in practice extended to heterospores: microspores are 'male', whereas megaspores are 'female'. When the sporophyte generation has 'separate sexes', i.e. when an individual sporophyte produces either micro- or megaspores, this is in turn referred to as *microsporophyte* or *megasporophyte*. In vascular plants, where the sporophyte generation is dominant, these two terms are replaced by the terms *male sporophyte* and *female sporophyte*, respectively, and a species with separate 'sporophyte sexes' is said to be *dioecious*. Accordingly, a species is called *monoecious* if the sporophyte produces both micro- and megaspores.

By combining the different types of spores and gametes that are produced, three main types of haplodiplontic cycles can be recognized (Figure 2.5):

- **Cycles with homospores and isogametes**. The sporophyte is sexually indeterminate and produces spores of a single type that give rise to a sexually indeterminate gametophyte that produces isogametes (Figure 2.5a). Spores, gametophytes and gametes may however express a specific mating type (Section 6.6). Two examples are the green algae *Cladophora vagabunda* and *Caulerpa*.

- **Cycles with homospores and anisogametes**. The sporophyte is sexually indeterminate and produces spores of a single type, each giving rise to a monoecious gametophyte that produces both male and female gametes (Figure 2.5b). This is the case in most ferns (Figure 7.11). Alternatively, the spores give rise to unisexual gametophytes (male or female) that will produce either male or female gametes exclusively (Figure 2.5c). This is the case in most liverworts.

- **Cycles with heterospores and anisogametes**. The sporophyte is sexually determinate (male, female or monoecious) and produces male and/or female spores that give rise to unisexual gametophytes producing either male or female gametes exclusively. This is the case in most spermatophytes, the sporophyte of which is generally monoecious (Figures 2.5d and 7.12), more rarely dioecious (Figure 2.5e).

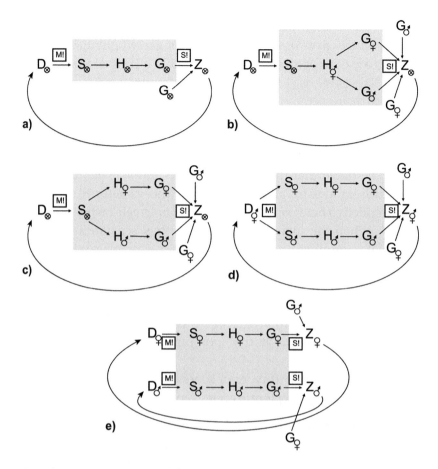

Figure 2.5 Schematic representation of the different combinations of spore and gamete formation processes in haplodiplontic cycles. (a) Cycle with homospores and isogametes: the sexually indeterminate sporophyte produces spores of a single type that give rise to a sexually indeterminate gametophyte that will produce isogametes. (b) Cycle with homospores, anisogametes and monoecious gametophyte: the sexually indeterminate sporophyte produces spores of a single type that develop into a monoecious gametophyte that will produce male and female gametes. (c) Cycle with homospores, anisogametes and unisexual gametophyte: the sexually indeterminate sporophyte produces spores of a single type that can develop into a male gametophyte that will produce male gametes, or a female gametophyte that will produce female gametes. (d) Cycle with heterospores, anisogametes and monoecious sporophyte: the sporophyte produces male and female spores (microspores and megaspores, respectively); these will give rise to male or female gametophytes (microgametophytes and megagametophytes, respectively) that will produce only male or female gametes, respectively. (e) Cycle with heterospores, anisogametes and unisexual sporophyte: the male sporophyte (microsporophyte) produces male spores (microspores) that will give rise to male gametophytes (microgametophytes) that will produce male gametes (sperm); the female sporophyte (megasporophyte) produces female spores (megaspores) that will give rise to female gametophytes (megagametophytes) that will produce female gametes (eggs). H, individual in the haploid phase; D, individual in the diploid phase; S, spore; G, gamete; Z, zygote; M!, meiosis; S!, syngamy; ⊗, sexually indeterminate. Haploid phases (n) on coloured background.

For the chromosomal sex-determination systems in the different types of haplodiplontic cycle see Section 6.1.1.

2.1.3 Haplodiplontic Cycles and Asexual Reproduction

In haplodiplontic organisms, asexual reproduction can alternate with sexual reproduction within the same cycle. Asexual reproduction can occur in the haploid phase (e.g. in mosses, *Sphagnum*), in the diploid phase (e.g. in most angiosperms) or in both phases (e.g. in some ferns, especially among the Hymenophyllaceae where, besides the sporophyte, the gametophyte can also reproduce by propagules, although less frequently). In addition to the haploid spores produced by the sporophyte by meiosis (*meiospores*), many gameto-phytes and sporophytes of the chlorophytes can produce spores by mitosis (*mitospores*), haploid and diploid, respectively, which therefore represent a form of asexual reproduction.

In these cases, asexual reproduction gives rise to multiple generations that not only share the same nuclear phase, but also present the same organizational form of the organism that reproduces. In the following sections we will see cases in which different organizational forms of the same organism (all diploid) and different types of reproduction are found within the same diplontic cycle.

2.2 Alternation of Sexual and Asexual Generations: Metagenetic Cycles

A **metagenetic cycle** is a multigenerational cycle in which exclusively asexual generations alternate with sexual generations, represented by distinct organizational forms (Figure 2.6). The metagenetic cycles are labelled as cycles with alternation of generations, and certainly they are multigenerational, but not in register with alternation of the nuclear phases. In a metagenetic cycle the individuals of different generations have the same ploidy level. Meta-genetic cycles are distinguished from other multigenerational cycles by the fact that there is at least one obligate asexual generation, morphologically and physiologically distinguishable from the sexual generations with which it alternates. Metagenetic cycles are found in many metazoans, including dicye-mids, cnidarians, digeneans, cestodes, polychaetes, naidid oligochaetes, cyclio-phorans and tunicates.

A classic example of a metagenetic cycle is seen in many species of cnidarian. The zygote develops into a larva which, after a pelagic phase, attaches to a substrate where it will metamorphose into a polyp. After a phase of growth, the polyp can reproduce asexually, generating a number of medusae by

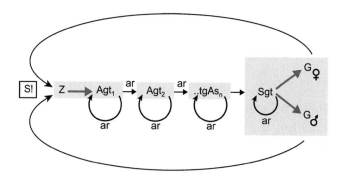

Figure 2.6 Schematic representation of a metagenetic life cycle. Thin black arrows represent reproductive phenomena: asexual reproduction (ar) and syngamy (S!). Thick grey arrows represent developmental phenomena, such as the development of the zygote within the first generation of the cycle and the production of gametes (G) in the last generation of the cycle. Developmental phenomena within each generation are not represented. Coloured rectangles contain several generations ascribable to the same type of generation, either asexual (Agt) or sexual (Sgt). Actual metagenetic cycles may deviate from this pattern in many respects.

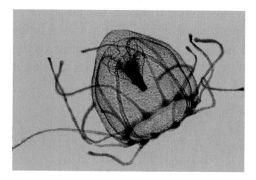

Figure 2.7 Medusa of *Turritopsis dohrnii*. This hydrozoan is able to reverse the direction of its life cycle, returning from medusa to polyp (Box 2.2).

budding or strobilation (Section 3.1.2.3), depending on the species. In turn, the medusa will grow and, when maturity is reached, it will reproduce sexually, producing gametes that through fertilization give rise to the zygotes of the next cycle (Figure 7.17). This case clearly shows that the alternation of generations is not necessarily accompanied by alternation of nuclear phases. In fact, both the polyp and the medusa, alternating within the diplontic cycle of many cnidarians, are diploid. In this group, however, there are many variations on this scheme (among which the one of the hydrozoan *Turritopsis* stands out; Figure 2.7, Box 2.2), some of which will be discussed later, in the context of the 'difficult boundaries' of metagenesis (see Box 2.3).

Among the chordates, asexual reproduction is obligate in pelagic tunicates such as salps and *Doliolum*, which regularly alternate sexual and asexual

Box 2.2 The 'Immortal Jellyfish'

In many hydrozoans, the alternation of generations between polyp and medusa coincides with an alternation between colonial and solitary organization (Section 2.5). The planktonic larva, called a *planula*, attaches to a substrate where it metamorphoses into a polyp that will start to reproduce asexually, producing a colonial aggregate of polyps. From this aggregate, medusae are generated asexually, and these constitute the next, solitary, phase of the cycle. Medusae lead a pelagic life until sexual maturity. From their fertilized eggs will develop the planulae of the next cycle.

Surprisingly, the hydrozoan *Turritopsis dohrnii* (previously identified as *T. nutricula*) is able to reverse the direction of its life cycle, returning to a phase with polyp organization (first a stolon, then a small colony of polyps), unable to reproduce sexually, after having reached sexual maturity, and possibly having reproduced, as a solitary medusoid individual (Piraino *et al.* 1996). In the laboratory, this phenomenon is observed in every *Turritopsis* medusa, but it has never been observed in nature, probably because the transition is so rapid.

This cycle inversion is obtained by altering the state of differentiation of some cells (cell transdifferentiation), a phenomenon generally associated with the regeneration of an organism's damaged or lost parts. Following a morphogenetic transformation that can proceed according to alternative developmental paths, the medusa attaches to a substrate, spawns its gametes (the last act of its life as a sexually mature individual), and produces stolons. Two days after the appearance of the first stolon, this starts producing the first polyps. Polyps will feed on zooplankton and will soon be able to generate new medusae.

Following this discovery, *T. dohrnii* became known as the 'immortal jellyfish'. This hydrozoan seems to have the ability to escape senescence (Section 1.5) – but, without diminishing the exceptional nature of the phenomenon, a more in-depth analysis, which necessarily involves taking a stance on the problem of what an individual is (Section 1.4), shows that this is not in fact a case of development without ageing.

The medusa regains polyp organization through a developmental process, but it was itself formed through a reproductive event. Moreover, the new polyp will not develop into a medusa, but it will generate several. In the life cycle of this species, it is not possible to come and go between polyp and medusa simply through development. The medusa and the polyp are two stages of development of the same individual in the transition from medusa to polyp, but not in the transition from polyp to medusa. In other words,

continues

Box 2.2 (cont)

reversion does not simply concern the developmental process, but actually involves the whole life cycle, including reproduction.

Let's describe the life cycle of *Turritopsis* from the perspective of a genetic concept of biological individual (Section 1.6.2). Since the genet reproduces only sexually, in cnidarians there would be no alternating generations, because the growth and proliferation of polyps as well as the production of medusae are aspects of development of the single genet. In this case we could actually say that in *T. dohrnii* the genet first ages, throughout its development into a medusa, and then rejuvenates, by returning to the stage (here, yes, it is a stage) of polyp. However, looked at in this way the life cycle of *T. dohrnii* is not so exceptional, resembling what is observed in any genet when asexual reproduction is interpreted as growth. The quaking aspen clone known as Pando (Section 1.6.2) has continued to age (in the development of each 'tree', or ramet) and to rejuvenate (every time a new tree is produced) for tens of thousands of years.

---

generations. These present morphologically distinct individuals that are called *blastozooids* and *oozooids*, respectively. For instance, in salps the zygote develops into a solitary oozooid (asexual generation), which lacks gonads but can produce a stolon by budding. These stolons give rise to chains of blastozooids (sexual generation), which possess gonads but cannot produce stolons (Nielsen 2012).

However, it is among the digeneans that the most complex metagenetic cycles are found. In these parasitic flatworms there is no such thing as a 'typical' life cycle, and, among metazoans, the diversity in their cycles rivals that of cnidarians. In most cases, a hermaphrodite adult (gonochoric in *Schistosoma*), called the *marita* stage, lives as a vertebrate parasite, and produces eggs that are fertilized and released into the environment, from which a free-living aquatic larva develops. This tiny larva, the *miracidium*, can infect a mollusc. In the body of the latter, the miracidium develops into a *mother sporocyst*. This reproduces asexually (or, according to some other interpretations, by parthenogenesis), giving rise to a second generation of sporocysts (*daughter sporocysts*), which, again by asexual reproduction, produce a first generation of *rediae*, the form that will mature into the adult. But even the rediae can reproduce asexually (or, according to some other interpretations, by parthenogenesis – hence the collective term *parthenita* for these generations), giving rise to new generations of rediae. Sooner or later, the redia produces a new type of larva, the *cercaria*, which sometimes turns into the non-mobile *metacercaria* before maturing into a *marita*, finally closing the cycle. If the reproduction of sporocysts and rediae is actually by parthenogenesis, rather than being asexual, this cycle should be described as heterogonic (Section 2.3) rather than metagenetic. But it is possible that uniparental reproduction follows different cytogenetic modes in different species, or even between different reproductive events within the same cycle. Thus a further 'difficult boundary' emerges, between metagenesis and heterogony, which adds to the even more difficult boundary between metagenesis and metamorphosis, anticipated in Section 1.6.1 and discussed in Box 2.3.

Box 2.3 Metagenesis vs. Metamorphosis

In the typical life cycle of cnidarians (a multigenerational cycle), the transition from polyp to medusa is interpreted as a change of generation, or *metagenesis*, which therefore includes a reproductive event, while in the typical life cycle of echinoderms (a monogenerational cycle) the passage from larva to adult (or juvenile) is described as a *metamorphosis*, i.e. as the transformation of an individual, without the interposition of a reproductive event. But is this distinction always so clear? What fixes the divide between metagenesis and metamorphosis? Returning to the theme of the 'difficult boundaries' introduced in Section 1.6, we examine here the value of some of the discriminating criteria that have been proposed.

A first criterion is rooted in the demographic meaning of reproduction: if there is reproduction, there must be an increase in the number of individuals. In scyphozoans and hydrozoans the detachment of one or more medusae from the parent polyp leads to an increase in the number of individuals. But what if the polyp disappears in giving life to a single medusa? In cubozoans, for instance, the polyp does not give rise to medusae by strobilation or budding, i.e. through a process that preserves the polyp as a parent (of one generation), distinct from the medusae (its offspring, belonging to the next generation). Instead, the polyp 'transforms' directly into a medusa. This seems to be a developmental process, and therefore a metamorphosis, rather than a reproductive event. Should we call the cubozoan polyp a larva and claim that the cycle of these cnidarians is monogenerational?

If the demographic criterion does not clearly separate metagenesis from metamorphosis in cnidarians, we could perhaps adopt a more restrictive criterion, considering that in asexual reproduction by budding the parent survives after the detachment of its descendants, while nothing survives

---

continues

Box 2.3 (cont)

metamorphosis except for the metamorphosed organism. But in the case of the cubozoans, no polyp survives the formation of the medusa, and in the hydrozoan *Eirene hexanemalis* the polyp is even planktonic (like a sea urchin larva) and produces by budding a single medusa that completely reabsorbs what remains of the polyp (Bouillon *et al.* 2006). This cycle can be clearly described as monogenerational, exactly as in sea urchins. In this case, the difference between the hydrozoan polyp and the sea urchin pluteus larva is no longer a difference between an adult and a larva, but between larvae with early (pluteus) or late (polyp) specification of the mass of cells that will give rise to the adult. On the opposite side, in the starfish *Luidia sarsi* the larva can continue to swim for three months after the juvenile that originated from it has detached (Williamson 2006). Should we say that the larva of this echinoderm reproduces asexually, and that its metagenetic life cycle includes two generations, like a cnidarian's?

Possibly, another way of clearly separating metagenesis from metamorphosis could be to enter into the details of the dynamics of the respective processes, arguing that in metagenesis reproduction occurs through buds that are only a part of the individual parent, while metamorphosis is a transformation of an entire individual. However, in the metamorphosis of many forms of marine invertebrates, most of the larval body is discarded or consumed and the young derives from a small number of founding cells, called *set-aside cells*. Indeed, the fish-parasitic larva of the freshwater bivalve *Mutela bourguignati* produces a true bud from which a juvenile develops (Fryer 1961). Are there two generations in the cycle of this bivalve?

Whenever evolution has preserved both polyp and medusa, cnidarian cycles are invariably described in terms of metagenesis, regardless of how the polyp-to-medusa transition occurs. On the contrary, mollusc and echinoderm life cycles that involve a larval phase are invariably described in terms of metamorphosis, regardless of how the larva-to-adult transition occurs. In many cases, a distinction between reproduction and metamorphosis is reduced to a lexical question, or to a question of taxon-specific tradition (Minelli 2009).

Finally, an observation that shows how the way we classify things can condition our perception of organisms' life cycles. Polyembryony, the generation of more than one embryo from a single fertilization event (Section 3.1.2.4), is usually considered a form of sexual reproduction with a partially clonal outcome (Avise 2008), since it generates multiple identical copies of the same genotype, even if this is different from that of the parents. However, polyembryony could instead be considered a form of asexual reproduction at a very early (embryonic) stage of development. From this perspective, the species that reproduce by polyembryony exhibit an alternation of sexual and asexual generations. Accordingly, we should also list among the species with a metagenetic cycle some species of flatworms, wasps and armadillos, where polyembryony is constitutive.

2.3 Alternation of Amphigonic and Parthenogenetic Generations: Heterogonic Cycles

In some eukaryotes, amphigonic reproduction alternates regularly with parthenogenesis (Section 3.6.2). These multigenerational cycles are called **heterogonic cycles** (Figure 2.8). Heterogonic cycles are found in some species of parasitic nematodes and, more famously, in most monogonont rotifers, cladocerans and aphids. In all these animals, the transition from parthenogenesis to amphigonic reproduction is regulated by the interpretation of specific cues from the environment, such as seasonal reduction in day length or an increase in population density. However, there are significant differences in how the environmental signal is received and subsequently transduced into the physiological response of the organism. These differences are in part related to the specific mechanisms of sex determination in these groups (Section 6.3).

In monogonont rotifers (Figures 2.9 and 7.18), the males are haploid, while the females are diploid and can produce two types of egg. During the

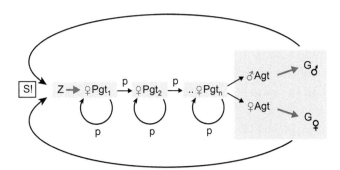

Figure 2.8 Schematic representation of a heterogonic life cycle. Thin black arrows represent reproductive phenomena: parthenogenesis (p) and syngamy (S!). Thick grey arrows represent developmental phenomena, such as the development of the zygote in the first generation and the production of gametes (G) within the last generation of the cycle. Developmental phenomena within each generation are not represented. Coloured rectangles contain several generations ascribable to the same kind of generation, either parthenogenetic (Pgt) or amphigonic (Agt). Actual heterogonic cycles may deviate from this pattern in many respects.

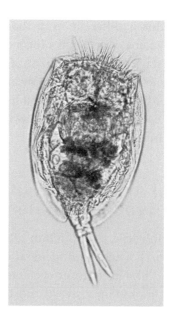

Figure 2.9 Monogonont rotifers (here, *Euchlanis*) have multigenerational life cycles with alternation between amphigony and parthenogenesis (heterogonic cycles).

favourable season, **amictic females** produce large diploid eggs by parthenogenesis, which have not undergone meiotic reduction (**amictic eggs**) and do not need to be fertilized. These eggs develop into females within 12–48 hours, and are therefore often called *subitaneous eggs*. Several parthenogenetic generations follow during the favourable season, until, typically in autumn, females develop with a slightly modified ovary, compared to that of amictic females. These **mictic females** produce haploid eggs, regularly obtained by meiosis, called **mictic eggs**. These, if fertilized, produce a thick shell and become resistant eggs (*resting eggs*, although, actually, the development is suspended at the embryonic stage with a few dozen cells; Boschetti *et al.* 2011), which will develop into mature (diploid) females the following season. If mictic eggs are not fertilized, in the same season they develop into (haploid) males, which can mate with their 'aunts'. Parthenogenesis allows rapid growth of the population, while the production of mictic eggs enables the rotifers to survive through the adverse season or shorter unfavourable periods. Generally, the production of mictic females by amictic mothers is induced by a specific chemical signal that diffuses in the water, produced by the amictic females in response to specific variations in relevant environmental parameters. Beyond a certain concentration threshold, this molecule stimulates the production in an amictic female of an internal chemical signal that is transmitted to the forming oocytes. A fraction of these will develop by parthenogenesis into mictic females after deposition (Snell *et al.* 2006). In some species there seems to be instead a sort of internal clock, and males appear after a certain number of parthenogenetic generations (Ricci 2001).

In cladocerans (Figure 7.19), males and females are diploid and the determination of sex is environmental. While females are regularly present in the aquatic environments in which they live, males appear only at certain times of the year, different according to species and latitude. Also in this group, reproduction by parthenogenesis alternates cyclically with amphigony, when males are present. Parthenogenetic females produce subitaneous eggs, which develop into females within 3–4 days. However, in response to specific environmental signals, unfertilized eggs can develop into males (Section 6.3). When males are present, the same parthenogenetic females can produce haploid eggs that, if fertilized, become resting eggs. These, protected singly or in pairs by special cases produced by the mother (*ephippia*; singular, *ephippium*) and/or by the exuvia released by the mother at the moult during which she lays the eggs, are able to face adverse environmental conditions. Since the production of eggs capable of parthenogenetic development is based on the non-completion of the second meiotic division (Section 5.2.3.3), the same female can produce both types of egg and, after amphigonic reproduction, resume reproducing by parthenogenesis. Some species perform one heterogonic cycle per year (*monocyclic species*), while others go through several cycles per year (*polycyclic species*).

Among the insects there are numerous species, belonging to different groups, that alternate more or less regularly between amphigonic and parthenogenetic reproduction (e.g. many species of cynipid hymenopterans and cecidomyiid dipterans). However, it is among the aphids, which have an X0 chromosomal sex-determination system (Section 6.1.1.2), that a notable diversity in heterogonic cycles has evolved (Figure 7.20). The cycle generally lasts one year (*holocycle*), but in some species it may extend over several years (*paracycle*), or the amphigonic generation may become sporadic or even disappear (*anholocycle*), as in some species living in warm climates. The cycles of aphids are further complicated by the alternation of winged dispersal forms and wingless sedentary forms, and by the possibility of developing on one or more host plants, or on one or more parts of the same plant. Schematically, in the annual cycle the following sequence is repeated: (i) a generation of **founders**, females that develop from the resting eggs that have passed through the winter season; these females are parthenogenetic, *virginoparous* (i.e. they generate daughters that are, in their turn, parthenogenetic) and generally wingless; (ii) one, but more often several generations of females, which take different names depending on whether they develop on the same host plant as the founders or on a different secondary host, also parthenogenetic and virginoparous, wingless or winged; (iii) a generation of **sexuparous** females (i.e. that generate amphigonic offspring), also parthenogenetic; depending on the species, these can be *amphiparous*, generating individuals of both sexes, or either *gynoparous* or *androparous*, generating individuals of only one sex, females or males respectively; and (iv) an **amphigonic generation**, generally winged. Whether the environmental signal that leads to the development of a sexupara is received and processed by the developing sexupara itself or by her virginoparous mother is still a controversial issue (Bickel *et al.* 2013). In most aphids the parthenogenetic generations are viviparous and only the amphigonic female is oviparous, but in some species all generations are oviparous.

Still among the insects, another example of a heterogonic cycle is provided by many species of gall wasps (cynipids), which have a haplodiploid sex-determination system (Section 6.1.3). These insects have two generations per year. Spring females are parthenogenetic, some androparous, others gynoparous. Their offspring constitute the summer generation of amphigonic males and females. From their fertilized eggs will hatch the parthenogenetic females of the following year (cycle). The individuals of the two generations may differ in morphology, in the shape of the galls they produce, and in the localization of the galls on the plant (Heming 2003). In *Biorhiza pallida*, males are winged, while females of the amphigonic generation have vestigial wings, and those of the parthenogenetic generation are completely wingless.

2.4 Alternation of Gonochoric and Hermaphrodite Generations: Heterogenic Cycles

Some nematodes show a particular alternation of gonochoric and hermaphrodite generations, sometimes referred to as a **heterogenic cycle** (Figure 2.10).

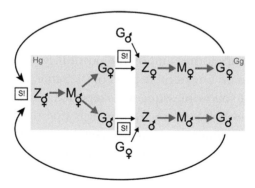

Figure 2.10 Schematic representation of a heterogenic cycle. Thin black arrows represent reproductive phenomena, thick grey arrows, developmental phenomena. Coloured rectangles contain the two different generations, hermaphrodite (Hg) and gonochoric (Gg). G, gamete; M, reproductively mature individual; Z, zygote; S!, syngamy.

Figure 2.11 Nematodes of the genus *Heterorhabditis* have multigenerational life cycles with alternation of gonochoric and hermaphrodite generations (heterogenic cycles).

In insect-parasitic nematodes of the genus *Heterorhabditis* (Figure 2.11), the infectious juveniles of the third stage are carriers of a bacterium in the lumen of the anterior tract of the intestine. When a suitable host insect is found, a young individual worm can enter it through the mouth, anus, spiracles, or directly through the cuticle. More individuals may enter the same host. A few hours after entering the host, the nematode moves into the haemocoel of the latter, where it releases the bacteria it carries. These multiply rapidly, killing the host in 1–2 days. In what remains of the host, the young nematode feeds on bacteria and develops into an adult about three days after the insect's death. This first generation of adults is hermaphrodite, and after mating they lay eggs in the insect's body cavity. The juveniles hatching from these eggs develop into the gonochoric adults of the second generation. The mating of the latter produces the eggs from which the infectious young of the next generation will develop. These emerge from what remains of the first host and disperse into the environment in search of a new host (Nguyen and Smart 1990).

2.5 Alternation of Solitary and Colonial Generations

Some multigenerational cycles are characterized by one or more phases of aggregation among the individuals that are generated. These are often labelled as **cycles with alternation of solitary and colonial generations**. Examples are found among sponges, hydrozoans, anthozoans, bryozoans, tunicates (e.g. *Botryllus*, Figure 2.12) and green algae (e.g. *Volvox*).

In Section 1.4 we explained the difficulties that colonial organisms pose for an unambiguous delineation of reproductive processes. The problem arises from the difficulty of finding an unequivocal criterion to establish what in a colony should count as an individual, or, in other words, what to place at the

Figure 2.12 A colony of *Botryllus schlosseri*. The 'petals' of the flower-shaped colony are mature zooids. This tunicate has a multigenerational life cycle with alternation of solitary and colonial generations.

centre of the reproductive event: the single member of the colony or the colony as a whole? The answer, setting aside personal preference, depends on what we intend to describe or investigate in a particular organism (e.g. ecology or evolution), as well as on the type of colony, with special regard to its level of integration and the degree of independence and the division of roles, including reproductive roles, among the members of the colony. Without adhering dogmatically to either of the two extreme viewpoints, we instead analyse the effects that these alternatives have on the description and interpretation of life cycles.

Let's start by interpreting a colony as an association of individuals, through the description of a generic life cycle that applies to many marine invertebrates. The planktonic larva of a founding individual finds an appropriate place to settle, metamorphoses and begins a sedentary adult phase. At a certain point it begins to reproduce asexually, typically by budding, generating a progeny of individuals called *zooids*, which in turn will continue to reproduce asexually in the same way. However, the zooids remain in anatomical connection with each other, producing a colonial aggregate. The primary zooid (the founding individual of the colony) may be morphologically similar to the secondary zooids that originate from it (as in the red coral, *Corallium rubrum*), but it may also differ considerably. In cnidarians, the primary polyp is often much larger than the secondary polyps, as in the case of the anthozoan *Pennatula*, where in the feather-shaped colony (from which the Latin name derives) the primary polyp constitutes the 'shaft' from which multiple series of secondary polyps branch off to form the 'vanes'.

After a phase of maturation, some or all members of the colony become competent to reproduce sexually. Depending on the species, there are colonies with only male or female zoids (e.g. *Corallium rubrum*), with zoids of both sexes (simultaneously, e.g. *Cladopsammia rolandi*, or at different times, e.g. *Stylophora pistillata*), and colonies with hermaphrodite zoids, as in most bryozoans. The fertile zoids produce gametes that fertilize to form zygotes that will go through a mobile phase of solitary development: the new generation of planktonic larvae. In this interpretation there is a real alternation between the single solitary generation of the individual founder and the colonial generations of the zoids that descend from it. However, unlike the metagenetic cycles of many cnidarians (hydroids), where the alternation between medusa and polyp coincides with an alternation of solitary and colonial phases, here the transition from solitary to colonial does not correspond with a change of reproductive mode, and often not even with the occurrence of deep morphological differences between the two phases. In these cycles, both the solitary founder and the many generations of secondary zoids of the colony can reproduce asexually, while some zoids switch to sexual reproduction, in a way that, depending on the species, can be more or less exclusive and/or reversible.

On the other hand, if we interpret the colony as a single individual in the generalized life cycle just described, the proliferation of zoids has to be seen as a growth phase of the individual-colony through the multiplication of its parts, rather than as a reproductive phase. The colony is thus understood as a modular organism, as a tree can be. And like a tree, which grows by adding multicellular modules (branches) and has multiple sexual organs (flowers), the colony grows by increasing the number of its modules (zoids) and reproduces through distributed sexual organs (fertile zoids). The organism-colony can also reproduce asexually, by fragmentation or detachment of propagules made of groups of zoids (as in the bryozoan *Discoporella*; Ryland 2005), but there is no alternation of generations.

2.6 Alternation of Unicellular and Multicellular Generations

When the individual-colony question is applied to aggregations of single-celled individuals, the problem translates into the question of whether to consider such an aggregation as a colony of unicellular organisms or as a true multicellular organism. As with the question discussed in the previous section, opinions differ, based on the level of integration and the division of roles between the cells of a multicellular aggregate. In some cases it is simply a matter of tradition: we see a single organism in the body of a sponge, despite the great autonomy and the modest differentiation of the cells that compose it, while, at the same time, an integrated aggregate of ciliates

Figure 2.13 Myxogastrids, or plasmodial slime moulds (here, *Fuligo septica*), have multigenerational life cycles with alternation of unicellular and multicellular generations. The photo shows the 'multicellular' phase, more correctly called *plasmodial*, because it is organized as a single multinucleate cell.

(e.g. *Zoothamnium*) is viewed as a colony, despite the tight connection between the cells that compose it, and also their differentiation.

The transition from unicellular to multicellular organization is obviously a subject of great interest in evolutionary studies. Bonner (2000) identifies at least 13 independent evolutionary transitions from unicellular to multicellular organization, distributed among eubacteria (e.g. myxobacteria), archaea (e.g. *Methanosarcina*) and eukaryotes (metazoans, fungi, social amoebae, embryophytes and brown algae, plus some groups within ciliates, foraminifera, green algae, red algae and diatoms). Let's briefly describe here two groups of organisms whose life cycle, with alternating unicellular and multicellular generations, places their organization on the border between colonial unicellularity and true multicellularity. These are two groups of mycetozoans, or slime moulds.

In myxogastrids, or plasmodial slime moulds, from the zygote, through mitosis not followed by cytokinesis, develops a plasmodial mass with many diploid nuclei that can reach a diameter of up to 1 m or more (Figures 2.13 and 7.5). This represents the *trophic phase* of the cycle, where the organism expands and moves across the ground, incorporating food particles by phagocytosis. Approaching the exhaustion of trophic resources, *sporangia* containing the nuclei from which haploid spores are produced by meiosis emerge from the mass. The spores, protected by a thick wall, are a phase of resistance: dispersed in the environment, on finding favourable conditions they break out of their wall, take an amoeboid or flagellate shape and undergo syngamy, forming an amoeboid zygote from which a new plasmodium will emerge.

In dictyosteliids, or cellular slime moulds (also known as social amoebae), there is an *asexual cycle* where the *trophic phase* is represented by haploid solitary amoeboid cells that feed on soil bacteria and reproduce asexually by binary fission (also called *vegetative cycle*; Figure 7.4). In *Dictyostelium discoideum* (a model species in developmental biology; Li and Purugganan 2011), when food is scarce, many cells aggregate, reciprocally exchanging chemical signals, to form an approximately discoidal mass and thus enter the *aggregation phase* (also called *social cycle*). The disc turns into a migrating mass (or *slug*, as it is roughly slug-shaped), which after a short time spent wandering stops and turns into a pedunculated fruiting body where some of the cells – those at the top of the structure – become resistant spores that are released and dispersed into the environment. Subsequently, under favourable conditions, the spore wall breaks apart, allowing the amoeboid cells to return to the trophic phase and to asexual reproduction. In this asexual cycle, reproduction occurs during the unicellular phase, since during the multicellular phase there is no mitosis, and some (lucky) individual cells simply become spores and may have descendants. There is therefore no multicellular generation in the strict sense, since the cycle is limited to an alternation between the trophic (dispersed) and the aggregation phases. However, free amoebae can also switch to a *sexual cycle*. Here two amoebae undergo syngamy and form a zygote that grows by attracting and engulfing solitary individuals of the same species (cannibalism). This giant zygote acquires a wall and becomes resistant. Within the wall meiosis occurs, followed by several mitotic cycles, so that from a zygote derive several amoeboid haploid cells ready to resume the trophic phase.

2.7 Alternation of Generations by Seasonal Polyphenism

The life cycles of organisms can be described with reference to several different traits of their biology. Among the life cycles of different organisms we find solitary and gregarious phases, sedentary and dispersal phases, trophic and non-trophic phases, benthic and pelagic phases, endogeic and epigeic phases, and more. In many cases, these different phases correspond to different segments of the life of the same individual, sometimes (but not always) corresponding to the same number of developmental stages, and have no effect on reproductive modes. However, in other cases these phases correspond to distinct generations within a multigenerational life cycle (for the difficulties in distinguishing transformations of the same individual from reproduction, see Section 1.6.1).

The ability of an individual to develop different phenotypes in response to specific environmental stimuli is called **phenotypic plasticity**. When these alternative phenotypic states are discrete, rather than distributed along a spectrum of continuous variation, this is referred to as **polyphenism**

Figure 2.14 Seasonal polyphenism: cyclomorphosis in *Daphnia retrocurva*. From spring (left) to late summer (right), there are several parthenogenetic generations of this cladoceran, with different phenotypes.

(Fusco and Minelli 2010). Environmental signals that evoke these responses are of various kinds, from the presence of a predator to interactions with conspecific individuals, but when these signals depend on seasonal variation in some environmental parameter, polyphenism can take the form of a regular alternation of generations, known as **seasonal polyphenism**.

For instance, in many freshwater rotifers, in some cladocerans and also in some algae, a progressive change in morphological traits is observed through the sequence of generations produced during the favourable season, so that it is possible to distinguish spring individuals from those of summer and autumn. This phenomenon is called **cyclomorphosis**. In some rotifers, for instance, throughout summer the lorica becomes more sculpted and rich in spines, while in the cladoceran *Daphnia* there is a progressive change in head size and shape (Figure 2.14).

In the life cycle of many species of butterflies from the equatorial regions there are two generations that reach the adult stage respectively in the dry season (autumn–winter) and in the wet season (spring–summer) (Brakefield and Zwaan 2011). The 'dry' and 'wet' generations of the African species of the genus *Bicyclus* differ greatly in the pattern of stripes and eyespots on the ventral surface of the wings, the one that is visible while the insect is at rest. The dry-season form, relatively inactive, has a darker colour, with a less marked wing pattern, which results in better cryptic camouflage against the background of dry leaves. By contrast, the more active wet-season form adopts a type of warning mimicry, capable of intimidating or disorienting potential predators by displaying conspicuous eyespots along the external edge of the wing. Among the European butterflies, striking seasonal polyphenism is observed in *Araschnia levana* (Figure 2.15).

In the cases mentioned above, reproduction occurs in a very similar way in the different generations of the life cycle, but this is not necessarily so. For instance, in the case of the aphids discussed above (Section 2.3), phenotypic plasticity, which here takes the form of an alternation of wingless and winged phenotypes, is closely associated with the heterogonic cycle.

Figure 2.15 Seasonal polyphenism: spring form (left) and summer form (right) of the European nymphalid butterfly *Araschnia levana*.

2.8 Cycles with Reproductive Options

The life cycle can be complex in different respects. As we have seen, a large number of morphologically distinct generations, separated by different reproductive phases, can follow one another during one cycle. In some species, however, a further contribution to the variety and complexity of cycles is provided by the possibility, at certain stages of the cycle, of taking one of two or more alternative options for reproduction or development. The 'choice' generally depends on the contingent state of the organism and/or the occurrence of specific environmental conditions. This is a form of phenotypic plasticity, which could be called *life-cycle plasticity*, through which developmental processes and/or the mode of reproduction can first diverge and then converge again in a subsequent stage, which can thus be reached through alternative paths within the same cycle. Many examples are found among parasites, nematodes and digeneans especially. Minelli and Fusco (2010) have hypothesized that the multigenerational cycles of many metazoans with alternating generations may have evolved from primitive cycles that presented optional paths, originally subject to control by environmental factors.

With regards to possible developmental options occurring within the same generation, we only mention the fact that in many species of nematodes, polychaetes, opisthobranch gastropods, insects and amphibians, alternative forms of larvae or juveniles eventually develop into adults of the same type. This phenomenon is known as **poecilogony**. However, for the subject of this book, reproductive rather than developmental options are more relevant.

Some of the cycles we have described fall into this category, for example that of *Dictyostelium* (Section 2.6), but in most of the cases attributable to this category a certain reproductive mode is qualified as *facultative* or *cyclical*, as opposed to *obligate* (or *constitutive*) (Ram and Hadany 2016). Parthenogenesis

Figure 2.16 Life cycles with reproductive options. The Komodo dragon (*Varanus komodoensis*) can facultatively reproduce by parthenogenesis.

is facultative in many molluscs, annelids and arthropods, and also in some vertebrates, among which is the Komodo dragon (*Varanus komodoensis*) (Avise 2008; Figure 2.16). Self-pollination is facultative in a number of plants, including various members of the legume, orchid and aster families; self-fertilization is facultative in several simultaneous hermaphrodite animals, among which are some pulmonate gastropods (e.g. the terrestrial snails of the genus *Rumina*; Prévot *et al.* 2013). Similarly, asexual reproduction is facultative in many organisms that usually reproduce sexually. In some animals, reproduction can be carried out either by the adult or by a juvenile stage (*paedogenesis*; see Sections 2.9 and 3.6.2.5), as in the larvae or pupae of some insects, among which is the beetle *Micromalthus debilis* (Heming 2003).

The 'choice' to take one of several possible reproductive options is generally influenced by specific environmental factors, which vary from species to species. In particular, species that exploit ephemeral food sources alternate different forms of reproduction in response to the variable availability of nutritional resources. For instance, many species of cecidomyiid dipterans reproduce as larvae by parthenogenesis (*paedogenesis*) when food is abundant, or as adults, by amphigony or parthenogenesis, when food is scarce. In *Heteropeza pygmaea* (Figure 2.17), when food availability is suboptimal, male and female larvae develop into adults, which reach new food sources by flying. There they mate and females lay eggs, either fertilized or amictic. From these eggs, owing to the abundance of nourishment provided by the newly colonized mushroom, viviparous gynoparous females develop and reproduce by parthenogenesis at the larval stage, generating other females. The larvae of the new generation develop in the body cavity (haemocoel) of the mother, where they complete embryonic development before the mother dies, at the

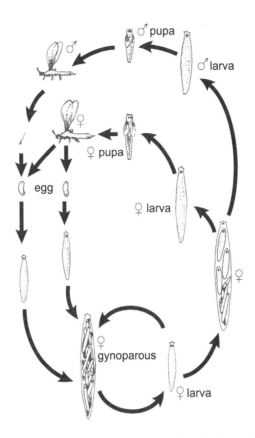

Figure 2.17 Life cycles with reproductive options. When food is abundant, the cecidomyiid midge *Heteropeza pygmaea* reproduces as a larva by parthenogenesis, and when food is scarce it reproduces as an adult by amphigony or parthenogenesis.

end of her last larval stage (semipupa phase). In turn, these larvae can be gynoparous, androparous or amphiparous (that is, generate only female or male offspring, or offspring of both sexes, respectively). Therefore, they either produce other female larvae that continue the paedogenetic cycle, or, if food resources become scarce, they generate male and female larvae that will develop into adults (Heming 2003; see also Section 3.6.2.5).

2.9 Distribution of Reproductive Phases Within One Generation

Individuals of many species, such as humans, reproduce repeatedly throughout their lives, while others, such as certain agaves (*Agave*, Figure 2.18) and all Pacific salmon species (part of the genus *Oncorhynchus*) reproduce sexually only once and then die. As for the number of times an individual reproduces, we distinguish two main modes of **parity**: in **iteroparous** species (in plants, most of

Figure 2.18 Many agaves, like *Agave americana*, bloom only once, then die. These plants, even if they can live more than 10 years, are therefore monocarpic (= fruiting only once), as are the annual plants.

the *perennial* species; see Section 2.10) the same individual experiences multiple reproductive seasons throughout its life, whereas in **semelparous** species (in plants, especially among the *annual* and *biennial* species; see Section 2.10) an individual has only one reproductive season in its life. Some specialists adopt a more restrictive definition of semelparity, to refer to the situation when an individual dies after its first and only reproductive episode. Among animals, generally the two definitions do not conflict with regard to females, but for males, things are different (Bonnet 2011). For instance, male mammals are considered semelparous if they die after only one breeding season, regardless of the number of partners and copulations they had, whereas male spiders are considered semelparous only if they die after a single copula (generally killed by the female, with extreme cases where sperm transfer requires the perforation of the male by the female's chelicera; Andrade 1996), but not if they die at the end of the only breeding season after having mated several times. In spiders of the genus *Tidarren*, the male self-amputates one of the two copulatory appendages (modified pedipalps) before using the other to inseminate the female. In *T. argo*, from Yemen, after mating, the females pull out this remaining pedipalp, so that the male can mate only once (*traumatic semelparity*, Knoflach and van Harten 2001). Similarly, in some land slugs of the genera *Limax*, *Ariolimax* and *Deroceras*, *apophallation* was often observed, namely the amputation of the penis – by the owner or by its partner – at the end of mating.

It is often impossible to define a species as semelparous or iteroparous, since males and females of the same species can exhibit different reproductive modes.

That being said, semelparous animals include the aforementioned salmon, eels, ticks, some spiders, solifuges, the scorpion *Bothriurus bonariensis*, some polychaetes and most of the coleoid cephalopods (i.e. cuttlefish, squid and octopus). Many miniaturized marine invertebrates that produce just a few large eggs and perform some kind of parental care are also semelparous (Chaparro *et al.* 2011). In mammals, semelparity is found in the tiny Australian marsupials of the genus *Antechinus* (with the exception of *A. swainsonii*) and in another marsupial, the South American didelphid *Monodelphis dimidiata* (Chemisquy 2015). In semelparous animals, adult life can be very short. In insects, where a moult separates the adult from the last juvenile stage, it is possible to quantify this duration precisely. Adult life is particularly short in ephemeropterans. The shortest-lived mayfly seems to be the female of *Dolania americana*, which completes its adult life within five minutes (Sweeney and Vannote 1982).

Other variations on the theme are the phenomena of **facultative iteroparity** and **facultative semelparity** (Hughes 2017). Many usually semelparous species can facultatively reproduce one or more times after the initial bout of reproduction. For example, mothers of the semelparous crab spider *Misumena vatia* (Thomsidae) typically lay and provision a single brood of eggs; however, in response to high food availability and/or unusually warm environmental conditions, they are capable of laying and caring for a second brood if sperm supplies stored in their body are not depleted. Other examples of facultative iteroparity are seen in the marsupial mouse *Antechinus stuartii*, the leech *Erpobdella octoculata*, the cephalopods *Loligo vulgaris*, *Sepia officinalis* and *Nautilus* spp., and the salmonid fish *Oncorhynchus mykiss* and *O. tshawytscha*. Conversely, examples of facultative semelparity are the clupeid fish *Alosa sapidissima*, the three-spined stickleback (*Gasterosteus aculeatus*), the lizard *Uta stansburiana* and the chaparral yucca (*Hesperoyucca whipplei*).

Besides annual and biennial plants, semelparous plant species, which in this case are called **monocarpic** (literally, which bear fruit only once), are also found among the perennial plants, some of which can live up to a century or more. These bloom only once and then die. Among these, in addition to the previously mentioned agaves and many species of bamboo (Section 4.3), there are some yuccas, some palms and the legume tree *Tachigali versicolor*, from Costa Rica and Colombia, which for this reason is known as the 'suicide tree'. In addition to the number of breeding seasons that a single individual can enjoy, living things differ in the age at which they reach reproductive maturity (Section 4.3). The time needed to reach this condition is extremely variable, also in relation to the fact that the life span of an individual ranges from a few minutes for a bacterium to over 5000 years for some specimens of the North American pine *Pinus longaeva* (Table 2.1). The shortest-lived vertebrate is the turquoise killifish (*Nothobranchius furzeri*), with a post-hatching life span of about three months in some populations (Cellerino *et al.* 2016); the longest-lived is the Greenland shark (*Somniosus microcephalus*), estimated to live for up to over 400 years (Nielsen *et al.* 2016).

In many organisms it is useful to distinguish **mature stages**, which are characterized by effective *reproductive maturity*, from **adult stages**, which, instead, are characterized by a definitive morphological condition, or *adult type* (Minelli and Fusco 2013). Mature stages and adult stages do not always coincide. Among arthropods, for instance, the adult condition precedes the mature condition in many species of myriapods that develop by *hemianamorphosis*, such as pill millipedes and lithobiomorph centipedes. These animals hatch with a smaller number of trunk segments than are found in the adult. During the early stages of post-embryonic development, at each moult they increase the number of trunk segments until the adult condition is reached. However, some further moults, during which no new segments are added, are usually needed before the animal becomes mature. Adult condition and reproductive maturity are achieved simultaneously in most holometabolous insects, where both conditions are reached with the moult that accompanies metamorphosis. Finally, reproductive maturity may precede the achievement of adult morphology. Ahead-of-time reproduction in a juvenile stage is called **paedogenesis** (Section 3.6.2.5). In extreme cases of paedogenesis, an individual can even begin to reproduce when it is still in the body of its mother. In some aphids, within a parthenogenetic female one can find her developing daughters, and within these their own developing daughters (granddaughters of the former). Like a Russian doll, these 'nested' generations are called *telescoped generations* (Figure 2.19). Telescoped generations are also found in some mites, in the parasitic flatworms of the genus *Gyrodactylus* (although here the phenomenon is associated with polyembryony, rather than with paedogenesis, see Section 3.1.2.4), among the stenolaematous bryozoans and in the colonial green alga *Volvox*.

Similarly, in the post-embryonic development of most plants an **adult vegetative phase** and an **adult reproductive phase** can be distinguished (Poethig 2003). Only the latter corresponds to sexual maturity in animals. The transition from the juvenile phase to the vegetative adult phase is often characterized by changes in vegetative structures such as leaf shape, phyllotaxis, or rooting capacity, while the advent of the adult reproductive phase is marked by the development of reproductive organs – in angiosperms, the flowers.

Because of the disparity of living beings, it is difficult to use rigid categories to classify organisms with respect to the mode and time at which reproductive maturity is achieved. Males and females of the same species can become mature at different ages. In many animals, however, there is a positive correlation between sexual dimorphism and sexual maturity, with the sex of larger size acquiring competence to produce gametes later than the smaller sex

Table 2.1 Approximate maximum age reached by some animals and plants (main sources Flindt 2003 and Fransson *et al.* 2017, integrated with some examples of exceptionally long-lived species)

	Years	Months	Days
PORIFERA			
Monorhaphis chuni	11,000		
CNIDARIA			
Sea anemone (*Cereus* sp.)	65		
Black coral (*Leiopathes* sp.)	4265		
SYNDERMATA			
Rotifers: many species		1	
RHABDITOPHORA			
Planarians: many species		14	
Tapeworms: many species	35		
MOLLUSCA			
Octopus (*Octopus vulgaris*)	2		
Cuttlefish (*Sepia officinalis*)	5		
Oyster (*Ostrea edulis*)	12		
Snail (*Helix pomatia*)	18		
Giant clam (*Tridacna gigas*)	100		
Ocean quahog clam (*Arctica islandica*)	500		
NEMATODA			
Caenorhabditis elegans			20
Large roundworm (*Ascaris* sp.)	5		
ARTHROPODA			
Fruit fly (*Drosophila melanogaster*)			45
Housefly (*Musca domestica*)			75
Bed bug (*Cimex lectularius*)		6	

continues

Table 2.1 (cont)	Years	Months	Days
Brown centipede (*Lithobius* sp.)	6		
Crayfish (*Austropotamobius pallipes*)	15		
American lobster (*Homarus americanus*)	100		
ECHINODERMATA			
Sea urchins: many species	7		
CHONDRICHTHYES			
Whale shark (*Rhincodon typus*)	70		
Greenland shark (*Somniosus microcephalus*)	400		
OSTEICHTHYES			
Salmon (*Salmo salar*)	13		
Herring (*Clupea harengus*)	20		
Goldfish (*Carassius auratus*)	40		
Eel (*Anguilla anguilla*)	88		
Sturgeon (*Acipenser* sp.)	150		
AMPHIBIA			
Tree frog (*Hyla* sp.)	22		
Common toad (*Bufo bufo*)	40		
Fire salamander (*Salamandra salamandra*)	43		
REPTILIA			
Wall lizard (*Podarcis muralis*)	8		
Green anaconda (*Eunectes murinus*)	30		
American alligator (*Alligator mississipiensis*)	65		
Tuatara (*Sphenodon punctatus*)	100		
European freshwater turtle (*Emys orbicularis*)	120		
Galápagos tortoise (*Chelonoidis* sp.)	150		

Table 2.1 (cont)	Years	Months	Days
AVES			
Eurasian wren (*Troglodytes troglodytes*)	6		
Hummingbirds: many species	8		
Great tit (*Parus major*)	15		
Barn owl (*Tyto alba*)	17		
Common swift (*Apus apus*)	21		
House sparrow (*Passer domesticus*)	23		
Rook (*Corvus frugilegus*)	23		
Canary (*Serinus canarius*)	24		
Penguins: many species	26		
Grey heron (*Ardea cinerea*)	37		
White stork (*Ciconia ciconia*)	39		
Griffon vulture (*Gyps fulvus*)	41		
European herring gull (*Larus argentatus*)	44		
Ostrich (*Struthio camelus*)	62		
Laysan albatross (*Phoebastria immutabilis*)	68		
MAMMALIA			
House mouse (*Mus musculus*)	4		
Squirrel (*Sciurus* sp.)	12		
Red fox (*Vulpes vulpes*)	14		
Bottlenose dolphin (*Tursiops truncatus*)	45		
Lion (*Panthera leo*)	30		
Giraffe (*Giraffa camelopardalis*)	34		
Cat (*Felis catus*)	35		
Brown bear (*Ursus arctos*)	47		

continues

Table 2.1 (cont)	Years	Months	Days
Gorilla (*Gorilla gorilla*)	60		
African elephant (*Loxodonta africana*)	80		
Right whale (*Balaena mysticetus*)	200		
GYMNOSPERMAE			
Larch (*Larix decidua*)	600		
Intermountain bristlecone pine (*Pinus longaeva*)	5060		
ANGIOSPERMAE			
Cyclamen (*Cyclamen* sp.)	80		
Hazelnut (*Corylus avellana*)	120		
Grape (*Vitis vinifera*)	130		
Aspen tree (*Populus tremula*)	150		
Dogrose (*Rosa canina*)	400		
Ivy (*Hedera helix*)	440		
Plane (*Platanus* sp.)	1300		
Sacred fig (*Ficus religiosa*)	2300		

(Stamps and Krishnan 1997). Examples are found among marine invertebrates, insects, spiders, fishes and mammals.

If an organism reproduces by different modes, for instance asexual and sexual, reproductive maturity can occur at different times for each mode. This happens, for instance, in many angiosperms, where vegetative propagation can occur long before the plant starts developing flowers. This scenario is further complicated by those organisms that change their reproductive mode or role during life, such as sequential hermaphrodite species, which change sex in the course of their lives (Section 6.2.2).

In addition, in some cases, reproductive maturity is not acquired once and for all with the achievement of the adult condition, but occurs instead in intermittent phases. In some animals there are two or more reproductive periods separated not just by a long temporal interval but also by a profound

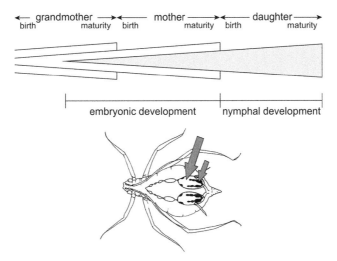

Figure 2.19 In aphids, paedogenesis gives rise to telescoped generations. In the ovaries of a parthenogenetic female are the embryos of her daughters (large arrow), and within the ovaries of the latter (small arrow), the embryos of the daughters of her daughters.

reorganization of the animal's body structures. Some ctenophores and polychaetes have a first breeding season when they are still in the larval stage and a second one as adults. This phenomenon is called **dissogony** (or **dissogeny**). In contrast, the males of some millipedes may have two or more reproductive adult stages, intercalated with stages where the copulatory structures regress and the animal cannot reproduce. This phenomenon is called **periodomorphosis** (Drago *et al.* 2011).

In fact, sexual maturity itself can be variably defined. Should it coincide with the maturation of the gonads, with the availability of mature gametes, or with the moment when an individual is ready to reproduce from the morphological, physiological and behavioural point of view? In the spider *Pholcus phalangioides*, the sperm cells are mature a couple of weeks before the moult to adult (Michalik and Uhl 2005). Similarly, among pterygote insects, in ephemeropterans, plecopterans and lepidopterans, sperm and eggs are ready before the metamorphosis to the adult stage (Minelli *et al.* 2006). On the other hand, in many dipterans and beetles maturation of gametes is dependent on the availability of energy resources provided by the adult female's first meals (Minelli *et al.* 2006). It seems, then, that the transition from *immature* to *mature* fully qualifies as another 'difficult boundary', to add to those we have already seen (Section 1.6) and those we will meet in the following chapters.

As a final note, Hughes (2017) reviewed ecological and molecular evidence for the continuity and plasticity of modes of parity and concluded that annual-semelparous and perennial-iteroparous life histories are better understood as endpoints along a continuum of possible strategies, which differ in the degree to which they disperse or concentrate reproductive effort in time.

2.10 Generation Times

In population biology and in demography, generation time is the average (or in some cases, the minimum) time interval between two consecutive generations. This time interval varies from minutes for a bacterium (even as few as 12 minutes), to tens of years for large animals (20–30 years in our species) and plants (30–40 years in the beech, *Fagus sylvatica*).

Generation time has effects on the number of generations (or the fraction of a generation) that can occur over a calendar year. Small- and medium-sized animals usually reach sexual maturity early (Section 4.3) and tend to have one generation per year, and they are therefore said to be **univoltine** (or **monovoltine**). The same is true of many herbaceous plants (see below). This reproductive rhythm is particularly widespread among the species that inhabit regions with marked seasonality. However, there are numerous animals with particularly short generation times, which have more than one generation per year, and they are said to be **multivoltine** (or **polyvoltine**). In the amphigonic generations of insects, the shortest generation time is as low as a week (in the mosquito *Psorophora confinnis*; Gunstream and Chew 1967), while some parthenogenetic generations are even shorter, with a minimum of 4.7 days observed in the aphid *Rhopalosiphum prunifolia* at 25 °C (Noda 1960). In contrast, terrestrial plants able to complete several generations per year in natural conditions are rare. Among these there are the European alien populations of *Galinsoga quadriradiata*, a member of the aster family native to South America, which can have 2–3 generations per year (Reinhard *et al.* 2003; Figure 2.20).

Figure 2.20 *Galinsoga quadriradiata* is one of the few land plants with more than one generation per year, in this case up to three.

More commonly, depending on the length of the life cycle, we distinguish between *annual* plants, which complete the entire cycle within a year, like many herbaceous plants; *biennial* plants, which in the first year develop an abundant photosynthetic apparatus, but bloom the following year, generally using the resources accumulated during the first year in bulbs, tubers or rhizomes; and *perennials*, which have longer life cycles. Annual and biennial plants have generation times of one and two years, respectively. In perennial plants the generation time depends on the age at which the plant reaches sexual maturity, which can vary from one year to several tens of years (Section 4.3).

Chapter 3: The Natural History of Reproduction

This chapter illustrates the ways in which new individuals are generated. As we will see, these modes are very diverse and depend on the structure of the organism, prokaryote or eukaryote, unicellular or multicellular, with modular or non-modular organization, plant or animal. The divide between asexual and sexual reproduction is discussed in Section 1.2, and the cytogenetic aspects of the different reproductive modes are discussed in Chapter 5. Unlike in other chapters, many topics are necessarily treated here with reference to individual groups of organisms.

3.1 Asexual Reproduction

The name of this reproductive mode derives from a delineation of phenomena in negative terms, signalled by the privative 'a' that precedes 'sexual': this is indeed non-sexual reproduction, i.e. reproduction that does not involve production of gametes, recombination and syngamy. Asexual reproduction, also called **agamic** or **vegetative**, generates individuals that are genetically identical to the parent, except for new mutations (Section 5.1). In the case of unicellular organisms, new mutations can occur during the DNA replication process that precedes the division of a cell into daughter cells. In multicellular organisms, different mutations may accumulate in the different cell lines forming the soma of the parent individual. If these mutations affect the cell lines involved in the production of buds or other types of propagules, they can be transmitted to the next generation.

In botanical and agronomic literature, a distinction is sometimes introduced between *agamic reproduction* in the strict sense and *vegetative multiplication*, reserving the first term to those cases in which reproduction is carried out through mechanisms that involve the differentiation of reproductive cells and sometimes also the production of fruiting bodies. The term *vegetative multiplication* is preferred instead when the process takes place by means of non-differentiated somatic cells or by detachment of portions of the parent organism. Here, to ensure uniformity of terminology, we have chosen to consider asexual, agamic and vegetative reproduction as synonyms, using further specifications (e.g. 'by fragmentation') when required.

Granted that the definition of asexual reproduction is based on aspects of transmission genetics, from the point of view of the mechanism through which it takes place, asexual reproduction can be seen as the process by which a portion of an individual organism is transformed into a new independent one: only one parent is involved and no special sex cells are recognized. Therefore, asexual reproduction is a form of *uniparental reproduction* (reproduction that involves one parent only), but not all modes of uniparental reproduction are asexual (see Section 3.6). For instance, self-fertilization, or those forms of parthenogenesis that involve recombination during the differentiation of the female gamete (which will not be fertilized), are forms of sexual uniparental reproduction which include some form of genetic remixing.

Asexual reproduction can occur through very different mechanisms, depending on the organism's structure and biology, in particular on its unicellular or multicellular condition and the phase of development or life cycle in which reproduction takes place.

3.1.1 Asexual Reproduction in Unicellular Organisms

Asexual reproduction in unicellular organisms in effect coincides with cell division.

3.1.1.1 CELL DIVISION IN PROKARYOTES

In prokaryotes, reproduction is always disjunct from sex (Section 5.2.1) and occurs by fission or (in the eubacteria, but not in the archaea) by sporulation. In **fission**, either **binary** or **multiple**, the dividing cell generates metabolically active daughter cells (vegetative cells), whereas in **sporulation** the division gives rise to one or more quiescent cells (spores) capable of resisting adverse environmental conditions for up to thousands of years. When the spores are produced inside the mother cell, the event is called *endosporulation* (Figure 3.1), whereas we speak of *exosporulation* when the spores are generated on the surface of the mother cell. According to some microbiologists (e.g. Pommerville 2011), when the production of a single spore is accompanied by the destruction of the mother cell, sporulation should not be interpreted as a reproductive process but rather as a form of differentiation. In *Bacillus subtilis*, for example, sporulation begins with an asymmetric division that leads to the formation of a small *prespore* and a large surviving mother cell. The prespore is subsequently internalized by the mother cell and matures as a spore, while the mother cell dies. On the other hand, a case of sporulation in which a mother cell produces two or more spores is unquestionably a

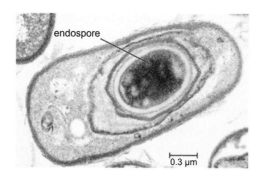

Figure 3.1 Endosporulation in the bacterium *Bacillus anthracis.*

reproductive process. This is for example the case in many Firmicutes bacteria, where there is regular production of multiple endospores within a parent cell.

Fission and sporulation are often found in the same bacterial species, and the switch from one process to the other is induced by changing environmental conditions. When these are favourable, the bacterium undergoes fission; when they become adverse, it turns instead to sporulation. In fact, these two reproductive modes can be much more similar than they might at first seem. A variant of the mechanism that usually leads to the formation of spores is known, in which the typical protective coating that ensures resistance is not produced; as a consequence, these cells function as normal active cells, rather than as spores. For instance, endocellular division of the bacteria of the genus *Epulopiscium*, which live in the intestine of some species of surgeonfishes, produces from 2 to 12 daughter cells per mother cell; in contrast to what normally happens in the sporulation processes, these cells are not dormant endospores, but active cells.

Simultaneous production of multiple daughter cells is known in cyanobacteria, proteobacteria and actinobacteria: this generates either active cells (multiple fission) or spores (multiple sporulation). The process involves a rapid succession of division cycles within a large cell, such as one of the specialized reproductive cells (*hormogones* or *baeocytes*) of the Pleurocapsales (a group of cyanobacteria). Some genera of these prokaryotes have also retained the ordinary binary division, which leads in this case to the formation of small clusters of cells, or small filaments, preceding the production of baeocytes, but in some groups of unicellular Pleurocapsales there is no binary fission at all.

The archaea also reproduce asexually by binary or multiple, symmetric or asymmetric division (budding), like the eubacteria, but in this group sporulation is unknown. The archaea show greater diversity than the eubacteria in the mechanisms of cell division. Following the discovery in the hyperthermophile archaean *Sulfolobus acidocaldarius* of a cell-division machinery based on proteins similar to those involved in the same function in eukaryotes (Lindås *et al.*

2008), a comparative genomic analysis showed that among the archaea there are at least three different systems that can form the ring structure (*constriction complex*) responsible for cell division: (i) a system similar to that of the eubacteria, where cell division is mediated by FtsZ filamentous proteins, (ii) a system based on Cdv proteins, homologous to the eukaryotic ESCRT-III proteins, and (iii) a system based on an actin-related protein peculiar to the archaea (Makarova *et al.* 2010).

3.1.1.2 CELL DIVISION IN UNICELLULAR EUKARYOTES

In unicellular eukaryotes, asexual reproduction consists of the division of the nucleus (**karyokinesis**) by **mitosis**, followed by the splitting of the cytoplasm (**cytokinesis**).

Binary fission is the most common reproductive mechanism among unicellular eukaryotes. At a certain stage of the cell cycle, for example when the cell has reached a certain size, mitosis occurs and the cell divides. Mitochondria and plastids, as well as flagella or cilia, can be replicated before or after the separation of the daughter cells. There may be a long interval between karyokinesis and cytokinesis, however, so in the meantime the cell remains binucleate.

If karyokinesis is repeated several times without interposed cytokinesis, there is no cell multiplication and therefore no reproduction of the unicellular organism, and a *multinucleate plasmodium* develops instead. However, the multinucleate condition may be transitory, a prelude to a multiple division into a large number of daughter cells, each of which normally inherits only one of the nuclei deriving from the previous nuclear divisions. This is **multiple fission**, in unicellular eukaryotes also called **schizogony**. It is frequent in the apicomplexans (Figure 3.2), in parasitic amoebae such as *Entamoeba histolytica*, in heliozoans and in radiolarians.

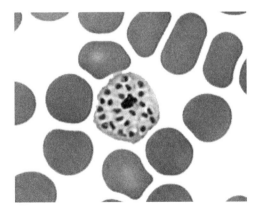

Figure 3.2 The apicomplexan parasite *Plasmodium vivax* dividing by erythrocytic schizogony (schizont within a human erythrocyte).

Cell division can be either *equal* (*isotomic*) or *unequal* (*anisotomic*), depending on whether the daughter cells are approximately the same size or not.

In unicellular eukaryotes, **budding**, as found in some ciliates and in some yeasts, is an extreme case of unequal cell division, either binary or multiple. Buds are generally formed on the surface of the parent cell (*exogenous budding*), but some species, for example of suctorian ciliates, reproduce through a form of *endogenous budding*. This specialized mechanism involves not only the formation of buds within the parent cell, but also a prolonged period of development within the shelter of a fold of the cell membrane functioning as a *brood pouch*. A particular form of endogenous budding, called *endodiogeny*, is found in the infectious stages of various species of coccidia and provides for the differentiation of two or more daughter cells within a mother cell whose cytoplasm is progressively and completely incorporated into the developing daughter cells. In ciliates, the term *budding* is reserved for those cases (e.g. the suctorian *Ephelota gemmipara*) in which the mother cell remains sessile, releasing one or more mobile daughter cells. These will undergo a kind of metamorphosis, turning into the adult sessile form.

Other mechanisms of cell division occurring in different lineages of unicellular eukaryotes are described in Section 5.1.5.

3.1.2 Asexual Reproduction in Multicellular Organisms

In multicellular eukaryotes, asexual reproduction occurs when a new individual originates from one or more somatic cells of the parent. The ways in which generation and detachment of new individuals from the parental body are prepared and carried out can be traced back to a limited number of models, which however do not represent rigid categories. These models refer to different characteristics of the reproductive process – for example, which part of the parent's body is involved in reproduction, or the stage of the parent's life cycle in which it occurs. Thus, the systems described below are not mutually exclusive categories, and different reproductive modes may have elements in common, e.g. the number of cells from which the new individual originates.

On the basis of the latter criterion, two main categories are distinguished:

- production of **unicellular propagules**, more or less strongly differentiated compared to the ordinary somatic cells of the parent

- production of **multicellular propagules** with their own characteristics

One of the features that separates sexual from asexual reproduction is the possible absence, in the latter, of the bottleneck represented by a single-cell stage (the meiospore in haplontic organisms, the zygote in diplonts, or both in haplodiplonts) that almost universally occurs following sexual exchange, marking the beginning of the development of an individual. Asexual reproduction in multicellular organisms can be **monocytogenous**, i.e. based on a unicellular propagule, a single parental cell acting as the founder cell of the new individual, or **polycytogenous**, i.e. based on a multicellular propagule. According to many authors (e.g. Grosberg and Strathmann 1998; Wolpert 2007), the absence of a single-cell phase between generations could represent a problem for the evolutionary maintenance of a multicellular organization. The reason is that by iterating this form of reproduction, the new individuals will derive from increasingly heterogeneous populations of founding cells, and sooner or later will suffer the consequences of competition, or at least a lack of integration, between genetically different cell lines. The same argument helps to explain the great rarity of multicellular organisms formed by aggregation of cells that are not members of the same clone (see the aggregation phase of the social amoebae of the genus *Dictyostelium*, Section 2.6), compared to the predominance of multicellular organisms of clonal origin that are formed from a single founding cell whose descendants remain united together, as normally occurs in the development of animals or plants.

3.1.2.1 UNICELLULAR PROPAGULES

In multicellular organisms, monocytogenous asexual reproduction occurs through unicellular propagules that do not have the value of gametes and to which the general term of **spores** applies (Box 3.1). The process producing them is sometimes called **sporulation**. In addition to representing a single-cell phase from which a new individual can originate without interposing sexual processes, very often the spores are also a resistance and dispersal phase. This explains the frequent presence of a robust external coating that will be abandoned at the time of germination.

The spores through which an organism reproduces asexually are produced by mitosis and are therefore called **mitospores**. These are haploid or diploid, depending on the ploidy of the individual that produces them, and should not be confused with the haploid spores produced by meiosis in cycles with alternating generations (*meiospores*), which are part of the processes of sexual reproduction of those organisms. Flagellated spores, able to move in a liquid medium (**zoospores**), such as those of brown algae (Figure 3.3) or those of the chytridiomycetes, unique among the fungi, contrast with those that lack flagella and thus are immobile (**aplanospores**), such as the asexual spores of the ascomycetes (called *conidiospores* or *conidia*), produced on specialized hyphae called *conidiophores*. Some algae, e.g. *Vaucheria* (Xanthophyceae), produce both zoospores and aplanospores.

Box 3.1 Spores

The term **spore** is used with many, more or less overlapping, meanings in the biology of unicellular and multicellular prokaryotes and eukaryotes. What is common to most of these spores is the fact that they are reproductive cells, of a sexual or asexual origin, which can develop into a new organism without merging with another cell, in this respect behaving unlike gametes. Moreover, the spore is often a form of quiescence and resistance, often in relation to a dispersal phase of the life cycle. The spores of certain fungi are so small and light that winds can take them a great distance. Spores have been found in air samples collected at 100 km altitude.

Spore terminology is very rich. The following overview covers the most commonly used terms.

Spores classified based on the production mechanism

Mitospore – a spore produced by mitosis (as found in some fungi and algae).

Meiospore – a spore produced by meiosis; a term used in fungi; but meiospores are also the spores of the embryophytes, as well as the flagellated **zoomeiospores** of many green algae, deriving by meiosis directly from the zygote; instead, the flagellated spores of other green algae formed by the meiotic products through an additional mitotic division are called **zoomitospores**.

Parthenospore – a spore derived from an unfertilized gamete (in green algae such as *Volvox aureus* and *Eudorina elegans*).

Zygospore – a quiescent zygote produced for example by gametangiogamy in zygomycetes or by isogamy in conjugate green algae; the latter and also some groups of fungi produce **azygospores** as well, i.e. quiescent spores which are not the result of a gamic process; the term could be used – but it is not – also in *Dictyostelium*.

Spores classified based on the structure of the spore

Zoospore – a spore with a flagellum (or multiple flagella).

Aplanospore – a spore without a flagellum.

Autospore – one of the aplanospores, all of the same shape as the mother cell, produced within the latter in a number equal to a power of two, as in many unicellular algae traditionally ascribed to the genus *Chlorella*.

Terms used in embryophytes in the presence of heterospory

Microspore – one of the meiospores, produced in groups of four from a sporophyte mother cell (microsporocyte), that will develop into a male gametophyte, or microgametophyte.

> ## Box 3.1 (cont)
>
> **Megaspore** – a meiospore, generally the only surviving product of a
> meiosis, produced from a sporophyte mother cell (megasporocyte), that
> will develop into a female gametophyte, or megagametophyte.
>
> ### Other terms used in different taxa
>
> **Basidiospore** – a spore carried on a basidium, in the fruiting body of a
> basidiomycete.
>
> **Ascospore** – a spore contained in an ascus, in the fruiting body of an
> ascomycete.
>
> **Teleutospore, aecidiospore, uredospore** – spores produced by a rust (a
> plant parasite of the basidiomycete group Uredinales) in the different
> phases of its life cycle; the first, diploid, produces a basidium on which
> meiosis will occur; the others, respectively produced in the aecidia and
> telia on the host plant, are dikaryotic.
>
> **Teliospore** – a spore produced by a smut fungus (Ustilaginales): initially
> dikaryotic, but in this spore karyogamy occurs, followed by meiosis right
> at the time of germination.
>
> **Carpospore** and **tetraspore** – spores respectively produced by the
> carposporophyte and the tetrasporophyte in the life cycle of a red alga.
>
> **Auxospore** – a diatom cell, as a rule (but not always) represented by the
> zygote, which increases in size until it is transformed into a vegetative cell,
> with the production of two valves.
>
> **Androspore** – a small zoospore produced by some green algae of the group
> Oedogoniales, of intermediate diameter between a sperm cell and an
> ordinary zoospore; the androspore attaches itself to an algal filament near
> the mother cell of the oogonium that has attracted it and germinates there
> as if it were a zoospore.

Figure 3.3 Zoospore of the seaweed *Fucus*, with the dimorphic flagella typical of the Heterokonta.

Mitospores (asexual spores) are found in many groups of algae and fungi
(although less widespread among the basidiomycetes), but not in embryophytes
and metazoans. In these groups, monocytogenous clonal reproduction occurs
instead by different forms of ameiotic parthenogenesis (Section 3.6.2).

In principle, from a genetic point of view, a monocytogenous origin of a new organism could also be obtained via multicellular propagules, for example buds (see Sections 3.1.2.2 and 3.1.2.3). If a bud corresponds to a single-cell clone, in terms of genetic homogeneity among its cells it would correspond to a spore germinated in contact with the parent. In fact, genetic homogeneity within a bud is probably a matter of degree rather than an all-or-nothing question. If the parent developed from a single cell, all of its cell lines (including those that will contribute to the buds) will be clones of the parent's founder cell. Therefore, the question of the clonal or non-clonal origin of a bud translates into a measure of the number of cell divisions to which the founding cell of the bud dates back, a cell that obviously could have also contributed to cell lines other than those in the bud. Otherwise, to determine if a bud has unicellular or multicellular origin it would be necessary to fix in a non-arbitrary way the moment at which a cell, or a group of cells, deserves to be called a bud, irreversibly committed to that fate. The answer to this question is generally unknown and probably, in many instances, impossible to determine (Section 1.3.3).

3.1.2.2 MULTICELLULAR PROPAGULES IN PLANTS AND FUNGI

In fungi and many algae, but also in mosses, there is asexual reproduction by **fragmentation** when small parts of an organism (the thallus of an alga, the gametophyte of a moss, the mycelium of a fungus) are detached from the rest and become independent, thus giving rise to new individuals. In vascular plants, an equivalent process occurs when a part of a plant that has been accidentally detached succeeds in taking hold by regenerating the missing parts. For example, the river current might rip a water-bathed twig off a willow tree, and this twig might take root downstream. In agricultural practice, *propagation by cuttings*, where a plant fragment is cut and planted, exploits these regenerative capacities in many plants of economic interest.

In addition to traumatic fragmentation, in embryophytes polycytogenous vegetative reproduction may take more specialized forms. This may imply the separation of a part of the individual at the level of a defined abscission zone, as in the release of bulbils, or a localized decay of tissues separating the surviving parts, e.g. radical buds. This is how stolons, rhizomes, bulbs and tubers separate from the mother plant (Figure 3.4).

Stolons (or **runners**) are specialized, horizontally developing stems in which long and thin internodes alternate with very short internodes capable of sprouting adventitious roots and therefore giving rise to new plants, which will acquire independence with the death of the stretch of stolon connecting them to the mother plant. This is the way strawberries (*Fragaria*) normally propagate.

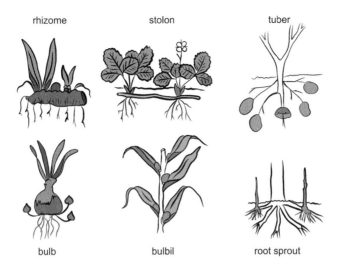

Figure 3.4 The most common types of plant multicellular propagules.

The **rhizome** is similar, but it is usually subterranean and stouter. Many ferns, many aquatic plants (e.g. water lilies) and many monocotyledons (e.g. the common reed *Phragmites*) have rhizomes. In some woody monocots (e.g. *Cordyline*), *aerial rhizomes* capable of rooting and developing into independent plants grow downwards from the main stem.

Bulbs are short stems with a vertical axis, more or less completely buried and with abundant reserve material localized in the foliar bases or in squami-form leaves. The buds that differentiate from the axils of these leaves develop into second-generation bulbs, which survive the rotting of the parent bulb and will eventually give rise to new plants (e.g. *Urginea*).

A **tuber** is a short, enlarged subterranean stem or an enlarged root with *adventitious buds* (i.e. buds which do not develop in the axilla of a leaf, as usual). The tuber, as in potato (*Solanum tuberosum*), provides a way for vegetative multiplication when the thin, long connection that initially unites it with the mother plant dries out.

Bulbils are small propagules that can develop in different places of the plant, for example on subaerial stems, in the leaf axils (*Lilium*). In *Allium*, bulbils often replace flowers inside the reproductive structure that would otherwise be an inflorescence. Bulbils are common among the Liliaceae and Amaryllidaceae, and also in ferns.

Characteristic of many aquatic plants such as *Utricularia* are the **turions** (or **hibernating buds**, or **hibernacles**), buds containing large amounts of reserve materials that detach from the mother plant and pass a period of quiescence during the unfavourable season, to give rise to new individuals when environmental conditions improve again.

Root sprouts are produced by many plants, mainly, but not exclusively, in response to damage to the roots. In some trees (e.g. *Populus, Liquidambar*), very large clones (*genets*; Section 1.6.2) can originate from root buds, whose 'individuals' (*ramets*) remain partially connected to each other through the roots. In herbaceous plants, the root connections among the individuals developed from these sprouts decay much more easily and quickly, turning the ramets into physically independent individuals.

In liverworts, the gametophyte can produce multicellular propagules, often on the tip of the leaflets or inside cup- or bowl-shaped structures. In mosses, the propagules, also multicellular, appear often as small, brown cell clusters similar to tiny tubers, which develop on the rhizoids from which they eventually detach and germinate, forming new plants, often after a period of rest; or, on the contrary, they may be green like the stems or the leaves on which they differentiate, ready to develop immediately as new seedlings.

3.1.2.3 MULTICELLULAR PROPAGULES IN METAZOANS

In the **binary fission** of metazoans, the entire parent is divided into two offspring individuals. Among the animals, and multicellular organisms in general, this mode is rarer than multiple fission, if we disregard accidental fragmentation of the body into two parts, followed by regeneration of the missing parts in each fragment. In **multiple fission** (or **schizotomy**; Schroeder and Hermans 1975), the whole parent individual is divided into a number, sometimes very high, of parts. Fragmentation can be induced by trauma of environmental origin, or controlled by the organism itself. In animals that reproduce by fission, two main modes, paratomy and architomy, can be distinguished (Figure 3.5).

In **paratomy**, a new complete individual is recognizable before its detachment from the parent, while **architomy** is the simple fission or fragmentation of the body before the whole organization of a new complete individual is

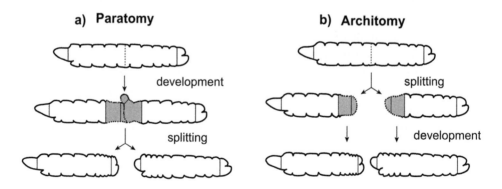

Figure 3.5 Asexual reproduction by (a) paratomy and (b) architomy in a hypothetical polychaete.

recognizable in each fragment. Architomy is thus associated with a regeneration process known as *morphallaxis*, where the regrowth of the missing parts is preceded by reorganization, de-differentiation and new differentiation of the propagule's tissues. There are also intermediate situations between paratomy and architomy. This class of phenomena also includes **strobilation**, a term that mainly applies to the subdivision of the polyp (a *scyphistoma* or *scyphopolyp*) of scyphozoans into a stack of tiny medusae (*ephyrae*) that progressively detach, but also to the articulation in *proglottids* of the body (the *scolex* excluded) of most cestodes (tapeworms in the broader sense).

These reproductive modes are widespread mainly in sponges, nemertines, flatworms (in particular, in the freshwater planarians), in some families of oligochaetes and in the starfishes. As a rule, but with exceptions, zoological groups that exhibit high regenerative capacities usually resort to reproduction by fission, and those that reproduce by fission also have high regenerative capacity (see Table 3.1).

Within the polychaetes, paratomy is known in several families (Syllidae, Serpulidae, Sabellidae, Ctenodrilidae and Spionidae; Figure 3.6) but architomy, documented in the Cirratulidae, Syllidae, Tomopteridae, Spionidae, Chaetopteridae and Dorvilleidae, is more widespread. *Dodecaceria* and *Zeppelina* can divide into many parts and continue to fragment until reducing to one-segment-long pieces. Of these, the pieces deriving from the mid-body segments can regenerate an entire individual. While all these polychaetes also reproduce sexually (Rouse and Pleijel 2001), sexual reproduction is exceptional in the Naididae, small freshwater oligochaetes which as a rule reproduce by transverse fission with paratomy.

In metazoans reproducing by **budding**, a portion of the parent's body grows through cellular proliferation and/or reorganization of already differentiated tissues to become a new organism that eventually detaches from the parent. We have already referred to analogous (and homonymous) mechanisms of asymmetric division in prokaryotes (Section 3.1.1.1) and unicellular eukaryotes (Section 3.1.1.2). A bud can be formed from somatic tissues with cells that have regained totipotency, or at least multipotency. However, this is not an absolute necessity, as exemplified by the case of *Hydra*, where the buds are formed by cells belonging to already differentiated germ layers (endoderm and ectoderm). Budding is frequent in placozoans, sponges, cnidarians (especially among the polyps of the hydrozoans and cubozoans), entoprocts, bryozoans and phoronids.

Similar to many plants (Section 3.1.2.2), some animals reproduce by **stolons**, elongated offshoots that remain connected to the parent organism, sometimes permanently. Groups of cells differentiate along the stolon, from which new individuals originate; the latter will separate by mechanical action

Table 3.1 Asexual reproduction and regeneration in different animal phyla or subphyla (from Minelli 2009, with modifications)

Phylum or subphylum	Asexual reproduction	Regeneration	Comments
Porifera	+	+	
Placozoa	+	?	Asexual reproduction is the dominant and perhaps exclusive form of reproduction, either by binary division or by production of tiny hollow-sphere propagules
Ctenophora	+	+	Asexual reproduction only in the benthic genera *Ctenoplana* and *Coeloplana*
Cnidaria	+	+	
Acoela	+	+	
Nemertodermatida	–	?	
Gastrotricha	–	+	Asexual reproduction only in *Turbanella*
Micrognathozoa	–	?	
Syndermata	–	–	
Gnathostomulida	–	?	
Catenulida	+	+	

Taxon			Notes
Rhabditophora	+	+	An asexual reproduction phase in the biological cycle of digeneans and some tapeworms (Cyclophyllidea). Asexual reproduction is known in Macrostomorpha and Tricladida, apparently absent in Polycladida and Lecithoepiteliata. Regeneration is known in Prolecitophora, Proseriata, Bothrioplanida and Rhabdocela
Cycliophora	+	?	
Bryozoa	+	+	Budding and statoblast production. Polyembryony in Stenolaemata
Entoprocta	+	+	
Dicyemida	+	?	
Orthonectida	–	?	
Nemertea	+	+	
Phoronida	–	**+**	
Brachiopoda	–	+	
Mollusca	–	+	Some instances of regeneration in Cephalopoda (arms) and Nudibranchia (dorsal processes = cerata)

continues

Table 3.1 (cont)

Phylum or subphylum	Asexual reproduction	Regeneration	Comments
Annelida, incl. Siboglinidae (= Pogonophora), Echiura and Sipuncula	+	+	Neither asexual reproduction nor regeneration in Hirudinea and other groups. In Sipuncula, tentacle regeneration and asexual reproduction limited to *Sipunculus robustus* (both by lateral budding from the posterior extremity and by transversal division in the posterior half) and *Aspidosiphon elegans* (by transversal division)
Priapulida	–	–	
Loricifera	?	–	
Kinorhyncha	–	–	
Nematoda	–	–	
Nematomorpha	–	–	
Tardigrada	–	–	
Onychophora	–	–	
Arthropoda	+	+	Asexual reproduction limited to the polyembryony of some hymenopterans and a

			few others. Instances of regeneration in different groups, but limited to the appendages
Chaetognatha	−	−	
Echinodermata	+++	+	
Xenoturbellida	?	?	
Enteropneusta	+	+	Asexual reproduction at least in *Balanoglossus australiensis*
Pterobranchia	+	?	
Cephalochordata	−	−	
Urochordata	+	+	Asexual reproduction at least in colonial ascidians
Vertebrata	+	+	Asexual reproduction limited to polyembryony (as regular reproductive mode only in the armadillos of the genus *Dasypus*)

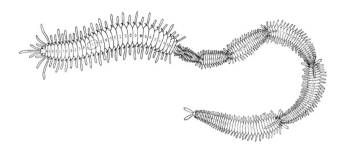

Figure 3.6 Paratomy in the syllid polychaete *Myrianida pachycera*, with a stolon of individuals in different degrees of maturation. The rearmost individual is ready for abscission.

or by degeneration of the interposed tissue, or remain united together to form a colonial group. Stolons are frequent in sessile marine invertebrates, e.g. among the cnidarians (colonial polyps), entoprocts, bryozoans, pterobranchs and tunicates (in the latter, not only in some benthic and sessile groups of Ascidiacea, but also in the pelagic Thaliacea).

3.1.2.4 POLYEMBRYONY AND LARVAL AMPLIFICATION

In the strictest sense of the term, **polyembryony** occurs when more than one embryo is obtained from one zygote. All individuals produced by polyembryony are genetically identical, but different from mother and father. Polyembryony could be regarded as a form of sexual reproduction with a partially clonal result (*intragenerational clonality*), since it generates multiple identical copies of the same genotype, even if different from that of the parents (Avise 2008). Alternatively, polyembryony can be considered in all respects a form of asexual reproduction of an individual still in a very early (embryonic) phase of development. According to the latter interpretation – as a special form of asexual reproduction by means of multicellular propagules – in the species in which this reproductive mode occurs regularly there is alternation of sexual and asexual generations (a *metagenetic cycle*, see Section 2.2).

Polyembryony is therefore a form of asexual reproduction that occurs when the organism is still at the embryo stage, and it is found especially in organisms where the embryo is protected by maternal structures or – in the case of parasitoid insects – by the body of the host.

Similar to polyembryony is the phenomenon of **larval amplification** (which some authors simply include within polyembryony), which occurs when the division into two or more genetically identical individuals occurs at the stage of larva rather than embryo. This is limited, however, to species where the larval stage in which asexual multiplication occurs has a lifestyle clearly different from that of the adult. Assimilation of larval amplification to polyembryony is motivated by the same adaptationist perspective that leads to assigning great significance to the single-celled bottleneck that usually accompanies the passage to a new generation in sexual reproduction (Section 3.1.2). In fact, the larval stage at which clonal multiplication occurs has not yet been subjected to the selective regime to which the adult will be exposed. Larval amplification is typical of the *sporocyst* phase of some parasitic flatworms (but see Section 2.2) and of the *bipinnaria* and *brachiolaria* larvae of some starfishes (Jaeckle 1994), but is also found in other metazoan groups (Table 3.2).

Rare or occasional cases of polyembryony occur in almost all major metazoan groups, while animals for which polyembryony is a common or even obligate form of reproduction belong to a much smaller set of taxa, listed in Table 3.2. The only vertebrates with obligate polyembryony are the armadillos of the genus *Dasypus* (Figure 3.7), where the number of embryos deriving from a zygote differs according to the species: 2 in *D. kappleri*, 4 or 8 in *D. sabanicola* and *D. septemcinctus*, 2–3 but more often 4 in *D. novemcinctus*, and finally 8–9, more rarely 12, in *D. hybridus*.

Still among the vertebrates, accidental polyembryony is known in domestic cattle, pigs, deer, cetaceans and various birds and fishes. In humans, monozygotic twins occur once in 300 births, and monozygotic triplets once in 50,000. In some parasitoid insects, the number of embryos that can derive from a single zygote may exceed 1000 (Table 3.3).

In the hermaphrodite flatworm *Gyrodactylus elegans*, after fertilization, a single egg begins to develop inside the uterus. However, before the embryo is released by the parent, a second embryo is produced within it, and soon a third embryo develops within the second and a fourth within the third, thus obtaining a pattern of telescoped generations (Section 2.9). However, the development of the second embryo stops until the first embryo is released by the parent worm.

The *substitutive polyembryony* described by Cavallin (1971) in *Carausius morosus*, a stick insect, may also be included among the phenomena of asexual reproduction. After a period of normal development that also includes germ-cell differentiation, the primary embryo undergoes degeneration and is replaced by a secondary embryo that differentiates from the serosa, the layer of blastoderm cells that remain to coat the yolk after the primary embryo has formed. This second embryo develops in turn a new germline.

In plants, the term polyembryony is often used with a different meaning, when multiple cells of the same megagametophyte are fertilized independently (*simple polyembryony*) and even when additional embryos develop from integuments (*integumentary polyembryony*) or nucellar tissue (*nucellar*

Table 3.2 Metazoans that reproduce regularly by polyembryony. Division into two or more individuals can occur at an embryonic (E) or larval (L) stage (data from Craig *et al*. 1997, with additions)

CNIDARIA		
Hydrozoa Trachilina	*Pegantha* spp. *Cunina* spp. *Cunochtantha* spp.	L
Hydrozoa Hydroida	*Polypodium hydriforme*	L
RHABDITOPHORA		
Gyrodactyloida	*Gyrodactylus elegans*	E
Eucestoda	*Echinococcus* spp.	L
Digenea	*Schistosoma* spp.	L
BRYOZOA		
Stenolaemata		E
CRUSTACEA		
Rhizocephala	*Loxothylacus panopaei*	L
INSECTA		
Hymenoptera Encyrtidae	*Copidosoma* spp.	E
Hymenoptera Platygastridae	*Platygaster* spp.	L
Hymenoptera Braconidae	*Macrocentrus* spp.	L
Hymenoptera Dryinidae	*Aphelopus theliae*	L
Strepsiptera	*Halictoxenos simplicis*	E
ECHINODERMATA		
Asteroidea	*Luidia* spp.	L
Ophiuroidea	*Ophiopluteus opulentus*	L
VERTEBRATA		
Mammalia	*Dasypus* spp.	E

Table 3.3 Polyembryony in parasitoid Hymenoptera: number of embryos or larvae per zygote (data after Segoli *et al.* 2010)

	Embryos or larvae	Hosts
Encyrtidae	2–1000+	Lepidoptera and Hymenoptera (eggs)
Platygasteridae	2–10	Diptera Cecidomyiidae (eggs)
Braconidae	2–50	Lepidoptera (larvae)
Dryinidae	up to 60	Homoptera: leafhoppers (nymphs)

Figure 3.7 Polyembryony is obligate in the armadillos of the genus *Dasypus*. In the nine-banded armadillo (*D. novemcinctus*, shown here) there are usually four embryos.

polyembryony) (Kishore 2014; Section 3.6.2.9). There may be more than one egg cell in an embryo sac and more than one embryonic sac in an ovule (e.g. *Citrus*, *Opuntia*). Synergid cells (Section 3.4.2.1) can occasionally also be fertilized. Obviously, none of these cases is an instance of asexual reproduction.

However, plants can also produce twin embryos (true monozygotic twins) by division of the original embryo. This process, called *cleavage polyembryony*, is common in gymnosperms (e.g. *Pinus*, *Tsuga*, *Cedrus*, Figure 3.8), but less frequent in angiosperms (e.g. *Erythronium americanum*, *Nymphaea advena*, *Nicotiana rustica*).

In any case, in plants polyembryony leads to competition for developmental resources among the embryos of the same ovule; the outcome, usually, is the elimination of all competitors except one.

Figure 3.8 Cleavage polyembryony is common in conifers of the genus *Cedrus* (here, *C. libani*).

3.2 Sexual Reproduction: Gametes and Syngamy

In Chapter 1 we defined **sexual reproduction** as a form of reproduction that generates new individuals carrying a genome obtained from the association and/or the reassortment of genetic material of more than one origin. In the most canonical form of sexual reproduction, the new genome is formed by the union of (partial) copies of the genomes of two parents through the fusion of two special cells, the *gametes*, into a single cell, the *zygote*. From the point of view of transmission genetics we can anticipate that sexual reproduction consists of two main processes: *recombination in the broad sense*, which occurs through the production of gametes whose genomes do not match either of the two chromosome sets of the parent organisms, and *syngamy*, the fusion of the genomes of two gametes into the zygote's genome. The cytogenetic aspects of sexual reproduction will be dealt with in Chapter 5, while in this section and the following ones in this chapter, attention will be focused on how syngamy is achieved, from the formation of gametes or gametic nuclei to their meeting and fusion.

Sexual reproduction is found in all multicellular eukaryotes and in most protists (but not, for example, in ciliates and euglenozoans); in any case its absence does not necessarily rule out other forms of sexual exchange. Despite the widespread occurrence of sexual reproduction (Aanen *et al.* 2016), its origin and maintenance through the course of evolution are still an unsolved enigma for evolutionary biology (Box 3.2). The oldest fossil record of sexual reproduction (presence of gametes) is provided by the red alga *Bangiomorpha pubescens*, dated about 1200 million years ago (Butterfield 2000), but in the

Box 3.2 The Evolutionary Enigma of Sex

The problem of the origin and maintenance of sexual reproduction is considered by many as the 'main problem of evolutionary biology', often labelled as the **'paradox of sex'**.

Sexual reproduction is widespread in all major eukaryotic groups, but it seems to present an insurmountable disadvantage compared to asexual reproduction. With the same reproductive investment (number of eggs), females that reproduce asexually can have twice as many second-generation descendants as females that reproduce sexually, simply because they do not waste resources generating males, which do not produce offspring by themselves. This is the so-called 'twofold cost of sex', but more correctly it should be called the 'cost of males', because it only applies in the case where sexual reproduction is not isogamous. This does not, however, alleviate the problem, since anisogamous reproduction has evolved repeatedly, and independently, in many eukaryotic clades. Furthermore, with regard to the genetics of hereditary transmission, sexual reproduction (anisogamous or not) can break apart favourable gene combinations that had been stabilized by selection in previous generations, or create deleterious or non-viable combinations of genes (e.g. due to genetic incompatibility).

Given these heavy costs of sex, it is assumed that sexual reproduction must provide some selective advantage, to an extent that at least compensates for these disadvantages. Many hypotheses have been formulated, generally based on the idea that despite the deficit in terms of number of descendants (low *fecundity fitness*), sexual reproduction can lead to an improvement in the quality of offspring (high *viability fitness*) in sexual populations.

Most of these hypotheses are variants of four main ideas: (i) sex facilitates adaptation to new environments by combining favourable genetic variants from different genomes (Fisher–Muller model); (ii) sex confers advantages to the host in coevolution with its parasites, through the negative frequency-dependent selection imposed by the latter (Red Queen model); (iii) sex maintains adaptation by removing deleterious mutations more effectively (deterministic mutational models); (iv) sex releases beneficial mutations from association with deleterious alleles in the genomes where they appear (mutational load models). Moreover, sexual reproduction in multicellular organisms favours the evolution of anisogamety, which in turn creates conditions for the advent of sexual selection, which accelerates the adaptive processes listed above. Different types of advantages could obviously operate in a synergistic way. More recent theoretical work suggests that occasional or conditional sex, involving facultative switching between sexual and asexual

continues

Box 3.2 (cont)

reproduction, is the optimal reproductive strategy. Therefore, the true 'paradox of sex' could turn out to be the prevalence of obligate sex (Burke and Bonduriansky 2017).

This enigma, 'why sex?' (or the 'paradox of sex'), is countered by the opposite problem, 'how to manage without sex?', a problem posed by so-called **'ancient asexual scandals'**. If the prevalence of sexual reproduction shows that it must necessarily have advantages over asexual reproduction, either those thus far hypothesized or others, how is it possible that there are groups of organisms that have exclusively reproduced asexually for millions of years? Among the 'ancient asexual scandals' in animals, which more precisely reproduce by ameiotic parthenogenesis (Section 5.2.3.3), there are the bdelloid rotifers, the darwinulid ostracods and several groups of oribatid mites (Table 3.4). The literature on the 'paradox of sex' and 'ancient asexual scandals' is vast, and the two problems are far from resolved. For a first level of in-depth analysis, refer to Stearns (1987), Agrawal (2001), Schön *et al.* (2009), Lehtonen *et al.* (2012), Neiman *et al.* (2017), Schön and Martens (2017) and Kondrashov (2018).

Table 3.4 Eukaryotes in which no clear forms of sexual reproduction are known (data after Schurko *et al.* 2009)

	Estimated age of the asexual clades (million years)	Residual evidence of sexuality
AMOEBOZOA: LOBOSA		
Entamoeba spp.		
METAMONADA		
Giardia intestinalis		Fusion of nuclei
Trichomonas vaginalis		
Trypanosoma brucei, T. cruzi		
APICOMPLEXA		
Toxoplasma gondii		Fusion of cells

Table 3.4 (cont)

	Estimated age of the asexual clades (million years)	Residual evidence of sexuality
FILICALES		
Vittaria spp.	10	
ASCOMYCOTA: SACCHAROMYCETINA		
Candida albicans		Fusion of cells
GLOMEROMYCOTA	400	Presence of sexual structures
BASIDIOMYCOTA		
Lepiotaceae cultivated by leaf-cutting ants	23	Presence of sexual structures
SYNDERMATA		
Bdelloidea	40–100	
BIVALVIA		
Lasaea spp.	3.0–7.9	
GASTROPODA		
Campeloma spp.	0.32–0.56	
Potamopyrgus antipodarum	ca. 0.5	Occasional presence of males; mating observed
NEMATODA		
Meloidogyne spp.	17–40	Occasional presence of males; mating observed
ACARI		
Oribatida (several parthenogenetic lines)	100–200	Occasional presence of males

continues

Table 3.4 (cont)

	Estimated age of the asexual clades (million years)	Residual evidence of sexuality
CRUSTACEA: ANOSTRACA		
Artemia spp.	3.5	Occasional presence of males
CRUSTACEA: OSTRACODA		
Darwinulidae	ca. 200	Occasional presence of males
Eucypris virens	2.0–4.5	Mating observed
Heterocypris incongruens	8–13	
PHASMATODEA		
Bacillus atticus	ca. 15	
Timema spp.	0.3–3.0	
HOMOPTERA: APHIDIDAE		
Rhopalosiphum padi	0.4–1.4	Occasional presence of males; mating observed
Tramini		Occasional presence of males
HOMOPTERA: SCALE INSECTS		
Aspidiotus spp.	ca. 1	
COLEOPTERA: CHRYSOMELIDAE		
Calligrapha spp.	0.3–3.1	
COLEOPTERA: CURCULIONIDAE		
Aramigus spp.	>2	
AMPHIBIA		
Ambystoma spp.	2.4–3.9	Egg activation by sperm required

light of genomic data (presence of meiotic genes) it is likely to have evolved with the earliest stages of eukaryote evolution, about 2 billion years ago (Schurko and Logsdon 2008; Speijer *et al.* 2015).

3.2.1 Isogamety, Anisogamety, Oogamety

Gametes that participate in a fusion (**–gamy**) process may have different degrees of morphological and functional similarity (Figure 3.9).

We speak of **isogamety** when the two gametes are morphologically indistinguishable. In this case they are called **isogametes**, the process that involves them is termed **isogamy**, and the species or the process of fertilization is said to be **isogamous**. Generally both isogametes are flagellated, as in most algae, but the amoeboid gametes of the dictyosteliid mycetozoans lack flagella.

Anisogamety (and also **anisogametes**, **anisogamy** and **anisogamous** fertilization) occurs when there are two types of gametes that are morphologically distinguishable. However, gamete dimorphism can take different forms. In the one we refer to as **anisogamety s.s.** (*sensu stricto*) the two types of gametes are similar in shape and function (e.g. being both flagellated and motile), but one of the two, the **macrogamete** or **female gamete**, is considerably larger than the other, which is called the **microgamete** or **male gamete**. This condition is widespread in the Apicomplexa and occurs also among the green algae, for example in a number of Volvocales. A more common and more extreme form of anisogamety is **oogamety** (with the associated terms **oogametes**, **oogamy** and **oogamous** fertilization), which is found in most of the sexually reproducing organisms. In oogamety, gamete dimorphism involves both size and shape. The largest gamete, or *female gamete*, immobile and generally full of reserve substances, is called, according to the different taxonomic traditions, **egg**, **egg cell**, **ovum**, **ovocell**, **oosphere** or **ovule** (but the last should not be confused with the *ovule* of seed plants; see Section 3.4.2.1). The gamete of the other type, or *male gamete*, is generally much smaller and mobile and is called, also according to the

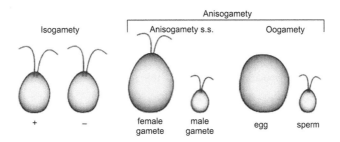

Figure 3.9 The three conventional degrees of differentiation between gametes of the same species. Biflagellate gametes are shown in the scheme, as typical of many green algae.

different taxonomic traditions, **sperm**, **sperm cell**, **spermatozoon**, **antherozoid**, **spermatozoid** or **spermatium**. In our own species, the egg has a volume about 100,000 times larger than the spermatozoon. Oogametes are discussed in more detail in Sections 3.4.1.3 and 3.4.2.2.

We refer to an individual's **sex condition** as its state with respect to sexual function, either *male* or *female*, but also both male and female (*hermaphrodite*) or neither male nor female (*sexually indeterminate*). Thus there are two sexes, but four sex conditions. In anisogamety, an individual's sex condition coincides with the type of gametes it produces: **male** if it produces male gametes exclusively, **female** if it only produces female gametes, and **hermaphrodite** if, simultaneously or at different times, it is able to produce both types of gametes. On the contrary, in isogamety, where gametes are not distinguishable as male and female, the ascription of an individual to a specific sex condition does not apply and we could designate it as **sexually indeterminate**. In many of these organisms, however, there are forms of reproductive incompatibility such that an individual, although not attributable to a specific sex, nevertheless belongs to a particular **mating type** which allows it to participate in sexual processes only with individuals belonging to mating types other than its own (see Section 6.6 and Box 3.3). The fundamental asymmetry between male and female gametes is therefore the basis of the distinction of sexual roles and the necessary condition for the evolutionary process of sexual selection (Parker 2014).

Box 3.3 How Many Sexes?

To establish the number of sexes in a species we must first define what is to be counted.

The simplest definition takes into account the number of morphological types of gametes that are produced in a given species. Sex is singular (or indeterminate) for a species with isogamety (i.e. with gametes of one type only), while there are two sexes in most species that reproduce sexually (with male and female gametes). On the other hand, by assimilating the notion of mating type to the concept of sex, sexes in one species can reach up to several thousand, as in certain fungi (Section 6.6).

However, counting the number of gamete types seems to ignore an essential aspect of sexual reproduction, that is, that generally a maximum of two individuals engaged in a sexual exchange share their genomes to form a new individual. According to this meaning, the number of sexes would be the number of types of gametes that must be joined to form a new fertile individual in the population. This number seems never to exceed two, so that a system of sexual reproduction basically appears to be a binary system.

> **Box 3.3** (cont)
>
> Nevertheless, if we broaden our perspective, attributing a value of individual, or 'superorganism' (Section 1.4.3), to the level of organization of complex societies such as those of eusocial insects, the number two does not seem to be an insurmountable barrier. In an American ant species of the genus *Pogonomyrmex*, of hybrid origin, two distinct gene pools coexist, which we can designate as 'blue' and 'yellow', each with fertile individuals (queens and males) that produce gametes (eggs and sperm) of the respective type (blue and yellow). In a colony with a blue queen, she can generate blue males by haploid parthenogenesis (Section 5.2.3.3) and new blue queens by mating with blue males. However, to generate workers the blue queen must be fertilized by male gametes of the yellow type. A reverse situation is found in the colonies with a yellow queen, where, to produce a sterile caste of workers, the contribution of blue male gametes is required. From the point of view of the colony seen as a superorganism, the queen (blue or yellow) must mate with two types of male (blue and yellow), i.e. she must receive both types of male gamete, to generate a self-sustaining colony. A colony with a fertile queen therefore has at least three parents, descending from three types of gametes. Furthermore, four types of gametes (male and female, blue and yellow) are necessary for the perpetuation of the hybrid system (Parker 2004).

3.2.2 Gametogamy, Gamontogamy, Autogamy

In the most common forms of sexual reproduction, the fusion of the chromosome complements of two distinct nuclei occurs through the union of two specialized cells, generally haploid, called the **gametes**. This event (**gametogamy**) involves the merging of the cytoplasms of the two gametes (including the cytoplasmic organelles and their genomes) to form the cytoplasm of the zygote (**plasmogamy**), followed by the fusion of the nuclei of the two gametes (**karyogamy**).

In many eukaryotes, cells functioning as gametes are recognized, but there is no process that could be described as the production of gametes by a parent organism, since the entire life cycle involves only unicellular phases. In these cases we have **hologamy**, that is, the union of two individual cells (e.g. *Chlamydomonas*). We refer instead to **merogamy** when unicellular gametes produced by multicellular individuals are involved.

Karyogamy, however, does not always involve the existence and the meeting of two gametes as distinct cells. Karyogamy can also occur between two **gametic nuclei** that are already present in the same cell. This process, called **autogamy**,

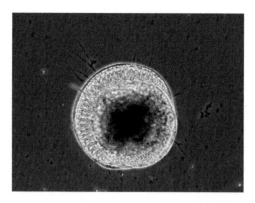

Figure 3.10 In the foram *Patellina corrugata* two or more individuals (gamonts) unite for sexual exchange before the formation of gametes, a form of reproduction called gamontogamy.

can for instance consist in the union of two nuclei within a single multinucleate cell, as in the foram *Rotariella*. More commonly, the term autogamy indicates the process of karyogamy between two meiotic products of the nucleus of a single cell, which in unicellular eukaryotes corresponds to self-fertilization (Section 3.6.1). In the ciliate *Paramecium*, autogamy, which can be induced by prolonged fasting, shows the same pattern of nuclear division found in conjugation (Section 5.2.5), with the production by meiosis of haploid nuclei, of which only one survives and divides once more by mitosis before karyogamy.

The temporary or permanent fusion of two or more individuals that will produce gametes (*gamonts*) sometimes instead precedes the formation of gametes. This is called **gamontogamy**, and it is found in gregarines, in some forams (e.g. *Patellina corrugata*, Figure 3.10) and in the amoebozoan *Sappinia*, a free-living amoeba that can accidentally become a parasite of humans. In the gregarines, apicomplexan protist parasites of insects, earthworms and other invertebrates, the union of two gamonts (*syzygy*) is followed by the formation of a common cyst, within which gametes are formed.

A particular form of gamontogamy, where there is only a sexual exchange without reproduction, is the **conjugation** of the ciliates (Section 5.2.5.1). In this process, between the two gamonts (called *conjugants* in ciliates) only a temporary union is formed, which allows an exchange of gametic nuclei, but no zygote is formed.

Fusion involving gametic nuclei rather than gametes is also found in fungi. The process is called **zygogamy** (or **isogamous gametangiogamy**) in the zygomycetes, where the multinucleate extensions (*gametangia*) of hyphae of different mating types merge and allow the meeting within a *zygosporangium* of several nuclei with gametic value. In the ascomycetes there is **anisogamous gametangiogamy**, because a specialized extension (*antheridium*) of a hypha of 'male' mating type joins a similar structure (*ascogonium*) carried by a hypha of 'female' mating type and transfers its nuclei into it. In the basidiomycetes, where no structure specialized for plasmogamy is present, the union (**somatogamy**) of two monokaryotic hyphae gives rise to a dikaryotic mycelium.

Admittedly, these categories are to some extent idealized. For instance, in the case of the metazoan egg the meiosis is often completed only after activation induced by the penetration of a spermatozoon, so it can be said that the latter actually enters a still diploid female cell (Section 3.5.2.1).

3.2.3 Cryptic Sex

There are numerous groups in which sexual reproduction has been lost for a long time, often leaving, however, a recognizable trace, or the possibility of an occasional resurgence of sexual reproduction, albeit as a very rare event. Groups that practise a form of obligate ameiotic parthenogenesis, such as the bdelloids among the rotifers, are also generally classified as asexual (Table 3.4).

For example, in many fungi only asexual reproduction is known. Lacking typical sexual reproductive structures such as asci or basidia (Sections 7.5.4 and 7.5.5), these fungi were traditionally relegated to an artificial class (deuteromycetes) from which they were retrieved if and when a sexual generation was found that allowed them to be assigned to one of the natural groups of the fungal kingdom. The possibility of determining the affinity of these fungi by examining their gene sequences has made the artificial group of deuteromycetes unnecessary. There is the possibility that some fungi have definitively lost the ability to reproduce sexually. However, a final demonstration that no trace of sexuality is present is very difficult to obtain. For example, just a few years ago it was possible to prove the occasional occurrence of sexual reproduction in a common mould such as the ascomycete *Aspergillus fumigatus* (O'Gorman *et al.* 2009). The yeast *Candida albicans* has long been considered an organism with no sex, represented only by diploid cells, but recently a sexual cycle has been discovered in this species, with a and α cells between which there is cytoplasmic and nuclear fusion, with the formation of a tetraploid cell. Also recent is the discovery of fusion events between 'identical' cells of this tiny ascomycete (Alby *et al.* 2009).

It was also previously thought that the glomeromycetes, fungi symbiotic with trees, with the roots of which they develop mycorrhizae, had been asexual for 400 million years (Jany and Pawlowska 2010). However, it turns out that in these fungi sexual processes are extremely rare but not completely absent. Evidence of recombination has been found (den Bakker *et al.* 2010), which implies some mechanism of reductional cell division; another clue (presence of meiotic genes) is mentioned in Section 7.5.3.

3.3 Distribution of Sex Conditions Within a Species or Population

Male and female gametes can be produced by distinct individuals (a *gonochoric* or *dioecious* species or population) or by the same individual (a *hermaphrodite* or *monoecious* species or population). In organisms with alternating generations (Chapter 2) it is necessary to consider the distribution of the sex conditions in the different generations separately. For example, in the life cycle of land plants there is alternation between a diploid phase (the sporophyte) and a haploid phase (the gametophyte). The sex condition of the gametophyte (male, female or monoecious) is defined by the type of gametes it produces (sperm cells, eggs or both, respectively). The sex condition (in a broad sense, as it does not produce gametes) of the sporophyte (male, female or monoecious), is instead defined on the basis of the type of spores (*heterospores*) it produces (*microspores*, *megaspores*, or both, respectively). Male gametophytes (*microgametophytes*) develop from microspores and female gametophytes (*megagametophytes*) from megaspores. The sex of the sporophyte is indeterminate if it produces a single type of spore (*homospores*) that develops into a gametophyte which in turn can be monoecious or dioecious (Section 2.1.2).

3.3.1 Gonochorism (Dioecy)

A species (or population) is said to be **gonochoric**, or **dioecious** in botanical terminology, when it includes distinct male and female individuals exclusively (Figure 3.11a). However, species are also considered gonochoric if (as developmental abnormalities or because of genetic mutations) *intersex* individuals

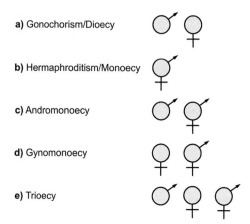

a) Gonochorism/Dioecy

b) Hermaphroditism/Monoecy

c) Andromonoecy

d) Gynomonoecy

e) Trioecy

Figure 3.11 Types of distribution of sex conditions in a population. Symbols refer to an individual's possible sex condition: male, female, hermaphrodite.

accidentally occur, i.e. ones with a mix of both male and female phenotypic characters (Section 6.1.1). In addition, species that reproduce exclusively by thelytokous parthenogenesis (Section 3.6.2.1) and therefore do not include males are also classified with the gonochoric taxa.

In the sporophytic generation of the flowering plants, only 6% of species are dioecious (Section 3.3.2.1). Dioecy is more common in tropical forests (e.g. 23% of the flora of Costa Rica) and in the floras of remote oceanic islands, such as Hawaii and New Zealand (Richards 1997). In a sclerophyll mountain flora of Chile, the percentage of dioecious species is 57% among the trees and 17% among the shrubs, but falls to 2% in perennial grasses and zero in annual species (Arroyo and Uslar 1993). In some cases, anatomically bisexual plants are functionally unisexual. In the mangosteen (*Garcinia*), for example, the flowers of individuals that function as males possess a sterile gynoecium with a nectariferous stigma capable of attracting pollinators. The opposite situation is exemplified by the Indonesian *Decaspermum parvifolium*, in which the flowers of the functionally female individuals produce sterile pollen, which provides a reward for pollinators.

In metazoans, gonochorism is the norm in cnidarians, priapulids, kinorhynchs, loriciferans, nematodes, nematomorphs, rotifers, molluscs (excluding some important clades of gastropods, many solenogasters and a certain number of bivalves), polychaetes including sipunculans, onychophorans, tardigrades, arthropods, brachiopods, echinoderms, hemichordates and chordates. Gonochorism can be accompanied by the development of **secondary sexual characters**. These phenotypic traits, which do not directly concern the reproductive system, are manifested with distinct forms in the two sexes, or are present in one sex only.

3.3.1.1 SECONDARY SEXUAL CHARACTERS IN PLANTS

Unlike the gonochoric species of many zoological groups, dioecious plants exhibit modest differentiation of secondary sexual characters. However, dwarf males are common in mosses (Haig 2016), where tiny males grow epiphytically on much larger females. Males are sometimes reduced to no more than a few leaves sheathing a single antheridium. Dwarf males are facultative in some species but obligate in others. Male dwarfism appears to have evolved many times independently.

According to Lloyd and Webb (1977), the only angiosperm in which it is possible to safely identify the sex of an individual without observing the flowers would be the hemp (*Cannabis sativa*), in which (in some cultivars, at least) the plants of the two sexes are recognizable for their different habitus: female individuals have a larger stem, a more developed root system and larger leaves. In several other dioecious species, less pronounced differences between males and females are observed in habitus, leaf morphology and other minor traits. Differences in plant size are observed in yew (*Taxus baccata*), ginkgo (*Ginkgo biloba*) and poplars (*Populus* spp.), where male plants are taller than females. In a number of perennial species, males surpass females in vigour and in the rate of growth or vegetative reproduction. In several short-lived species including hemp, spinach and some species of *Silene*, females are larger than males. In *Asparagus*, male plants are larger than full-grown females, but the individual shoots of female plants are larger. In some long-lived species, the greater survival rate of males contributes to a numerical predominance of males in the population, but in some species of *Silene* and in *Rumex acetosa* males have a higher mortality rate than females. Males and females sometimes occupy different microhabitats, as in *Mercurialis perennis* and *Rumex acetosella*, and this seems to be associated with the existence of a distinct environmental optimum for the two sexes. Male plants often start to flower at a younger age than females, flower more frequently and produce a larger number of inflorescences with a larger number of flowers per inflorescence (Lloyd and Webb 1977).

3.3.1.2 SECONDARY SEXUAL CHARACTERS IN ANIMALS

The main roles of secondary sexual characters in animals are found in the interactions between the two sexes (recognition of the partner, courtship), in within-sex competition for mating, and in the different parental investment in the production of gametes or in the fulfilment of parental care. Therefore, as a rule, we do not expect to find sexual dimorphism in animals with external fertilization.

A sensational case of extreme sexual dimorphism was revealed by a re-examination, based on molecular characters, of the relationships between representatives of three putative families of fishes (Megalomycteridae, Cetomimidae, Mirapinnidae), which in fact turned out to be the male, the female and the larva, respectively, of the same species (Johnson *et al.* 2009).

The different scenarios in which sexual dimorphism occurs in animals correspond to the many forms that sexual selection can take (Ghiselin 1974a):

1. *Choice of partner by the female.* In this context, well exemplified by birds such as peacock, pheasants and birds of paradise, the male has a much more conspicuous plumage than the female. Most of these animals have a polygynous mating system (a harem of females for every male that can reproduce; see Section 3.5.6).

2. *Competition between males through direct confrontation.* This form of sexual selection is present for example in deer, whose males possess weapons (the antlers) used in ritualized confrontations and actual fights between rival males.

3.   *Competition between males based on their efficiency in reaching and inseminating one or more females.* This form of sexual selection re-enacts at a macroscopic level the dualism that exists, at a microscopic level, between the male gamete, small and mobile, and the female gamete, large and immobile. The females of many animals are much more corpulent and less mobile than the males of the same species. In different kinds of fireflies (lampyrid beetles), for example, the male is winged, while the female is wingless. In this situation, the production of large masses of eggs was favoured in the female by natural selection, even at the cost of a progressive reduction in the animal's mobility. By contrast, in the male it proved to be more advantageous to develop the agility accompanying smaller size and those physiological and behavioural adaptations that allow males to readily find females at the time they begin to be available for mating.

4.   *Seizure of the female by the male.* Once insemination has occurred, it is important for the male that the female does not mate again, in the same (or only) breeding season, with other males – which could render useless the first male's reproductive effort. Multiple insemination, in fact, would trigger a *sperm competition* in which it is generally more probable that the spermatozoa of the last male mating with the same female will prevail; this is as much as certain in the case of animals (e.g. many insects, millipedes etc.) in which the male has copulatory organs with which he can remove the sperm cells resulting from previous inseminations. One of the possible strategies that allow the male to avoid this risk consists in remaining attached to the female for a sufficient time after insemination, in a position that hinders insemination attempts by other males. This behaviour is observed in several isopods and amphipods, in which, in relation to this strategy, the male is considerably bigger than the female, contrary to the prevailing trend in the animal kingdom.

5.   *Long-lasting physical integration between male and female.* While in the previous cases the morphological and behavioural traits that correspond to greater mobility are accentuated in the male, in some cases this divergence between the two sexes is less pronounced, and even an inversion of this general trend is possible, with selection favouring males that are less mobile than the females of the same species. The most striking among these cases, involving the abyssal fishes known as seadevils and their relatives (Ceratioidei) is discussed later (Section 3.5.2.1). Let's focus here instead on *Blastophaga psenes*, the little wasp which is the sole pollinator of the common fig (*Ficus carica*). This tiny insect develops inside a fig inflorescence; more precisely, each individual feeds as a larva on one of the female flowers deep inside the fig. Fertilization takes place within the same inflorescence in which male and

Figure 3.12 A male sea horse (*Hippocampus comes*) showing a prominent ventral pocket (marsupium) containing the eggs deposited therein by the female.

female have developed. The male, which has no wings and is very little mobile, seeks female wasps, fertilizes them and dies inside the fig. In contrast, the female, after insemination, leaves the fig where she has grown up and flies to a new fig (possibly of the next flowering cycle), which she will penetrate to leave her pollen load and lay eggs. Almost larviform, wingless males alongside females with normally developed wings are also found in some bark beetles such as *Xyleborus dispar* and *Ozopemon* spp. (Jordal *et al.* 2000).

In some species or small clades, **sex-role reversal** with respect to the usual conditions is observed. In the seahorses (*Hippocampus* spp., Figure 3.12), incubation is entrusted to the male, which is equipped with an *incubation pouch*. In the spotted hyena (*Crocuta crocuta*) the females are larger than the males and have external genitalia that look superficially male: this condition, correlated with an unusually high level of androgens in the blood, makes fertilization of a non-consenting female practically impossible. The external genitalia of the female are normal in the other hyena species, where the male is larger than the female, as almost universally in mammals. In the South American frog *Rhinoderma darwinii*, the male spends three or four weeks guarding the eggs laid by the female among dead leaves; when they are close to hatching, he picks them up and then hosts the tadpoles, until metamorphosis, in his vocal sacs – whose walls produce material on which the offspring feed. In almost all spiders the male is smaller (sometimes much smaller) than the female, but in the wolf spider *Allocosa brasiliensis* the male is larger than the female and sedentary, while the female goes in search of a partner and starts courtship, risking falling victim to the cannibalistic behaviour of the male (Aisenberg and Peretti 2011). An inversion of mating roles is observed in the insect genus *Neotrogla*, as described in Section 3.5.2.1.

At times, a parasitic attack can have serious effects on the development of the victim's gonads: this leads to *parasitic castration*, as usually results from the parasitism of a rhizocephalan crustacean such as *Sacculina* on a decapod crustacean. In other cases, however, the prolonged presence of a parasite can selectively attenuate or erase some secondary sexual character. In the midges of the genus *Chironomus*, males parasitized by nematodes have normal male genital appendages, but their forelegs and antennae tend to resemble those of the females. Similar effects of parasitism by nematodes and gordiaceans have been observed in insects of numerous orders. Conspicuous effects, both on the development of the gonads and on secondary sexual traits, have also been observed when homopterans (leafhoppers or Delphacidae) are parasitized by dryinid wasps, when various insects (mayflies, stick insects, earwigs, homopterans, heteropterans and hymenopterans) are parasitized by dipterans of the Tachinidae, Pipunculidae and Phoridae, and when silverfishes, orthopterans, heteropterans, homopterans and hymenopterans are parasitized by strepsipterans (in this last case, the parasitic attack syndrome is called *stylopization*).

In some animals, including a number of insects, sexual dimorphism coexists with the presence, in the same population, of two or more distinct phenotypes within one of the two sexes. Almost always this is a form of polyphenism, i.e. phenotypic differences not based on genetic differences, but due to diet, photoperiod or other environmental influence. Polyphenism that affects the male sex (**poecilandry**) is seen in some beetles: there may be two or more male phenotypes differently 'armed' with larger or smaller jaws (many stag beetles), or males may differ in the development of cephalic or prothoracic horns (many scarabaeoids, e.g. species of the genus *Onthophagus*; Figure 3.13). Polyphenism in the female sex (**poecilogyny**) is exemplified by many species

Figure 3.13 Two males of the scarabaeid beetle *Onthophagus nigriventris*, one with prominent horns, the other without. This form of polyphenism, restricted to the male sex, is called poecilandry.

of damselflies (e.g. the European *Ischnura elegans* and *Pyrrhosoma nymphula*), and also by some vertebrates (e.g. about 30% of the many species of the lizard genus *Anolis*; Paemelaere *et al.* 2011).

More complex and weakly related to the topic discussed here is the situation of many eusocial hymenopterans, especially ants, which have at least two female phenotypes (queen, worker), and not infrequently even three or four, but only one of them (the queen) includes fertile individuals, and it is only those females that should be compared with males in any assessment of possible sexual dimorphism. Different again is the nature of female poly-morphism in the African butterfly *Papilio dardanus*, whose various phenotypes represent mimics of many distasteful butterflies belonging to other genera.

3.3.2 Hermaphroditism (Monoecy)

An individual is said to be **hermaphrodite**, or **monoecious** in botanical terminology, when, at the same time or at different stages of its life it possesses functioning male and female sexual organs, and can therefore produce both male and female gametes. Correspondingly, a species (or a population) exclu-sively composed of hermaphrodite individuals is said to be *hermaphrodite* (or *monoecious*) (Figure 3.11b). Hermaphroditism should not be confused with *intersexuality*, where an individual (male, female, or reproductively dysfunc-tional) presents phenotypic characteristics of both sexes (Section 6.1.1). Herm-aphroditism is the sex condition typical of many seed plants and some groups of invertebrates, but it can also occur, as a developmental anomaly, in many other species, even among the vertebrates, including humans. A modern monograph on hermaphroditism, from which several examples in this section are taken, is Avise (2011).

3.3.2.1 MONOECY IN LAND PLANTS

As mentioned above, the biological cycle of terrestrial plants involves alterna-tion between a diploid phase (sporophyte) and a haploid phase (gametophyte). As shown in Table 3.5, the structures that produce male gametes (e.g. the antheridia of the liverworts) and those that produce female gametes (e.g. the archegonia of the same plants) are often borne on distinct gametophytes, but they do not necessarily differ in their morphology. Unisex-ual gametophytes are found in all seed plants (both gymnosperms and angio-sperms) and most liverworts and lycophytes; in other plants (in particular, in most mosses and almost all ferns), male and female gametes are produced on the same (monoecious) gametophyte. This distribution of sex conditions in the gametophytic generation is not necessarily correlated with the distribution of sex conditions (more precisely, the homosporous or heterosporous condi-tions) in the sporophytic generation of the same plant species (see Section

Table 3.5 Land plants: distribution of sexes in the gametophytic and sporophytic phases (several sources)

	Number of species	Gametophyte	Sporophyte
Hepaticae	7500	Dioecious in 68% of species	Homosporous
Musci	9000	Monoecious in 95% of species	Homosporous
Anthocerotophyta	200	Monoecious in 60% of species	Homosporous
Lycophytina	1300	Dioecious in 65% of species	Heterosporous, monoecious in the species with dioecious gametophyte (*Selaginella*, *Isoetes*), homosporous in the others (Lycopodiaceae)
'Filices'	9000	Monoecious in 99% of species	Nearly always homosporous; heterosporous and monoecious in Salviniaceae and Marsileaceae (aquatic ferns with dioecious gametophyte)
Cicadophyta	300	Dioecious	Heterosporous, dioecious
Ginkgophyta	1	Dioecious	Heterosporous, dioecious
Gnetophyta	80	Dioecious	Heterosporous, dioecious

continues

Table 3.5 (cont)			
	Number of species	Gametophyte	Sporophyte
Coniferophyta	1050	Dioecious	Heterosporous, monoecious in 99% of species
Magnoliophyta	ca. 300,000	Dioecious	Heterosporous, monoecious in 94% of species

2.1.2). Only a small number of conifers and 6% of flowering plants have unisexual (i.e. heterosporous dioecious) sporophytes. Moreover, focusing again on the sporophytes, in the angiosperms the distribution of the sexes within the population is very diverse, both for the existence of unisexual and bisexual (or hermaphrodite) flowers, and for the possible heterogeneity of the distribution of sex conditions in the population. For an overview of the wide range of conditions and the cumbersome corresponding nomenclature, see Section 7.4.5.

The vast majority of flowers open at maturity, exposing stamens and stigmas, or in any case allowing pollinators to get access to these. This condition of open-flower reproduction (**chasmogamy**) contrasts with **cleistogamy**, in which the flower does not open, although it reaches maturity, and self-pollination occurs. Some plants are always cleistogamous, like the four-leaf allseed (*Polycarpon tetraphyllum*) and another 70 or so species, but as a rule the two conditions coexist in the same species. The production of cleistogamous flowers can be simultaneous with the production of chasmogamous ones, as in *Acanthus*; at other times it precedes it, as in *Ononis*, or follows it, as in *Oxalis* and *Viola*. Cleistogamy often occurs in low light conditions. Cleistogamous flowers are generally small and inconspicuous, with more or less reduced petals and pollen-poor anthers.

Forms of sequential hermaphroditism, where an individual changes sex over its lifetime, similar to what happens in some animals (see next section), are also found in plants. In *Arisaema triphyllum*, each year one half of the individuals change sex and every individual can change sex several times in the course of its life. In *Eurya japonica*, the frequency of sex change depends on the combination of physiological parameters such as rate of growth and environmental factors including the amount of light to which the plant is exposed (Wang *et al.* 2017).

Many substances and conditions are known which are capable of modifying a plant's sexual phenotype. The conversion of a male into a female plant is favoured by abscisic acid, auxin, carbon monoxide, ethylene, cytokinins, methylene blue, by an excess of nitrogen, by potassium, but also by age, the presence of gall-producing eriophyid mites, high light intensity, abundant fertilization, short-day photoperiod and pruning. The opposite conversion, with transition from the female to the male condition, is instead favoured by an excess of boron, by the removal of storage organs, by arid soil, by a long-day photoperiod. The answer to the same stimulus, however, can be opposite in different plants. For instance, gibberellins favour transition from male to female in *Cleome*, *Cucumis* and *Ricinus*, but the opposite transition in *Cannabis* and *Spinacia* (Freeman *et al.* 1980).

3.3.2.2 HERMAPHRODITISM IN METAZOANS

Hermaphroditism is estimated to occur in 5–6% of animal species (and almost one-third of non-insect species), with over 70% of animal phyla containing at least one hermaphrodite species (Jarne and Auld 2006). In the following phyla or subphyla of metazoans, all or almost all species (in total, an estimated ca. 65,000 species) are hermaphrodite: sponges, ctenophorans, acoels, nemertodermatids, flatworms in the strict sense (catenulids and rhabditophorans), gnathostomulids, gastrotrichs, placozoans, entoprocts, bryozoans, chaetognaths and tunicates; in addition, some cnidarians, many annelids (in particular clitellates, such as earthworms and leeches) and molluscs (solenogasters, many opisthobranchs and all gymnomorphs and pulmonates – e.g. terrestrial slugs and snails – among the gastropods, many bivalves), some nematodes, phoronids and vertebrates and a very reduced number of arthropods, brachiopods, nemerteans, tardigrades and echinoderms. There are no known hermaphrodites among the syndermata (rotifers and acanthocephalans), cycliophorans, kinorhynchs, loriciferans, nematomorphs, priapulids, onychophorans and hemichordates.

In most hermaphrodite bryozoans, all zooids in a colony are hermaphrodite, but in some species the colony is formed by unisexual zooids, those that produce only eggs and those that produce only spermatozoa. Within the crustaceans, the cephalocarids and the remipeds, as well as *Apseudes hermaphroditus* among the tanaidaceans and a certain number of isopods, are hermaphrodite. In the isopods, hermaphroditism is practically limited to some families whose representatives are parasites of other crustaceans (e.g. Bopyridae, Hemioniscidae and Cryptoniscidae) or of fishes (e.g. Cymothoidae). Among the homopterans, three species of scale insects of the genus *Icerya* are hermaphrodite; rudimentary accessory ovaries are a characteristic of the males of some species of the plecopterans *Perla* and *Dinocras*. In fishes, about two species in a hundred are hermaphrodite.

Table 3.6 Temporal and functional modes of hermaphroditism in animals
Hermaphroditism

Simultaneous (simultaneously male and female)
 	
 • **Sufficient** (self-fertilization possible)
 • **Insufficient** (cross-fertilization as a rule)
   – **with reciprocal insemination** (mutual insemination of partners during a mating)
   – **with non-reciprocal insemination** (each partner takes only one sexual role during a mating)

Sequential (male and female at different stages of life)   • **Protandrous** (male first, then female)   • **Protogynous** (female first, then male)   • **Alternating** (alternation between the two sexes in the course of life)

In some animals (Table 3.6) the individual matures as male and female at the same time (**simultaneous hermaphroditism**), whereas in others it matures as male and as female at different times (**sequential hermaphroditism**).

Simultaneous hermaphroditism is said to be **sufficient** when self-fertilization is possible, **insufficient** when cross-fertilization normally occurs. In the latter case, insemination can be **reciprocal** (when during the same mating each of the two partners acts both as male and as female), or **non-reciprocal** (when an individual takes only one sexual role at each exchange of gametes with the partner).

Sufficient simultaneous hermaphroditism, in which self-fertilization is possible (Section 3.6.1), is widespread in some zoological groups (ctenophorans, parasitic flatworms such as tapeworms and flukes, pulmonate gastropods (Figure 3.14), pyrosomid thaliaceans, and a few bony fishes among the Serranidae and Sparidae), whereas it occurs only in some species of nematodes, nemerteans, phylactolaematous bryozoans, phoronids and gnathostomulids. Self-fertilization is usually optional, and the species that practise both cross-fertilization and self-fertilization, alternating the two modes more or less regularly, are said to have a **mixed mating system** (although, this should more appropriately be called a *mixed fertilization system*). For example, the rhabditophoran *Macrostomum hystrix*, a sufficient simultaneous hermaphrodite, can resort to both cross- and self-fertilization. The latter is practised in a very unusual way: the animal uses its very thin copulatory organ

Figure 3.14 The freshwater snail *Lymnaea stagnalis*, a simultaneous hermaphrodite, is able to practise self-fertilization.

to inject spermatozoa into the front half of its own body (Ramm *et al.* 2015) (Section 3.5.2.1).

Kryptolebias marmoratus (also known as *Rivulus marmoratus*), a small cyprinodontid fish of the mangrove environments along the Atlantic coasts from Florida to southeast Brazil, is the only vertebrate that reproduces routinely by self-fertilization. Its hermaphrodite gonad (*ovotestis*) produces eggs and sperm that meet in the common segment of the genital tract. Cross-fertilization can occur occasionally between individuals of distinct clones. Male individuals show up at a low frequency. Males are of two types: primary males, in which only functional testicular tissue is present since birth, and secondary males, which were simultaneous hermaphrodites first and subsequently lost ovarian function. Cross-fertilization (in this case, as external fertilization) can occur when occasionally a hermaphrodite releases unfertilized eggs that come into contact with the sperm of these males. In nature, cross-fertilization is very rare or absent in some populations, more frequent (up to 20%) in others. Thus, strictly speaking, *K. marmoratus* is not a hermaphrodite reproducing exclusively by self-fertilization, but an androdioecious species (see following section), another trait that makes this species unique among the vertebrates (Devlin and Nagahama 2002).

Insufficient simultaneous hermaphroditism with reciprocal insemination, in which at each mating the two partners behave simultaneously as both male and female, and each of them fertilizes the other's eggs, is known in digeneans, oligochaetes (Figure 3.15), pulmonate gastropods, gnathostomulids and in the chaetognath *Spadella cephaloptera* (which exchanges spermatophores; see Section 3.4.1.3). This does not exclude the possibility that during one reproductive season the hermaphrodite will first

Figure 3.15 Earthworms are insufficient simultaneous hermaphrodites with reciprocal insemination. Here, two paired individuals are inseminating each other: the sperm of one will fertilize the eggs of the other.

Figure 3.16 Leeches (here, *Hirudo* sp.) are insufficient simultaneous hermaphrodites that as a rule do not practise reciprocal insemination: during one mating event, each partner assumes only one of the two sexual roles.

mature the gametes of one sex, then those of the other, as happens for example in the earthworms, where the sperm mature first. In these annelids, after the sexual exchange the spermatozoa of the partner are stored until the animal's own eggs are mature. This is a form of *seasonal protandry* that should not be confused with the protandry of sequential hermaphroditism (see below), where there is an actual inversion of sex.

In leeches (Figure 3.16) and in some pulmonate gastropods, on the other hand, **insufficient simultaneous hermaphroditism with non-reciprocal insemination** is usually observed.

Within **sequential hermaphroditism**, three modes are distinguished: **protandrous** (or **proterandrous**) hermaphroditism, in which the individual is first male, then female; **protogynous** (or **proterogynous**) hermaphroditism (first female, then male); and **alternating** hermaphroditism (at least two changes of sex in opposite directions).

Protandrous hermaphroditism is widespread in sponges, rhabditophorans, entoprocts, gastrotrichs (marine species), gnathostomulids (some species), nemerteans (some species), leeches and gastropods (e.g. limpets), in the sea cucumber *Pentactella laevigata* (sometimes cited under the name *Cucumaria laevigata*) and in numerous shrimps. An example among the latter is the genus *Hippolyte*, where a diatom-based diet during post-larval life advances the transition to the female phase. A particular form of protandry, with sex change from male to simultaneous hermaphrodite, is known in another shrimp genus (*Lysmata*), in the large land snail *Achatina fulica* and in the polychaetes of the genus *Capitella* (Premoli and Sella 1995). Protogynous hermaphroditism is less common than protandrous and is found among the sponges, in some species of tapeworms and in salps and doliols among the thaliacean tunicates; in another group of thaliaceans, the colonial pelagic genus *Pyrosoma*, the species that produce colonies that can grow rapidly, reaching maturity at a modest size, are usually protogynous, whereas those that produce larger colonies are mostly protandrous.

In the protandrous gastropod *Crepidula fornicata*, several individuals live stacked on top of each other (Figure 6.15). The larger and older individuals, which occupy the lowest positions, are female and the youngest, in upper positions, are male; females, however, are often fertilized by vagrant males without fixed position within the group. Some of the individuals in a stack are pure males that will never become female; others instead change at some point during development into females. Some individuals of this species lead a solitary life and may belong to one or the other sex (Section 6.2.2).

Among the fishes, sequential hermaphroditism is more common than simultaneous hermaphroditism. Protandry is documented in several orders – Anguilliformes, Clupeiformes, Cypriniformes, Stomiiformes and Perciformes – while protogyny occurs in Anguilliformes, Cypriniformes, Perciformes (e.g. the rainbow wrasses of the genus *Coris*) and Symbranchiformes. Less widespread is alternating hermaphroditism, known for example in the Serranidae of the genus *Hypoplectus*. Many fishes with simultaneous hermaphroditism at the histological level are actually sequential hermaphrodites from a functional and behavioural point of view. In *Serranus fasciatus*, older individuals may lose female function and become males *de facto*. A list of hermaphrodite fish species is given in Table 3.7.

The protandrous clownfishes (*Amphiprion* spp.) live in groups formed as a rule by a female, a large breeding male and several smaller males that do not reproduce as long as they remain subordinate. When the female dies or leaves the group, the breeding male becomes female and the largest among the other males takes its place as the breeding male.

Social control of the sex condition of the individual is also found, with reversed roles, in protogynous fishes, for example in the bluestreak cleaner wrasse (*Labroides dimidiatus*; Devlin and Nagahama 2002). In *Thalassoma*

Table 3.7 Examples of hermaphrodite fishes (data mainly after Devlin and Nagahama 2002)

COBITIDAE		
Spined loach	*Cobitis taenia*	Protandrous
RIVULIDAE		
Mangrove rivulus	*Kryptolebias marmoratus*	Androdioecious
POECILIIDAE		
Green swordtail	*Xiphophorus helleri*	Protogynous
SERRANIDAE		
Dusky grouper	*Epinephelus marginatus*	Protogynous
Mottled grouper	*Mycteroperca rubra*	Protogynous
Comber	*Serranus cabrilla*	Simultaneous
Painted comber	*Serranus scriba*	Simultaneous
SPARIDAE		
Bogue	*Boops boops*	Protogynous
Pink dentex	*Dentex gibbosus*	Protogynous or protandrous
Annular seabream	*Diplodus annularis*	Protandrous
White seabream	*Diplodus sargus*	Protandrous
Sand steenbras	*Lithognathus mormyrus*	Protandrous
Axillary seabream	*Pagellus acarne*	Protandrous
Common pandora	*Pagellus erythrinus*	Protogynous
Red porgy	*Pagrus pagrus*	Protogynous
Salema	*Sarpa salpa*	Protandrous

Table 3.7 (cont)

Gilthead seabream	*Sparus aurata*	Protandrous
Blotched picarel	*Spicara maena*	Protogynous
Picarel	*Spicara smaris*	Protogynous
GOBIIDAE		
Zebra goby	*Lythrypnus zebra*	Simultaneous
Okinawa rubble goby	*Trimma okinawae*	Protogynous or protandrous
POMACENTRIDAE		
Yellowtail clownfish	*Amphiprion clarkii*	Protandrous
Tomato clownfish	*Amphiprion frenatus*	Protandrous
Damselfishes	*Dascyllus* spp.	Protogynous
LABRIDAE		
Mediterranean rainbow wrasse	*Coris julis*	Protogynous
Cuckoo wrasse	*Labrus mixtus*	Protogynous
Ornate wrasse	*Thalassoma pavo*	Protogynous
Bluestreak cleaner wrasse	*Labroides dimidiatus*	Protogynous
Ballan wrasse	*Labrus bergylta*	Protogynous
Five-spotted wrasse	*Symphodus roissali*	Protogynous
Bluehead	*Thalassoma bifasciatum*	Protogynous
Pearly razorfish	*Xyrichtys novacula*	Protogynous
Peacock wrasse	*Xyrichtys pavo*	Protogynous
SCARIDAE		
Common parrotfish	*Scarus psittacus*	Protogynous
Stoplight parrotfish	*Sparisoma viride*	Protogynous

bifasciatum, all individuals initially have yellow skin; some of them are male, others female, but all of them can later pass to an exclusively male polychrome phase with blue, white, black and green. The gobiid *Trimma okinawae* lives in groups usually formed of a large male and one or more smaller females. In this species, 22 cases of sex inversion from female to male condition were observed, along with three cases of the opposite change, from male to female. Two individuals have been recorded to switch from female to male and then to female again (Manabe *et al.* 2007).

Alternating hermaphroditism is even less common, and is known for some polychaetes, some bivalves (including the common oyster, *Ostrea edulis*) and some fishes. Depending on species, sex inversion can occur in response to signals from conspecific individuals, or in particular physiological conditions, or as a result of mechanical trauma. Overall, alternating hermaphroditism is known for 12 genera in six fish families (Munday *et al.* 2010). In *Lythrypnus*, tiny fishes living along the Atlantic and Pacific coasts of the American continent, all possible sexual phenotypes between pure male and pure female are found. These are hermaphrodites with a prevalent female gonadal component and hermaphrodites with a prevalent male component, as in simultaneous hermaphrodites, but at any time the individual functions only as male or as female. In five species of this genus bidirectional patterns of sex change have been observed. In the presence of another female, a female can be converted into male within two weeks, and the same individual can change sex many times, depending on the environmental conditions to which it is exposed, in particular the sex ratio in the local population.

Different forms of **rudimentary hermaphroditism** have been described, both in functionally gonochoric species belonging to clades where the hermaphrodite condition is primitive and in any case dominant, such as the trematodes, and in species belonging to strictly gonochoric clades. Within the Digenea, traces of female reproductive organs are sometimes observed in the males of *Schistosoma*, one of the very few gonochoric genera of flatworms. More complex is the case of *Wedlia bipartita*, a parasite of birds, in which the permanently associated partners are both rudimentary, but functionally complementary hermaphrodites: the smaller individual has a complete male apparatus, but only rudiments of the female one, while the larger individual functions as a female, its male apparatus being atrophied (Matthes 1988). A different form of rudimentary hermaphroditism, which reveals a perhaps unexpected bi-potentiality of the vertebrate gonad (fully expressed in a number of fishes, as seen above), is the presence in the males of toads (*Bufo*) of a rudimentary ovary (*Bidder's organ*), located near the testis, which becomes functional in case of excision or regression of the male gonad.

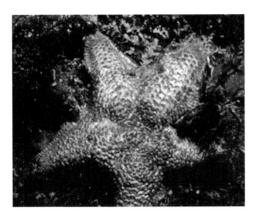

Figure 3.17 The starfish *Asterina gibbosa* is a protandrous sequential hermaphrodite, but there is individual variation in the duration of the male phase, to the point that some individuals actually remain male throughout their lives. This phenomenon is called unbalanced hermaphroditism.

Sometimes, reproductive modes vary considerably between populations of the same species. For example, the Roscoff (Brittany) population of the daisy anemone *Cereus pedunculatus* is a protogynous and viviparous hermaphrodite, while in the Mediterranean, the Naples population of this anthozoan is gonochoric, amphigonic and larviparous, and the Livorno population is gonochoric, parthenogenetic and viviparous. The marine gastropod *Patella caerulea* is most commonly a protandrous hermaphrodite, but there is individual variation in the duration of the male and female phases, and a few males never reach the female phase, while a few females do not pass through a male phase: this **unbalanced hermaphroditism** is also known in other gastropods and in bivalves, polychaetes, *Hydra*, the land isopod *Philoscia elongata* and the starfish *Asterina gibbosa* (Figure 3.17).

Freshwater snails belonging to the Basommatophora among the pulmonate gastropods are generally simultaneous hermaphrodites with a mixed mating system, with prevailing non-reciprocal cross-fertilization, plus occasional self-fertilization. In the land snail *Rumina decollata*, however, cross-fertilization can occur, but self-fertilization prevails.

In other gastropods, the reproductive behaviour depends on the particular anatomical condition of the individual. In some snails there is polymorphism for the length of the copulatory organ, with *euphallic, hemiphallic* and *aphallic* individuals: the latter reproduce only as females or practise self-fertilization.

3.3.3 Androdioecy, Gynodioecy and Trioecy

Among both plants and animals there are species in which hermaphrodite individuals are present within a population, together with males (**androdioecy**, Figure 3.11c) or with females (**gynodioecy**, Figure 3.11d) or with individuals of both sexes (**trioecy**, Figure 3.11e). These three distributions of individual sex conditions within a species (or population) are sometimes called **mixed breeding systems** (contrasting with the simple breeding systems represented by gonochorism and hermaphroditism).

Androdioecy is known in 50 plant species (see Table 7.1) and 36 species of animals, including the nematode *Caenorhabditis elegans*, several crustaceans among the cirripeds, anostracans, conchostracans and decapods, and the bivalve *Montacuta phascolionis*. In most cases, androdioecy seems to have evolved from dioecy, but in the case of barnacles it has evolved from the hermaphrodite condition. Dwarf males of the barnacle *Scalpellum scalpellum* are located on the rim of the mantle cavity of their hermaphrodite partner and mate by extending their penis, which is four times longer than their body (Dreyer *et al.* 2017).

Some sea anemones and sponges and more than 250 genera of flowering plants are gynodioecious (see Table 7.1).

Among the shrimps, the distribution of sex conditions in the population can be very diverse, sometimes among the species of the same genus or even within one species (Bauer and Newman 2004). In the genus *Pandalus*, some species are protandrous hermaphrodites, but others are gynodioecious. *Thor manningi* is trioecious (see below), but other species of the same genus (*Th. dobkini* and *Th. floridanus*) are gonochoric. Some species in the genus *Lysmata*, which pass from an initial male phase to a phase of sufficient hermaphroditism, have been categorized as *protandrous simultaneous hermaphrodites* (Bauer 2000).

Trioecy is very rare. In plants, this condition is known for example in the common ash (*Fraxinus excelsior*), in two cacti (*Opuntia robusta* and *Pachycereus pringlei*) and in some populations of annual mercury (*Mercurialis annua*; Perry *et al.* 2012). A condition close to trioecy is known in some conifers. In *Pinus johannis*, the vast majority of individuals are unisexual, with occasional production of some strobili of the other sex. *P. edulis*, however, is generally monoecious, but some individuals produce male and female strobili in different proportions. In these plants there is also a certain variation in the sex condition of the individual tree, depending on age (Flores-Renteria *et al.* 2013). In many tanaidacean crustaceans, sex is subject to very complex and little-understood environmental control. In some species there are different kinds of males: *primary males* that were already male when juveniles and *secondary* and *tertiary males* that are in fact protogynous hermaphrodites that in their female phase have generated one or two broods, respectively (Larsen 2005). In some of these species there are also *primary females*, destined to remain female throughout life. These species are therefore trioecious, while others are androdioecious, because the female condition is only found in protogynous hermaphrodites. In *Thor manningi*, a decapod crustacean, there are primary males, primary females and protandrous hermaphrodites (Bauer 1986).

Conditions similar to androdioecy and gynodioecy are found in some monoecious plants, if we focus on the sexual function in the different flowers of the same individual plant. There is *andromonoecy* when bisexual flowers and male flowers coexist on the same individual, as in *Solanum carolinense*. In this species, the basal flowers of each inflorescence are bisexual, while the terminal flowers are functionally male, having rudimentary carpels. On the contrary, there is *gynomonoecy* when bisexual flowers and female flowers coexist on the same individual. In *Aster*, within each flower head, the ray flowers are female, the disc flowers bisexual (see Table 7.1).

3.4 Germ Cells in Sexual Reproduction

In multicellular organisms, as in unicellular organisms with advanced colonial organization (e.g. in *Volvox* and other genera of green algae), there is a distinction between **germ** (or **germline**) **cells**, which are the potential genetic founders of the next generation, and **somatic** (or **somatic line**) **cells**, which will not bring any genetic contribution to the next generation.

In haplontic and diplontic organisms, and in the gametophytic phase of haplodiplonts, the germ cells give rise to the gametes, while in the sporophytic phase of haplodiplontic organisms and in the fungi the germ cells produce (meio)spores.

In addition to the cell differentiation that characterizes the development of germ cells, these often occupy a distinct position within the body (or colony); this separation (or *segregation*) from the other cells becomes more precise during ontogeny. At the time the individual reaches reproductive maturity, the production of gametes (**gametogenesis**) or spores (**sporogenesis**) is commonly carried out in specialized body districts or organs, different from group to group, called **reproductive organs**. An early segregation of the germline in the course of development is found only in some groups of metazoans (Box 3.4), but see Lanfear (2018) for possible evidence of early setting aside of germ cells in plants.

3.4.1 Production of Gametes in Animals

In metazoans, regardless of the early or late segregation of germline cells, the production of gametes is generally (but not always) localized in reproductive organs that are characteristic of the different taxonomic groups.

Box 3.4 Germline Segregation in Animals

The notion, introduced by Weismann (1892), of a very early separation between germ and somatic cell lines, in the embryonic development of metazoans, is often taken for granted. However, this is an unjustified generalization (Extavour and Akam 2003, Whittle and Extavour 2017). A very early separation between germ cells and somatic cells is found in fact only in the following taxa: Rotifera, Digenea, Cephalopoda, Bivalvia, Nematoda, Collembola, Chaetognatha, Anura and Archosauria (crocodiles and birds). A later differentiation, in the course of development, of the progenitor cells of the future gametes has been described instead for Porifera, Ctenophora, Cnidaria, Acoela, Gastrotricha, the flatworms excluding the Digenea, Bryozoa, Nemertea, Phoronida, Brachiopoda, Aplacophora, Polyplacophora, Kinorhyncha, Myriapoda, Xenoturbellida, Hemichordata, Crinoidea, Asteroidea, Holothuroidea, Agnatha, Dipnoi, Urodela, Lepidosauria, Testudinata and Mammalia. Both conditions occur in gastropods, annelids (only preformation in the leeches, epigenesis in echiurans and sipunculans), Onychophora, Insecta, Crustacea, Chelicerata, Echinoidea, Cephalochordata, Tunicata, Osteichthyes and Chondrichthyes.

It is not always easy, however, to decide whether or not a cell belongs to the germline. The problem is most difficult in cases of uniparental reproduction, in which the cell from which the new individual originates could be an egg that has not been produced by meiosis (an amictic egg) and which has not been fertilized (one would describe this as parthenogenesis), or a somatic cell from which a new individual originates by asexual reproduction. This is the case, for example, in some phases in the complex life cycle of the digeneans (Section 2.2). In these parasitic flatworms there is still uncertainty surrounding the nature of the reproductive process through which sporocysts or rediae give rise to other sporocysts or rediae, or to cercariae destined to grow up into maritae, i.e. into adults of the next sexual generation. In any case, no evidence of meiosis has been found so far in these animals (Galaktionov and Dobrovolskij 2003).

3.4.1.1 REPRODUCTIVE ORGANS IN ANIMALS

In some animal groups, e.g. sponges and nemetodermatids, gametogenesis is *diffused*, as it does not occur in a dedicated place or organ. In others, it is limited to clusters of cells attached to an epithelium, for example to the wall of the coelomic cavities, as typically occurs in annelids. The lack of true reproductive organs is often coupled with the lack of specialized pathways for the

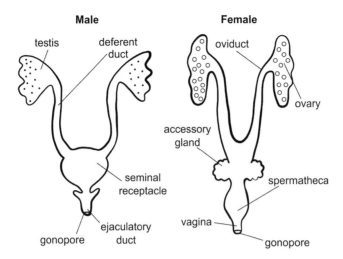

Figure 3.18 General scheme of male and female reproductive systems in metazoans.

release of gametes, which occur instead for example through the coelomoducts or (in some polychaetes) by simply breaking through the body wall.

Most animals, however, have distinct **reproductive organs**, collectively part of the **reproductive system** (or **reproductive apparatus**), among which there are the **gonad** (female gonad or **ovary** and male gonad or **testis**) where germ cells are found accompanied by specialized somatic tissues, a canal (the female **oviduct** or the male **deferent duct**) and an opening (**genital pore** or **gonopore**) through which the gametes reach the outside world (Figure 3.18).

Since in the different animal groups germline differentiation occurs at different stages of embryonic development (see Box 3.4), in some cases the germ cells differentiate before the somatic tissues of the ovary, while in others the opposite occurs. Differences in this sense can also occur within relatively circumscribed taxonomic groups, as in the Diptera, where Chironomidae, Cecidomyiidae, Sciaridae, Tipulidae and Psychodidae show a precedence of germline differentiation with respect to the formation of the gonad, while the opposite occurs in Drosophilidae, Calliphoridae and Muscidae.

The reproductive system is often complemented by other organs. For example, in the female, in addition to the ovary there can be: (i) a **vitellarium**, inside which are formed vitelline cells in which nutritional reserves for the embryo are stored (Section 4.4.3); (ii) a structure (**spermatheca**) for the storage of the spermatozoa received from the partner, where these will be kept alive until they are used in the fertilization of the eggs; (iii) **accessory glands** producing material that will form a protective coating around the egg before it is released outside; (iv) copulatory structures (e.g. a **vagina**) designed to accommodate the intromittent organ of the partner; (v) structures responsible for the digestion of excess sperm and/or seminal fluid (a function that in many flat-worms seems possible thanks to a genitointestinal duct that connects the female genital tract with the intestine); and (vi) structures capable of hosting the offspring for a more or less long period, in the case of viviparous species. In the male, the following organs are often found: (i) **seminal receptacles** in which mature sperm cells are stored waiting to be transferred to the partner; (ii) an **ejaculatory duct**, connecting the receptacle to the gonopore; (iii) a copu-latory or intromittent organ, the nature and name of which are different in different taxa (e.g. **penis**, **aedeagus**, **male gonopod**), which may or may not correspond to the genital opening; and (iv) glands, also differing according to taxon. In several taxa, a single organ, especially in the female reproductive system, performs more than one function.

In the case of hermaphrodite animals, male and female structures are usu-ally distinct, but in some groups (hermaphrodite molluscs, some fishes), there is only one gonad with both functions, called an **ovotestis**. In many herm-aphrodite animals, also, male and female genital openings are distinct, whereas in others there is only one opening, common to both kinds of gametes. Both conditions are known in the flatworms. In some animals, such as the cicadas and other homopterans, but also in the tantulocarids among the crustaceans, the vagina that receives the male copulatory organ is distinct from the genital opening through which the eggs are released. In marsupials, the young are born through a path (*pseudovagina*) different from those passed by the spermatozoa to reach and fertilize the eggs (*lateral vaginae*).

There is not always a correspondence between the structural complexity of the reproductive system and the complexity of other body structures. Con-sider, for example, the contrast between the extreme complexity of the repro-ductive system of most flatworms (both parasitic and free-living ones like the planarians) and its simplicity in polychaetes. The latter have no gonads (gametes are formed on the walls of the coelomic cavities), and distinct gono-ducts are often lacking: in this case, gametes are delivered through the excre-tory organs or simply following rupture of the body wall.

3.4.1.2 GAMETOGENESIS IN ANIMALS

In sperm formation (**spermatogenesis**, Figure 3.19a), the diploid germ cells, or **spermatogonia**, differentiate into **primary spermatocytes**. Through the first meiotic division, these produce two **secondary spermatocytes**, each of which divides into two haploid **spermatids** at the completion of the second meiotic division. Each spermatid then matures into a **spermatozoon**.

In the formation of the egg (**oogenesis** or **ovogenesis**, Figure 3.19b), diploid germ cells differentiate as **oogonia** and finally as **primary oocytes**.

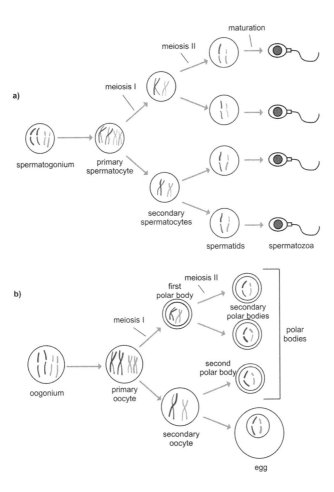

Figure 3.19 Gametogenesis in animals: (a) spermatogenesis; (b) oogenesis.

Meiosis then proceeds with the asymmetric division of the primary oocyte into a **secondary oocyte** and a **first polar body**. Next, at the second meiotic division, the secondary oocyte divides, asymmetrically again, into a haploid **egg** and a haploid **second polar body**, which is therefore the meiotic sister product of the egg cell. The first polar body can also complete the second meiotic division, forming two haploid polar bodies that are not sisters of the egg cell, called **secondary polar bodies**. Polar bodies have no role in reproduction except in some forms of parthenogenesis (Section 5.2.3.3).

In metazoans, maturation of oogonia and spermatogonia is preceded by a phase of multiplication of these germ cells, during which, as a rule, the sequence of mitotic divisions is not accompanied by a significant increase in the total mass of cells, which therefore become smaller and smaller. This phase is not necessarily simultaneous for all germ cells in a gonad. In particular, in animals that repeatedly produce considerable masses of eggs at different times, distinct batches of oogonia in the same gonad complete the multiplication phase at different times. On the contrary, in the females of mammals and rotifers, the multiplication phase is completed, or almost completed, during the embryonic life of the individual.

In women, the population of primary oocytes reaches a maximum of about 7 million at the twentieth week of fetal life. Most of them are destined to degenerate (*follicular atresia*), but the rest (about 400,000–500,000) will mature up to the stage of secondary oocytes. Approximately every 28 days from puberty to menopause there is the ovulation of a secondary oocyte (for a total of 400–500 during the whole life), which matures as an egg only in the case of fertilization. Instead, in human males, as in the males of other mammals, spermatogenesis is an almost continuous process extending into advanced age.

3.4.1.3 GAMETES IN ANIMALS

The egg is generally a large or very large cell, owing to the accumulation of storage materials (*yolk*) which will sustain the development of the embryo at a time during which the animal does not have organs for the intake of nourishment from outside.

Despite the almost universal presence of yolk, egg size is usually less than 1 mm, both in animals with external fertilization and in those with internal fertilization. Echinoderms, for example, typically produce eggs with a diameter of 0.1 mm, while those of mammals (except those of monotremes, which are laid) vary from 0.07 to 0.25 mm. Larger eggs, however, are not uncommon in teleost bony fishes (1–6 mm) and especially in elasmobranchs (15–100 mm) and even more so in oviparous land vertebrates. Among the birds, the smallest eggs are those of the hummingbirds, the smallest of which are just 10 mm in length, weighing 0.2 g; the largest are those of the ostrich (*Struthio camelus*), about 16 cm long, with an average diameter of 13.8 cm and a weight in the order of 1.6 kg. Examples of birds with egg size between these extremes are the chaffinch (*Fringilla coelebs*; 19 × 15 mm, 2.1 g), the carrion crow (*Corvus corone*; 40 × 30 mm, 19 g), the black-headed gull (*Chroicocephalus ridibundus*; 52 × 35 mm, 38 g), the white stork (*Ciconia ciconia*; 77 × 53 mm, 118 g), the mute swan (*Cygnus olor*; 11.5 × 3.8 cm, 340 g) and the emperor penguin (*Aptenodytes forsteri*; 131 × 86 mm, 450 g). By weight, an ostrich egg represents 1.7% of the animal that produces it, but this percentage rises to 13% in the Eurasian wren (*Troglodytes troglodytes*), 20% in kiwis (*Apteryx* spp., Figure 3.20) and 22% in the European storm petrel (*Hydrobates pelagicus*).

The egg can be coated with one or more *envelopes* additional to the cell membrane: (i) primary envelopes produced by the oocyte in the ovary, (ii) secondary envelopes produced by follicular cells, also in the ovary, and (iii) tertiary envelopes, produced in the oviduct or the uterus. Only a vitelline

Figure 3.20 X-ray of a female kiwi (*Apteryx* sp.) showing the enormous proportions of the egg.

membrane surrounds the eggs of many marine invertebrates such as sponges and cnidarians, and also those of sea urchins, where however this membrane is thicker. In some fishes, a secondary envelope (*chorion*) is added to the vitelline membrane. A chorion also surrounds the eggs of the arthropods. In mammals, the primary vitelline membrane, subtle and transitory, is replaced by a secondary envelope, the *zona pellucida*. Tertiary envelopes are the jelly coating of the eggs of amphibians, which is produced in the oviduct, and the egg white (albumen), the egg-shell membranes and the membranous or calcareous shell of the eggs of reptiles, birds included. Many eggs are protected by secretions from the accessory glands of the female reproductive system: there are examples in molluscs, insects, clitellate annelids and flatworms; analogous is the origin of the silk cocoon that protects the eggs of spiders.

The primitive animal **sperm** is a small cell in which three parts are typically recognizable: an ovoid *head* that contains the nucleus and is very often provided with an anterior *acrosome*; a *midpiece* containing mitochondria; and a *tail*, corresponding to a flagellum, often of a length in the order of 50 µm. However, deviations from this model are frequent, especially in animal groups that practise internal fertilization.

The **acrosome** is a structure that facilitates penetration by the spermatozoon through the protective envelopes of the egg, the vitelline membrane in particular. In many marine animals, this occurs as a result of the explosive discharge of the *acrosomal filament* on the egg. The *acrosome reaction* (Section 3.5.4.1) leads to the fusion of the acrosome membrane with the plasma membrane of the egg, thus allowing penetration by the male nucleus.

A typical acrosome is missing in sponges and cnidarians, but it is present in the ctenophorans. The acrosome has been lost in several groups, for example in the aflagellate sperm of dicyemids, rotifers and acanthocephalans, entoprocts, some bryozoans and kinorhynchs. In arthropods, a primitive spermatozoon is found only in the xiphosurans. In this phylum the structure of the *axoneme*, the cytoskeleton of the tail, shows considerable variation. In addition to the 9 +2 formula (nine microtubule doublets on the periphery, plus two microtubules in the centre) there are axonemes without central microtubules (9+0), axonemes with a number of central microtubules different from two and also variable in the same species, and even more aberrant formulae, for example 12 +0. In the pycnogonids of the genus *Nymphon* the central microtubules are missing, while the number of peripheral doublets is different from species to species: 9 in *N. rubrum*, 12 in *N. leptocheles*, 18 in *N. gracile*. A unique level of morphological diversity is found in the sperm tail of gall midges (Cecidomyiidae), where the number of microtubule doublets can be as high as 1000 (*Diplolaboncus*) and even 2500 (*Asphondylia ruebsaameni*) (Dallai 2014). Among the crustaceans, only mystacocarids, cirripeds and branchiurans produce flagellated spermatozoa. Nematode sperm cells are devoid of both flagellum and acrosome. There are also animal sperm with two flagella. This condition is very widespread in the flatworms (with a 9+1 axoneme) but is also found in the polychaete genus *Tomopteris* (with a 9+0 axoneme).

The spermatozoon of the pelagic tunicate (appendicularian) *Oikopleura* is exceedingly small, just 30 μm. By contrast, some small invertebrates produce very long spermatozoa (Smith *et al.* 2016). Ostracods of the superfamily Cypridoidea have some of the longest sperm in the animal kingdom, second only to a few insects (Vogt 2016). Sperm length in cypridoidean species ranges from 268 μm to 11.8 mm, and from 0.33 to 4.3 times the carapace length of the male that produces them. For example, the 3.5 mm long ostracod *Australocypris robusta* produces spermatozoa that reach 12 mm. Among the insects, there are sperm cells of exceptional size in some species of Coleoptera (up to 10 mm), Lepidoptera (up to 12.5 mm) and Hemiptera (up to 16.5 mm), up to the extreme value of 58 mm reached by the spermatozoa of *Drosophila bifurca* (Figure 3.21), 20 times longer than the whole animal. Also noteworthy is the spermatozoon of *Zorotypus impolitus*, an insect belonging to the tiny order Zoraptera, measuring 3 mm in length, just over the length of the entire animal (Dallai *et al.* 2014). Also relatively huge, compared to the size of the animal, are the spermatozoa of the macrodasyid gastrotrichs, with a length of 120–180 μm in an animal of 160–320 μm.

The spermatozoa of the Decapoda Reptantia, which include freshwater crayfish, lobsters and crabs, are aflagellate and non-motile '*explosion sperm*': they can achieve sudden short-term motility by a special acrosomal reaction

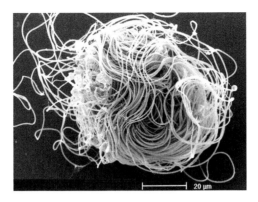

Figure 3.21 A single 60 mm spermatozoon of *Drosophila bifurca*.

characterized by an abrupt eversion of the acrosome, which causes a leap-like forward movement.

Some of the spermatozoa produced by an animal may have a function other than as gametes. In scorpions, masses of spermatozoa are left on the female's genital opening as a copulatory plug, reducing the risk of competition from subsequent inseminations. In some Hemiptera, such as the bed bug (*Cimex lectularius*), a part of the sperm is digested and turns into a source of nourishment that the female can exploit in the production of eggs (see also Section 4.1.1).

Several gastropods produce two types of spermatozoa, some (*euspermatozoa* or *eupyrene spermatozoa*) with a normal haploid set of chromosomes and others (*paraspermatozoa* or *apyrene spermatozoa*) in which the nucleus is reduced or even lacking. Dimorphic sperm cells are also known for the bivalve *Montacuta tenella*, the polychaetes of the genus *Siboglinum*, the centipedes (except for the geophilomorphs) and some representatives of the copepods and lepidopterans.

In some gastropod species of the Littorinidae and Scalidae, and in *Janthina*, paraspermatozoa consist of a fibrous plate which extends into a long peduncle to which many euspermatozoa are attached, thus forming a complex called a **spermatozeugma** (or **sperm bundle**). With different mechanisms, spermatozeugmas are also produced by other animals, especially by species with internal fertilization, such as the teleost fish *Poecilia reticulata*, where each bundle includes about 30,000 sperm cells (Figure 3.22; Evans *et al.* 2004).

In several groups – some polychaetes (e.g. Siboglinidae, formerly Pogonophora), leeches, onychophorans, arthropods, urodeles, kinorhynchs, phoronids, cephalopods, gastropods, bivalves, and even amphibians and cyprinodontid fishes among the vertebrates – the spermatozoa are grouped in **spermatophores**, which are deposited by the male on the ground or, in aquatic environments, on the seabed or on a suitable submerged support, or directly delivered to the female. The spermatozoa are enclosed in a capsule of

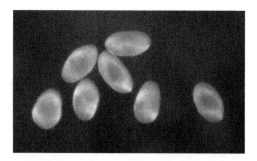

Figure 3.22 Oval sperm packages (spermatozeugmata) of the fish *Poecilia reticulata*. Each package is 0.15–0.25 mm long and includes up to 30,000 spermatozoa.

albuminoid or pseudochitinous nature, or agglutinated around an axile support, similar to a plume, to be used in fertilization (almost always internal). Spontaneously, or induced by the male, the female collects the spermatophore and brings it or at least its contents into contact with her genital opening.

3.4.2 Production of Gametes and Spores in Plants

In embryophytes, and in plants in general, germ cells do not separate from somatic cells during early development. However, plants usually have reproductive organs that are well identified and characteristic of the various taxonomic groups.

3.4.2.1 REPRODUCTIVE ORGANS, GAMETOGENESIS AND SPOROGENESIS IN LAND PLANTS

In land plants there is alternation of generations, and sexual reproduction is found (although in different forms) in both the gametophyte and the sporophyte (Section 5.2.2.1), which therefore have distinct reproductive organs.

The gametes are produced by the gametophyte in structures that have the value of reproductive organs, called **gametangia**. The gametophyte can be monoecious (or hermaphrodite), that is, carrying both male (**microgametangia**) and female gametangia (**megagametangia**), or dioecious (or gonochoric) and thus producing a single type of gamete.

In land plants gametangia are always multicellular and the germ cells (which will give rise to the gametes) are always surrounded by a coating of *sterile cells* (analogous to the somatic cells of animal gonads) with a protective function. Except in the angiosperms, the microgametangium is called the **antheridium** and the megagametangium is called the **archegonium** (Figure 3.23a). Some plants also produce larger reproductive structures of which the gametangia are parts: examples are the pedunculated *antheridiophores* and *archegoniophores* of the liverworts.

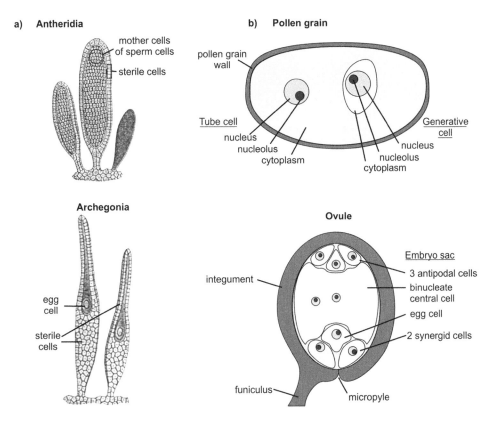

a) Antheridia

mother cells of sperm cells
sterile cells

Archegonia

egg cell
sterile cells

b) Pollen grain

pollen grain wall

Tube cell
nucleus
nucleolus
cytoplasm

Generative cell
nucleus
nucleolus
cytoplasm

Ovule

Embryo sac
3 antipodal cells
binucleate central cell
egg cell
2 synergid cells

integument

funiculus
micropyle

Figure 3.23 Schematic structure of the reproductive organs of the gametophyte in land plants. (a) Male (antheridia) and female (archegonia) gametangia (shown here as they are in mosses). (b) Pollen grain (microgametophyte) and embryo sac (megagametophyte) within the ovule of an angiosperm, where the reproductive organs coincide with the whole individual in the gametophytic phase.

The gametophytes of the seed plants are very small, and the whole gameto-phyte takes on the role of a reproductive organ. The male gametophyte develops from a microspore (see below) and includes only the few cells of a pollen grain. This has a different structure in the different groups. To simplify, within a pollen grain we generally recognize a **sterile cell** (none in the Cupressaceae and the flowering plants) and a **vegetative cell** (also called the **tube cell**, because it will produce the pollen tube), both mononucleate, and a **generative cell** which, after a mitosis without cytokinesis, becomes binucleate. In most gymnosperms, this binucleate cell takes on the function of a sperm cell (with two **sperm nuclei**) (Figures 3.23b and 3.24). Instead, in the Cupressaceae and in the flowering plants the karyokinesis of the generative cell is followed by cytokinesis to form two **sperm cells** complete with membrane and wall. In the Pinaceae two *prothalliar* cells complement sterile, tube and generative cells to form a five-cell mature pollen grain (Fernando *et al.* 2005).

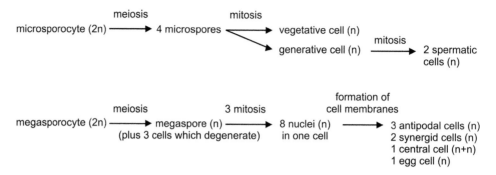

Figure 3.24 Sporogenesis and gametogenesis in angiosperms (*Polygonum* type), schematic.

The structure of the female gametophyte of the seed plants, which develops from a megaspore (see below), differs more extensively from group to group than does the microgametophyte. In the gymnosperms the megagametophyte is initially syncytial, with from 256 (*Taxus baccata*) up to about 8000 nuclei (*Ginkgo biloba*); in a subsequent phase, 1–25 archegonia, but usually 2–4 (depending on species) differentiate, each of them with an **egg cell**. In the flowering plants, the megaspore develops through three successive mitotic divisions into a megagametophyte (the **embryo sac**). The eight haploid nuclei obtained from these mitoses are found at first within a single cell. Subsequently, membranes and cell walls are formed which separate three **antipodal cells**, a binucleate **central cell**, two **synergid cells** and an **egg cell**, for a total of seven cells (Figures 3.23b and 3.24).

Spores are produced by the sporophyte in structures called **sporangia**. In heterosporous species, sporangia producing female spores are called **megasporangia**, and those producing male spores are called **microsporangia**. In the megasporangia, diploid **megasporocytes** undergo meiosis, producing four haploid spores, of which only one will become a **megaspore**, while the others degenerate. In the microsporangia, diploid **microsporocytes** produce, by meiosis, four haploid **microspores** each (Figure 3.24). Sporangia can in turn be borne on specialized structures called *sporangiophores* (for example in horsetails) or *sporophylls* (for example in ferns and seed plants), often of two kinds, *megasporophylls* and *microsporophylls* (e.g. in the flowering plants), or in even more inclusive structures, such as the **strobili** of gymnosperms (e.g. pine cones) and the **flowers** of angiosperms. In seed plants, the complex of a megasporangium plus the nearest tissues of the sporophyte that surrounds it is called an **ovule** (Figure 3.23b). Following the development of a megaspore into a megagametophyte and the subsequent fertilization of the egg cell contained in the latter, the ovule will develop into a **seed** (Section 4.6.3). In gymnosperms, ovules are borne on the surface of the

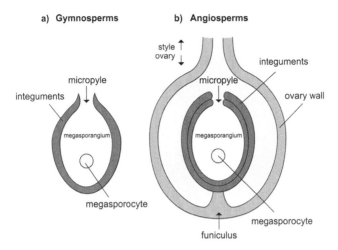

Figure 3.25 Schematic structure of female reproductive organs in the sporophyte of land plants: (a) gymnosperms; (b) angiosperms. At the developmental stage shown here, all the tissues are diploid.

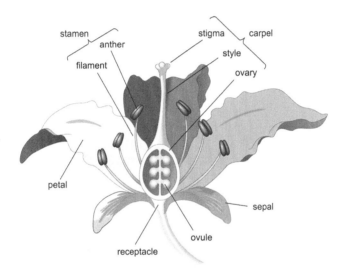

Figure 3.26 Schematic structure of a flower.

sporophylls, while in angiosperms they are instead enclosed in an **ovary** formed by the female sporophylls (**carpels**, Figure 3.25).

In angiosperms, the reproductive structures are grouped in the **flower** and are represented by male (microsporophylls or **stamens**) and female sporophylls (megasporophylls or **carpels**, Figure 3.26). In a typical stamen, a sterile proximal portion (the **filament**) is distinguished from a fertile distal portion (**anthers**) with the mother cells of the microspores. The carpels of a flower are generally fused to form a **pistil**, whose dilated lower part (**ovary**) houses the

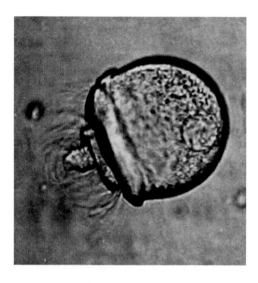

Figure 3.27 The sperm cell of *Ginkgo biloba* has about 1000 flagella.

mother cells of the megaspores. The upper part of the pistil includes the **style** and the terminal **stigma**, the structure which receives the pollen.

In monoecious angiosperms, the flower can be **bisexual** (or **perfect**), i.e. bearing both stamens and carpels, or **unisexual** (or **imperfect**), that is, carrying only male organs (**stameniferous** flower) or only female organs (**pistilliferous**, or **carpellate** flower). Dioecious flowering plants clearly have only unisexual flowers. For the distribution of sex conditions within the population in angiosperms, see Section 7.4.5.

3.4.2.2 GAMETES IN LAND PLANTS

The oldest and most conservative embryophyte lineages have flagellated male gametes (*antherozoids*, or *spermatozoids*), which reach the immobile female gametes (*oospheres*) by moving through the film of water in which a plant can, at least temporarily, be covered. The antherozoids of the bryophytes and lycopodiophytes have two flagella; those of the horsetails and ferns have many flagella, as have the antherozoids of two groups of gymnosperms (cycads and ginkgo, Figure 3.27). In the other groups, including all flowering plants, there are no flagellated male gametes, but only cells with the value of gametes, or sperm cells, which are part of the tiny male gametophyte (Palevitz and Tiezzi 1992).

In some gymnosperm genera (*Gnetum, Ephedra*) the male gametophyte produces a single binucleate sperm cell; through a pollen tube similar to the one produced by flowering plants, the two nuclei of the sperm cell reach two ovules and fertilize them, producing two twin zygotes (false twins; *simple polyembryony*, see Section 3.1.2.4). Also in most conifers and in *Welwitschia* only one binucleate sperm cell descends along the pollen tube, but here only one nucleus fertilizes an egg cell, while the other degenerates. In the Cupressaceae and in the flowering plants, instead, the male gametophyte produces two sperm cells (exceptionally more than two, up to 14 in *Cupressus arizonica*), with one nucleus each. In cypress trees, the two sperm cells can fertilize two distinct egg cells of the same ovule, but generally only one of the two zygotes develops into an embryo. In the flowering plants one sperm cell fertilizes the egg, forming a zygote, while the other fertilizes the central cell, forming the secondary endosperm, in a process called *double fertilization* (see Section 3.5.4.2).

The female gamete of the land plants is an immobile cell, sometimes easily recognizable, as in the case of the **oosphere** produced inside specialized structures (**archegonia**) of the gametophytes of bryophytes, pteridophytes and gymnosperms. In the angiosperms, the egg cell, flanked by the two synergid cells, is located at the end of the megagametophyte where the *micropyle* is located, an opening in the integuments of the ovule through which pass the two sperm cells carried by the pollen tube.

3.5 Biparental Sexual Reproduction (Amphigony)

This is sexual reproduction *par excellence*, where two distinct sexually compatible individuals (parents) undertake a sexual exchange that leads to the generation of new individuals with a genetic constitution obtained from the association and/or the reassortment of their genomes (Figure 3.28a). The key event in this mode of reproduction, also called **amphigony**, is the fusion of the two gametes produced by the parents to form a zygote (*syngamy*). In the following sections we will deal with the ways the two gametes can meet and merge. For the cytogenetic aspects of syngamy see Section 5.2.3.1.

3.5.1 Encounter and Fusion of Gametes

The **release** and/or **transfer** of the spermatozoa should not be confused with *fertilization*, which strictly speaking coincides with *syngamy*, the merging of the male with the female gamete. The meeting of the two gametes can take place in very different ways, regardless of whether, in the end, *fertilization* is *external* or *internal*. The release of gametes into the external environment is known as **spawning**; when male gametes are released in the immediate vicinity of the eggs, whether internally or externally, the process is called **insemination**.

External fertilization occurs when gametes meet freely in the environment, generally in water or another fluid where, in most instances, at least the male gametes can swim. In *Plasmodium* spp., unicellular parasites many of which cause malaria in humans and other vertebrates, fertilization occurs in the intestine of the intermediate host, a mosquito, filled with blood sucked

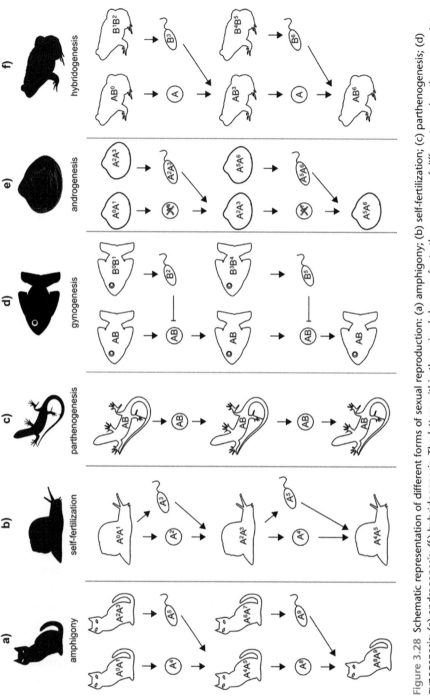

Figure 3.28 Schematic representation of different forms of sexual reproduction: (a) amphigony; (b) self-fertilization; (c) parthenogenesis; (d) gynogenesis; (e) androgenesis; (f) hybridogenesis. The letters within the animal shapes refer to the genomes of different species, the superscript numbers to the set of chromosomes. In a gamete, a number that differs from those of the individual that produced it indicates recombination (e.g. crossing over). To exemplify parthenogenesis, a form of ameiotic parthenogenesis of hybrid origin is shown; to exemplify androgenesis, a form with diploid sperm cells.

Figure 3.29 The males of most anuran species (here, the European common frog, *Rana temporaria*) practise a form of external insemination, releasing the spermatozoa directly over the eggs just laid by the female (visible under water surface).

from a warm-blooded vertebrate. External fertilization can be obtained in different ways; more often, and more simply, through the release and free dispersal in the water of gametes of both sexes (as in algae with isogamety or anisogamety, but also in many marine invertebrates), or by **external insemination**. The latter way is obtained when the male releases sperm directly over the eggs just laid by the female (as in most of the anurans; Figure 3.29), but also in the case where sperm cells are transferred by means of spermatophores (Section 3.4.1.3) if the female (as in symphylans) receives the spermatophores not in the genital tract but in another receptacle, and releases the sperm onto the eggs at a later time, when these are laid.

Release of gametes into the water, followed by external fertilization, is a primitive strategy, simple but burdensome in terms of energy, although functional in the case of sessile organisms that live in dense populations, like many algae and a number of marine invertebrates. But many small mobile aquatic invertebrates do not practise it. Auxiliary systems that increase the chances of meeting between gametes have evolved independently in different animal groups. These include: (i) synchronization in the release of gametes among all individuals of the population, in relation to a specific phase of the lunar calendar and/or a specific tidal condition (sponges); (ii) gatherings of reproductive individuals (many medusae); (iii) stability of pair bonds for the whole reproductive period (females seized by males in isopods and anurans); (iv) construction of nests where eggs are laid (many fishes, horseshoe crabs); (v) release of eggs only in the presence of sperm (sipunculans); (vi) physical contact between the partners (nemerteans).

Fertilization is said to be **internal** when the male gamete merges with the egg while the latter is still inside the body of the female parent, or protected by it in body folds or cavities. In gonochoric animals and in those with insufficient hermaphroditism, internal fertilization presupposes transfer of sperm from the male to the female. This usually involves active **internal insemination** (Section 3.5.2.1), or the production of **spermatophores** (although, as mentioned above, sperm transmitted by spermatophores do not always end up fertilizing the eggs inside the female's body). In some orthonectids (tiny animal parasites of other marine invertebrates), the whole male enters the female during fertilization.

In many cases, however, internal fertilization is accomplished by the release into the external environment of male gametes that individually swim to reach the egg cells retained by the females. This form of fertilization, a sort of halfway house between external and internal fertilization, is also called **in-situ fertilization**. This is practised by many algae, mosses and ferns, and by sessile aquatic animals such as many sponges, corals, entoprocts and bryozoans.

3.5.2 Modes of Gamete Encounter

The ways in which the gametes come into contact differ very much from group to group, and in many cases even between closely related species. Here we discuss, in general terms, what is observed in animals and in land plants.

3.5.2.1 SPERM TRANSFER IN ANIMALS

In terrestrial animals there are various modes of transfer of spermatozoa (Proctor 1998):

1. In **completely dissociated transfer** there is no physical or chemical interaction between male and female, and the latter stumbles by chance across a spermatophore released by a conspecific.

2. In **incompletely dissociated transfer**, the male interacts chemically or by touch with a female before depositing one or more spermatophores, but does not court a particular female.

3. In **mating with indirect transfer** the male courts a particular female before, during, or after depositing a spermatophore, and often directs the partner towards the spermatophore it has just released.

4. In **direct transfer** (or **copulation**) the male releases the ejaculate directly into the female structure suitable for receiving sperm. The male usually has an *intromittent organ* (*copulatory organ*), but copulation may also consist of the simple juxtaposition of the genital openings of the two partners. This behaviour, common among polychaetes and in some nemerteans and typical of earthworms, is called **pseudocopulation**.

In crustaceans, the male transfers spermatozoa or spermatophores directly to the female; only one species of hermit crab (*Pagurus prideaux*) attaches the spermatophore to the substratum (the shell in which it lives) and the female collects it from there. Hexapods have a greater variety of modes: in the pterygotes (winged insects) sperm transfer is direct, but it is indirect in proturans and diplurans. More diverse modes of sperm transfer have evolved in collembolans, and also in the centipedes. In millipedes, the transfer of sperm is mostly direct, but not so in the pincushion millipedes (Pselaphognatha) and pill millipedes (Glomerida). Symphylans and pauropods also have dissociated sperm transfer.

Xiphosurans and pycnogonids have external fertilization, while the arachnids have internal fertilization. Spiders, harvestmen, solifuges and ricinuleans transfer sperm by copulation, as also do ticks and many other mites. The transfer is indirect in the other arachnids, with association between the partners in scorpions, whip spiders, whip scorpions, many pseudoscorpions and some mites. The two partners remain dissociated, however, in other representatives of the pseudoscorpions and mites. Indirect transfer and copulation coexist in some groups, as in the collembolans, and even within one genus (e.g. in *Eylais* and *Unionicola*, two genera of aquatic mites).

Devices to facilitate sperm expulsion (e.g. muscular pumps) are present in the penis of mammals, in the aedeagus of insects and in the palps of spiders. In comparison, spermatophores may seem to offer just passive support, but this would be an incorrect assessment. Spermatophores of scorpions and some pseudoscorpions (Atemnidae and Withiidae) contain mechanical devices comparable to levers, and those of other pseudoscorpions, crickets and cyprinodontid fishes use osmotic systems. Spermatophores of ticks open in a very special way: when the male inserts the neck of the spermatophore into the genital tract of the female, a chemical reaction takes place in the bulb; this produces CO_2, causing explosive expulsion of the contents.

In some animal groups, the copulatory organ is located at a considerable distance from the male genital opening. In these cases, before copulation it is necessary for the male to transfer the sperm from its genital opening to the copulatory organ. This is the case in dragonflies and damselflies, spiders and cephalopods.

In the dragonflies and damselflies (Odonata), the male copulatory organ is located on the ventral side of the second abdominal segment, while the genital pore opens on the ventral face of the ninth segment (the same position in which the female genital pore is located). This peculiar anatomical arrangement explains the curious positions assumed by the two partners during mating: the male clings to the female by grasping the first segment of her thorax, using his terminal abdominal appendages. The female, meanwhile, flips her abdomen forward, eventually bringing her genital opening into contact with the secondary copulatory organ of the male: this results in the characteristic heart-shaped posture of the two partners (Figure 5.14).

In spiders, the copulatory function is entrusted to the male pedipalps, often very profoundly modified with respect to the other appendages. In two genera of Theridiidae (*Tidarren* and *Echinotheridion*), the sub-adult male amputates one of his two pedipalps. As an adult, he therefore possesses only one copulatory organ. During mating, the female snatches the latter from the body of the male, which becomes food for her (Knoflach and van Harten 2000). The torn pedipalp remains inserted for hours in the female's genital tract, continuing to release its sperm load (Michalik *et al.* 2010) and, most likely, preventing sperm competition from other males.

In most cephalopods, one arm, sometimes strongly modified, has copulatory functions. Sometimes it ends up detached from the male's body, remaining attached to the female; initially believed to be a parasite, this is still known as an *ectocotylus*, originally attributed to it as the taxonomic name of an alleged genus of parasitic worms. In some animals, the male copulatory organ is not introduced into the female's genital tract, but is used for the **insemination by hypodermic** (or **traumatic**) **injection** through the wall of a body part devoid of openings, although frequently specialized, as in the females of the bed bugs (*Cimex*) and relatives (Figure 3.30), in which a specialized organ (*spermalege* or *organ of Ribaga*) has the dual function of limiting damage from traumatic insemination, being covered by cuticle with lower resistance to penetration and which heals more easily, and of intercepting the path of the sperm to the eggs to be fertilized, in the meantime converting a considerable amount of sperm into a source of nourishment. Even in some animals with non-hypodermic insemination, for example in many pulmonate gastropods, most spermatozoa transferred to the partner end up serving a nutritional function (Section 4.1.1).

Hypodermic insemination is also practised by some opisthobranch slugs such as *Elysia crispata*, which has an extroflectible penis with a pointed stylet, by means of which the animal injects sperm into the partner's coelomic cavity. Hypodermic insemination has also been recorded in some nematodes

Figure 3.30 In the bed bug (*Cimex lectularius*), insemination occurs through a specialized area (Ribaga's organ) of the female body wall and is therefore called traumatic insemination.

(*Auchenacantha, Citellina, Passalurus*), the acanthocephalan *Pomphorhynchus bulbocolli*, some rotifers (*Brachionus* spp., *Asplanchna brightwelli*), various free-living flatworms, strepsipterans and the spider *Harpactea sadistica*.

Similar to insemination by hypodermic injection is **insemination by dermal impregnation**. Spermatozoa deposited on the exterior of the female penetrate independently into the body of the latter. This mechanism is known in polychaetes, chaetognaths, onychophorans, nemertodermatids and acoels, and in the marine planarian *Sabussowia dioica* (Mann 1984; Tekaya *et al.* 1997).

In some animals, the two partners implement, in different ways, a permanent union. Examples are found among gonochoric animals with very strong sexual dimorphism. In some cases the male is larger than the female, and hosts his companion within a *gynecophoric groove* (digeneans of the genus *Schistosoma*). It is more common, however, to find a dwarf male living inside the female's body, as in the echiurid annelid *Bonellia viridis*. In the spionid polychaete *Scolelepis laonicola*, the dwarf male lives as a parasite inside the female; epidermis and cuticle of the two partners are continuous in the contact zone and anastomoses are formed between the blood vessels of male and female (Vortsepneva *et al.* 2008). Dwarf males attach externally to the female in the Ceratioidei, a group of abyssal fishes classified within the Lophiiformes together with the monkfish. In the case of *Ceratias holboelli*, the female is 60 times longer and half a million times heavier than the male! Many species in this group include males that attach themselves to the female temporarily or permanently; in three genera, males are facultative parasites of the female, while those of seven other genera are obligate and permanent parasites, whose tissues merge with those of the partner (Pietsch and Orr 2007). The dwarf males of the chondracanthid copepods, parasites of fishes, also live attached to the females. Finally, joined together in a permanent manner are the tissues of the two partners in the monogenean *Diplozoon paradoxum*, a flatworm parasitic of frogs, which is an insufficient hermaphrodite.

From the moment of insemination, the life span of a spermatozoon (which, being now in the body of an individual other than the one which produced it, is also called an **allosperm**) is often short, from a few hours to a few days (as in most mammals), but in some animals sperm cells remain viable much longer in the female body, protected inside a **spermatheca** by the secretions of specialized glands.

The life span of allosperms is in the order of a few weeks in many birds and insects, several months in many snails, in some fishes and some salamanders, and up to some years in a few snakes and turtles. Consequently, a single insemination allows the fertilization of eggs that will mature over a long time: in bees and ants, for example, the sperm stored in the course of the nuptial flight will remain viable for the entire life of the queen, from three to five years in the honey bee (*Apis mellifera*). Some viviparous animals can use sperm stored in the female's body to fertilize eggs that will give rise to the embryos of many successive gestations, up to five in many scorpions (Warburg 2011).

Finally, we should mention the apparently unique strategy of the cave psocopterans of the genus *Neotrogla*, where females possess an intromittent organ (a *pseudopenis*) called a *gynosome*, while males have no external copulatory organs but only a genital chamber (*pseudovagina*). The female penetrates the genital tract of the male during a long copulation, extracting spermatozoa and the nutrient-rich seminal fluid. Females compete for access to males, so roles in sexual selection are reversed (Yoshizawa *et al.* 2014).

3.5.2.2 TRANSFER OF GAMETES IN EMBRYOPHYTES

Bryophytes and pteridophytes practise a form of *in-situ fertilization*, with male flagellated gametes reaching the immobile female gametes by swimming in a film of water. However, recent experimental studies have shown that in some moss species the probability and effective distance of cross-fertilization are increased by small soil arthropods such as springtails and mites (Cronberg *et al.* 2006). The moss *Ceratodon purpureus* even emits volatile substances capable of attracting different species of these small arthropods (Rosenstiel *et al.* 2012). These, moving through the moss carpet, act as vectors for the antheridia, thus playing in cross-fertilization a role similar to that of the pollinators in the zoogamous pollination of spermatophytes (Section 3.5.3.3).

In seed plants, the female gamete is always represented by an immobile cell, solidly integrated into the megagametophyte (see Section 3.4.2.2). Furthermore, in all angiosperms and in most gymnosperms the male gamete is led to fertilize the egg cell through the pollen tube developing from the microgametophyte. A free, mobile male gamete is found only in a few gymnosperm genera (cycads and *Ginkgo*). In *Cycas* (Figure 3.31), for example, the male gametophyte (a pollen grain) is carried by the wind until it comes into contact with the ovule, which generally contains two archegons, each with a huge egg cell (the largest female gametes by volume in the entire plant kingdom). Here the male gametophyte develops a pollen tube that contacts the megagametophyte, absorbs nourishment from it and then dissolves, releasing two flagellated antherozoids that reach the egg cell by swimming. Generally, only one of the two embryos that may be formed from two distinct fertilization events in the same ovule completes development into seed.

The completion of male gametophyte development is often postponed until shortly after its arrival on the female gametophyte, or on a receptive structure such as the stigma of a flowering plant, from which the pollen tube can be extended. In conifers, the time between pollination and fertilization varies

Figure 3.31 In cycads (here, a female individual of the sago palm, *Cycas revoluta*) the pollen grain releases two mobile sperm cells that reach the egg cell by swimming.

from a few weeks (as in most Cupressaceae and Pinaceae) up to one year in *Pinus* and some Araucariaceae (Fernando *et al.* 2005).

The diversity of ways the male gametes are transferred to the female reproductive organs corresponds to a diversity of *pollination* modes by which the microgametophyte (a pollen grain) is brought to the vicinity of the megagametophyte. Strictly speaking, this process corresponds to the encounter between two sexual partners, i.e. the two gametophytes of opposite sex (see Section 3.5.3.3).

3.5.3 Strategies that Promote Encounters Between Gametes

Syngamy requires physical contact between the two gametes. Different strategies have evolved that increase the probability of a meeting between the gametes, or the potential partners that produce them. These strategies differ according to the kind of life cycle, the organism's habitat and the modes by which it produces, disperses or transfers the gametes.

3.5.3.1 STRATEGIES OF EXTERNAL FERTILIZATION

In many algae and marine animals that resort to external fertilization, the meeting between the gametes is promoted by the frequently aggregated spatial distribution of adults (e.g. in the case of sponges, cnidarians or sessile molluscs), but also by the precise **synchrony** with which the whole population releases the gametes into the water.

In the algae, the release of gametes can be induced by different factors such as photoperiod or light intensity in excess of a certain threshold (Agrawal 2012). In the dioecious *Monostroma angicava*, a marine green alga living on the Pacific coasts of the Japanese island of Hokkaido, gametes are released every two weeks from February to June, during the low daytime tide of the

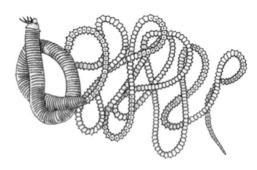

Figure 3.32 The synchronous swarming of the epithokous (posterior) segments of a whole population of the polychaete *Palola viridis* facilitates external fertilization.

first day of the new moon or full moon (Togashi and Cox 2001). On the other hand, a link with the lunar or tidal cycle does not seem to exist in the case of numerous green algae of the Caribbean area, in which there is massive release of gametes before dawn, with close synchrony between the individuals of the same species (the male gametes, however, are often emitted a few minutes before the female ones) and a clear temporal separation between the gamete releases by different but related species (Clifton 1997). Well known is the periodic appointment (*swarming*), in the sea waters around the island of Samoa, for the main breeding season of a polychaete, the palolo (*Palola viridis*, once known as *Eunice viridis*; Figure 3.32), in the night of the second or third day after the third quarter moon of October or November.

At short range, the encounter in water between gametes of the opposite sex is often facilitated by forms of chemical communication, with the emission of attraction substances (*pheromones*) acting as a booster. This form of communication is widely used by animals, including by the adult individuals that produce the gametes, as described more fully in the following section.

3.5.3.2 MEETING OF PARTNERS IN ANIMALS WITH INTERNAL FERTILIZATION

In animals, especially in those with internal fertilization, reproduction is preceded or accompanied by *movements*, for different reasons and of different extent, and/or by specific forms of *communication* between individuals of opposite sexes. The two kinds of actions are often associated, as for instance in **active partner search**, where the movements of (usually) male individuals are guided by attraction signals emitted by the females. These can culminate in forms of **courtship**, sometimes complex and more or less ritualized. These behaviours are generally mediated by various types of signal, e.g. chemicals or sounds. The adaptive value of these forms of communication is multiple: in addition to serving as a call that facilitates meeting between conspecifics of the opposite sex, they also represent an important mechanism of precopulatory reproductive isolation between closely related species, by means of the specificity of the signals that are used for the purpose.

As for **movements**, the reproductive success of an individual can depend both on those carried out during the breeding season and on those performed previously, sometimes starting in the larval or even embryonic phases. The reproductive success of an individual, in fact, depends to a large extent on the number of gametes it produces that will participate in syngamy with conspecific gametes of the other sex. However, reproductive success also depends on the genes carried by the gametes and the environmental resources available to the individuals of the new generation. Therefore, it is easy to understand the advantages associated with a dispersal strategy capable of minimizing the probability of inbreeding, while ensuring access to suitable places or substrates not immediately close to those in which the parents have grown. Relevant, therefore, are (i) the dispersal (including passive dispersal, e.g. as a component of plankton) of tiny eggs and larvae, as is the case of many invertebrates and fishes, especially in the sea, and (ii) the active movements carried out in the course of larval or juvenile life (e.g. by the first post-embryonic stage of scale insects, whose females in most cases become irreversibly sessile afterwards and then mate), and finally (iii) the movements of adults. These last movements are sometimes short-range and end when a partner is encountered, but can be of a greater distance and much longer duration.

Long-distance shifts usually have the character of seasonal **migration** from a region in which reproduction occurs, in a defined season, to another region that offers better trophic conditions during another part of the year. In short-lived animals, these two journeys are performed by individuals of successive generations, as is the case with the monarch butterfly (*Danaus plexippus*). In addition to a number of sedentary or almost sedentary populations, this species includes populations that perform regular cycles involving up to four generations per year: the adults of the last generation of the year migrate south (up to 3000 km) to the wintering area; at the beginning of the new year they reproduce, starting a series of two or three generations that follow each other during the summer. These generations move stepwise (only during the adult phase) to the north, to finally give life to the new generation that will migrate south before the next winter.

In animals in which life expectancy is more than one year, one individual takes both trips. In the Atlantic salmon (*Salmo salar*), which spends most of its life in the sea but reproduces in the fresh and oxygen-rich waters of mountain streams, a few years pass between the descent to the sea as an immature and the ascent of freshwater rivers where adults will reproduce (*anadromous behaviour*). Movements in the opposite direction (*catadromous behaviour*) characterize the migrations of eels (the European *Anguilla anguilla* and other species), which are born in the sea, but spend a few years in freshwater environments before returning to the sea.

In these fishes, the individual performs only one migratory cycle in its life. In other animals whose life spans several years, however, the individual repeats the journeys between the *breeding area* and what is generally defined as a *wintering area*. This is the case in many birds, some of which fly long distances every year. The record holder is the arctic tern (*Sterna paradisaea*), which in the northern summer lives in the Arctic and Subarctic regions of North America and Eurasia, while it spends the austral summer on the coasts of the Antarctic continent. It has been estimated that these birds fly over 70,000 km each year, with a documented case of 91,000 km, the longest migration so far recorded for any animal.

In animals where adults are not social, or at least gregarious, the forms of **communication** that allow the identification of a partner of their own species and of the opposite sex are very important, sometimes also using forms of distance call or dialogue. Communication can occur through a range of different sensory channels. A few examples follow. For more a complete treatment see Bradbury and Vehrencamp (2011).

Vocal communication, widespread among terrestrial vertebrates, is also well known in some groups of insects such as orthopterans (crickets, grasshoppers) and homopterans (cicadas), but it is actually much more common than our ears can perceive: indeed many insects emit sounds in the precopulatory phases.

The widespread **chemical communication** involves the release of an **attraction pheromone** (or **sex pheromone**) by the female (or, in a few cases, the male) and the presence in the male of specialized receptors, often sensitive to single pheromone molecules. In the male of the silkworm moth, on the antennae of which there are 17,000 chemical receptors sensitive to bombykol, to trigger the insect's search reaction it is sufficient that one receptor in a hundred be stimulated by a single molecule of the pheromone. Sex pheromones are very diverse, but are generally characterized by a relatively low molecular weight, which represents an acceptable compromise between two opposite needs: (i) the specificity necessary to maintain the message at the level of communication between conspecifics, while also reducing the probability that it becomes an easy attractive cue for a predator or a parasite; and (ii) the ease of dispersal, in water or air, necessary to carry the message far enough to guarantee a good chance of meeting with a male, or a male gamete. Some attraction pheromones are made up of a single substance (such as bombykol, or E-10,Z-12-hexadecadienol, the sex pheromone of the silkworm), but very often they are mixtures of a number of different molecules.

Sex pheromones are a central topic in behavioural ecology, but this cannot be adequately developed here. We refer the reader to the specialized literature

Figure 3.33 The showy bipectinate antennae of this male saturniid moth carry numerous receptors for the sex pheromones released by the females.

(e.g. Wyatt 2014). As an example, however, we briefly illustrate some features of sex pheromones in moths. In Saturniidae, Noctuidae, Tortricidae, Lymantriidae and other families, the female produces a mixture of several attractive substances; the males, on their often conspicuous and double-comb antennae (Figure 3.33), possess different receptors for the different pheromone molecules. In *Panolis flammea*, detection of (Z)-9-tetradecenyl acetate, the most abundant substance in the pheromone mixture produced by the female, induces the male to fly against the wind. This response has a clear adaptive value, as the origin of the wind indicates the direction from which the chemical signal is most likely to come. As the male flies into the wind, it is not only likely that this signal will increase in intensity, but it is also possible that the initially perceived molecule, more abundant or more volatile, will be joined by others, less volatile or produced in lower quantities. Thus, in the case of *P. flammea*, a male continuing to fly into the wind will sooner or later also detect the presence of other components of the pheromone, (Z)-11-tetradecenyl acetate and (Z)-11-hexadecenyl acetate, which trigger its landing reaction and the start of a new exploratory phase that culminates in the physical encounter with the female (Bradshaw *et al.* 1983).

The female of *Lymantria monacha*, a robust lymantriid moth, produces a mixture of the two optical isomers of a substance (*disparlure*), in a 9:1 ratio in favour of the (–) isomer, for which the males of the species do not have antennal receptors. A possible adaptive explanation for this apparent waste of phero-mone is found when considering the behaviour towards the two disparlure isomers shown by the males of *L. dispar*, a similar species often cohabiting with *L. monacha*. The *L. dispar* males have antennal receptors for both isomers, but only the (+) isomer has an attractive effect on them, while the (–) isomer has the opposite effect. Therefore, the abundant production of (–) disparlure by the females of *L. monacha* has the effect of keeping away the males of *L. dispar*, which are instead attracted by conspecific females, whose pheromone mixture consists almost exclusively of the (+) isomer (Hansen 1984).

Curiously enough, female Asian elephants, to signal that they are ready to mate, use in their urine the same pheromone ((Z)-7-dodecenyl acetate) that is used by more than 100 species of butterflies and moths, among them the cabbage looper (*Trichoplusia ni*). There are also several other cases in which phylogenetically very distantly related animals use similar, or identical, compounds for communication (Kelly 1996), and they provide beautiful examples of convergent evolution, offering important clues for the study of the evolution of pheromone signalling pathways.

In a number of insects, there is evidence that the detection of female pheromones only informs the male of the direction along which it may be profitable to search for a female. The male's directional searching behaviour along a gradient of increasing concentration of the pheromone eventually brings the two potential partners within close range, and **tactile communication** between them can then begin. This further exploration eventually results in mating.

Communication by light flashes is practised by fireflies (lampyrid beetles). In this family, the males are always winged and fly when in search of a partner. Females are winged in some genera (even if less active and mobile than the males), but are often larviform and remain on the ground or, at most, on low herbs and grasses. In these beetles, both sexes have an organ that allows the emission of flashes of light; dialogues between them allow both the localization of a prospective partner and the identification through a sort of Morse code of the species and sex of the other individual.

3.5.3.3 POLLINATION IN SEED PLANTS

In seed plants, meeting of gametes can only be achieved if the male (pollen grain) and female (embryo sac) gametophytes are brought in close proximity, a condition that often occurs after a pollen grain's long journey. The transfer of pollen from the microsporangia where it is produced up to the vicinity of the megagametophyte is called **pollination**. This can take place in three main ways: by wind, water or animals.

In **anemogamous pollination**, pollen is carried by the wind. This is common among the gymnosperms but also by no means rare among the angiosperms, in particular among the monocots (e.g. grasses). This is perhaps the most primitive form of pollen transport. Since the wind is not a specialized vector, **anemophilous** plants entrust the success of pollination to abundant pollen production. Moreover, many species have evolved specific morphological adaptations that facilitate the dispersal of pollen and increase the probability that this is intercepted by the reproductive organs of conspecific plants. In the flowers of anemophilous angiosperms, a large quantity of pollen grains is produced in anthers carried by long and flexible filaments, while the stigmas, often long and feathered, have a shape that allows easy pollen capture. Pollen grains of anemophilous species are light, of small size (and therefore often allergenic) and sometimes equipped with devices that favour suspension in the air (such as the *air sacs* in conifer pollen grains). The flowers of the anemophilous species are generally small and inconspicuous, lacking attractions for pollinators such as coloured petals, nectaries, etc.

Hydrogamous pollination is mediated by water and is limited to some freshwater (e.g. *Najas*) and marine plants (e.g. *Posidonia*, Figure 3.34). The approximately 40 species of the cosmopolitan genus *Najas* are predominantly monoecious with unisexual flowers and live completely submerged. The pollen is carried by the current in the body of water (i.e. under the surface) to the stigmas of the female flowers. Similarly, in *Posidonia oceanica*, a monoecious plant with bisexual flowers that forms vast submarine meadows in the Mediterranean, the pollen is dispersed by sea currents. In contrast, in brackish-water species of the genus *Ruppia*, the pollen is transported on the water surface.

Figure 3.34 A submerged meadow of the seagrass *Posidonia oceanica*, a plant with hydrogamous pollination.

There are also forms of 'water-assisted' pollination, different from true hydrogamy. For example, in *Vallisneria spiralis*, a dioecious species of slow-moving waters, male flowers develop within a spathe that opens only at maturity. The flowers abscissed from the spathe are carried on the water surface and, like minuscule boats, float with the anthers upwards until they come into contact with the female flowers, also emerging on the surface, which remain attached to the plant. The pollen leaves the anthers when the male flower bumps into a female flower, so the pollen actually runs along the anther-to-stigma route outside of water. Other aquatic plants with flowers floating on the water surface, such as water lilies (*Nymphaea*), rely instead on entomogamous pollination (see below).

In **zoogamous pollination**, the transfer of pollen is mediated by the action of an animal, which thus takes the role of pollinator. There are a number of animal species that act as pollinators, and pollination is thus said to be (i) **entomogamous** when carried out by insects, especially lepidopterans and hymenopterans, but also dipterans and coleopterans, (ii) **ornithogamous** when birds are involved (hummingbirds and some parrot species in the Americas, honeyeaters (Meliphagidae) in Australia, sunbirds (Nectariniidae) mainly in Africa but also in Southern Asia and Australasia), (iii) **chiropterogamous** when the pollinators are bats, and (iv) **malacogamous** when pollinators are molluscs. The plants that entrust these carriers with pollen are referred to as **entomophilous, ornithophilous, chiropterophilous** or **malacophilous**, respectively. Other animals can occasionally act as pollinators. Among these are some nocturnal primates and some species of opossum, which open the flowers in search of nectar, thus getting their hair dusted with pollen, and some species of lizards, geckos and skinks, whose muzzles become smeared with pollen when they lap the nectar.

The diversity of pollinators that ensure pollination of different plant species is generally reflected in specific flower traits evolved as adaptations that facilitate recognition by the potential pollinator, sometimes at a distance, or the transfer of pollen from the stamens of a flower to the animal that will carry it to the stigma of another flower.

The shape and colour of the flower often serve to attract pollinators, for example the bright red of many tropical flowers pollinated by small birds, or the star shape (sometimes reinforced by radial lines on the petals, converging towards the reproductive organs of the flower) which is attractive to bees and other insects. The flowers of many entomophilous plants attract pollinators by releasing perfumes – or at least strong scents, including the smell of putrefied flesh emitted by the inflorescences of many plants of the arum family, which is attractive to the flies that act as their pollinators. Also significant is the widespread presence of floral nectaries, whose secretions are an important source of

Figure 3.35 The lower lip (or labellum) of the flower corolla in the mint family (here, *Salvia pratensis*) provides a 'landing strip' for the visit of pollinators.

food for many pollinators, like the pollen itself, which is often produced in quantities greatly in excess of the amount necessary for the reproduction of the plant. Moreover, many flowers show particular morphological adaptations that make the interactions with their usual pollinators more precise, while excluding other animals. For example, the nectaries of a large flower, placed in the deepest point of a campanulate corolla, can only be reached by an insect (more often a lepidopteran, sometimes a hymenopteran or a dipteran) with mouthparts shaped like a very long straw. In other flowers, in particular those of many Papilionaceae, Lamiaceae and Orchidaceae, a particularly developed petal (or tepal) may serve as an airstrip for the landing of an insect, onto which the pollen-filled anthers will release their load. The pollen is often poured onto the insect by means of a lever mechanism operated by the weight of the insect (as in the sage, *Salvia*; Figure 3.35) or a spring, whose elastic force is freed by the advancing insect, which displaces the petals that were keeping the stamen filaments in a folded position (as in some legumes). Zoogamous pollination also requires a fairly precise agreement between the dates and times of day when the flowers are open and dates and times when the potential pollinator is active. This concordance is often evident, as in the case of many flowers pollinated by moths, which open their corollas and/or begin to emit their scent only in the evening; an example is the hoary stock (*Matthiola incana*).

Some plants have **mixed pollination systems**, having structures suitable for both the visit of insects and the release of pollen into the atmosphere. Heather and cyclamen, for example, behave as entomophilous at the beginning of the flowering season and as anemophilous later on.

3.5.4 Fertilization

The fusion of the two gametes in fertilization involves cellular processes that, although sharing common traits, can have a highly diversified course in different taxa. Here we mention what happens in metazoans and spermatophytes.

3.5.4.1 FERTILIZATION IN ANIMALS

At the time of the penetration of the spermatozoon, the female cell is not always a true egg, that is a germ cell that has completed meiosis. In sea urchins the process is already finished by this stage, including the elimination of polar bodies, but in amphibians and mammals the future egg (which properly *is not yet an egg*) is still in metaphase II, and it is precisely fertilization that triggers the completion of the second meiotic division. In ascidians, meiosis is no further than metaphase I when fertilization triggers the completion of the first meiotic division, immediately followed by the second. In still other cases, meiosis has yet to begin (Table 3.8).

As long as they remain in the male genital tract, even typical flagellated sperm are usually immobile, but they become mobile when in contact with sea water or the alkaline environment of the female genital tract.

There are several molecules that come into play in the process of fertilization; those produced by the egg are called **gynogamones**, those produced by the spermatozoon, **androgamones**. Depending on the function, we distinguish a *gynogamone I* or **antifertilisin**, which attracts and activates the spermatozoon (e.g. echinochrome A, B and C of the sea urchin, astaxanthin of the rainbow trout), and a *gynogamone II* or **fertilisin**, which enables the spermatozoon to adhere to the surface of the egg. Likewise, an *androgamone I*, which is the antifertilisin of the spermatozoon, is distinguished from an *androgamone II*, which dissolves the gelatinous coating of the egg. When the spermatozoon comes into contact with the egg, the enzymes contained in the acrosome are released, while one or more *acrosomal filaments* are projected forward (**acrosomal reaction**).

A regular zygote is produced provided that only one spermatozoon enters the egg. This is guaranteed by specific mechanisms (**cortical reaction**) that prevent *polyspermy*, i.e. the potential penetration of more than one spermatozoon. This involves either a modification of the membrane potential of the egg surface, with intake of sodium ions, or, as in the sea urchin, the formation of a fertilization membrane. However, a physiological polyspermy is frequent among ctenophores, insects, selachians and birds. In this case, entry of extra sperm cells into the egg does not lead to the union with the egg nucleus of more than one sperm nucleus. Instead, once a normal diploid zygotic nucleus has been formed, the supernumerary male nuclei are brought to the periphery of the cell, where they degenerate.

Table 3.8 Condition of the female germ cell at the time it is fertilized, with examples from the metazoans

Stage	Examples
Immature oocyte I	*Otomesostoma* (Rhabditophora) *Dinophilus, Saccocirrus, Histriobdella* (Polychaeta)
Mature oocyte I	*Grantia* (Porifera) *Dicyema* (Dicyemida) *Nereis, Myzostoma* (Polychaeta) *Spisula* (Bivalvia) Many crustaceans
Oocyte II at the start of the first meiotic division	*Cerebratulus* (Nemertini) *Chaetopterus* (Polychaeta) *Dentalium* (Scaphopoda) Most sea squirts (Ascidiacea)
Oocyte II at the end of the first meiotic division	*Amphioxus* (Cephalochordata) Most vertebrates
Egg	Cnidaria Sea urchins (Echinoidea)

3.5.4.2 FERTILIZATION IN SEED PLANTS

The different courses of fertilization in gymnosperms have already been presented in Section 3.4.2.2.

A process of **double fertilization** has evolved in flowering plants. This consists in the fertilization, by the two sperm cells carried by the pollen tube, of two distinct cells of the same ovule: the egg cell and the central cell. The diploid zygote obtained from the fertilization of the egg cell, which contains only one haploid nucleus, will develop into the embryo of a new sporophyte. From the fertilization of the central cell, which is binucleate, and from the subsequent fusion of the three haploid nuclei, a triploid tissue will develop, called the *secondary endosperm*, that will provide nourishment to the embryo (Sections 4.6.2 and 4.6.3).

Two sperm cells thus move through a single pollen tube, but polyspermy is avoided, and the two cells end up merging with two different cells in the ovule. The details of how polyspermy is prevented are not yet fully understood, but the existence of two distinct mechanisms to block polyspermic fertilization, one for the nucleus of the egg cell, the other for the two nuclei of the central cell, has been demonstrated, at least in the model plant *Arabidopsis thaliana* (Scott *et al.* 2008; Maruyama *et al.* 2013).

3.5.5 Reproductive Incompatibility

The genetic exchange associated with amphigonic sexual reproduction cannot always occur between any pair of individuals in a population. There are many forms of **reproductive incompatibility**. The most obvious limitation depends on the fact that an individual belongs to a sex or to a particular mating type which allows it to mate only with individuals of the opposite sex or of a compatible mating type. Other forms of reproductive incompatibility are those that prevent self-fertilization in hermaphrodite individuals (**self-incompatibility**, or **self-sterility**). In animals, many forms of hermaphroditism (e.g. sequential hermaphroditism) are irreconcilable with self-fertilization. In the haplodiplontic organisms with monoecious sporophytes, including most flowering plants, self-fertilization, strictly speaking, cannot occur, because male and female gametes are produced by two separate individuals (gametophytes), although generated by the same sporophyte.

In monoecious plants, either with unisexual or with bisexual flowers, cross-fertilization, i.e. fertilization of the ovules by flowers produced by another plant (**xenogamy**) is not guaranteed. Ovules could be fertilized either by pollen from other flowers of the same plant (**geitonogamy**), or by pollen produced by the same flower (**autogamy**, but this term is also used with other meanings, e.g. as a generic synonym of self-fertilization; see Section 3.6.1).

The mechanisms that prevent or at least circumvent self-fertilization (either geitonogamy or autogamy) are of two types: mechanical and physiological.

In some species, a *mechanical* barrier to autogamy is provided by the spatial separation of stamens and stigma within a bisexual flower (*hercogamy*, or *herkogamy*). *Heterostyly* is a form of intraspecific flower polymorphism that involves hercogamy. Heterostylous plants produce bisexual flowers of two or three different types (conditions called *distyly* and *tristyly*, respectively), which differ from each other in the relative length of the style and the stamen filaments (Figure 3.36). In the case of distyly, within the population there are individuals with *pin* (or *brevistylous*) flowers, where stamens are about half the length of the pistil, next to individuals with *thrum* (or *longistylous*) flowers, in which stamens are about twice the length of the pistil. In the case of tristyly there is an additional form with a long pistil and stamens of intermediate length. The two (or three) classes of floral morphology are neatly distinct, without intermediates. Each individual plant produces only flowers of the

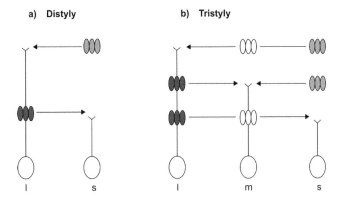

Figure 3.36 Mechanical barriers to self-pollination. Heterostyly with (a) two or (b) three alternative phenotypes. Anthers (small ovals) are carried by filaments of different height (not shown); correspondingly, the V-shaped stigma rises over the ovary (large oval) borne on a style of different lengths, long (l), medium (m) or short (s). Possibilities for crossing are represented by the arrows.

Figure 3.37 The common primrose (*Primula vulgaris*) is an example of a plant producing two different types of bisexual flowers, which differ in the relative length of stamen filaments and style (distyly).

same type, but the two (or three) types are all represented in the population. Pollination leads to fertilization only if it occurs between flowers of two different types. The different length ratios between the male and female parts of the flower make it very difficult to exchange pollen between two flowers of the same type, and the transfer of pollen from a thrum flower to a pin flower generally depends on the activity of a pollinator other than the one that could mediate pollination in the opposite direction. Examples of distylous plants are many species of *Primula* (Figure 3.37), *Linum*, *Lythrum*, *Cryptantha* and *Amsinckia*. Other species of *Lythrum*, such as the common *L. salicaria*, and

Oxalis pes-capreae are tristylous. In *Narcissus* distyly and tristyly can coexist, resulting in incomplete self-sterility and asymmetric incompatibility among the morphs.

More widespread are the *genetic–physiological* mechanisms of self-incompatibility, which function in the same way as the mating types in fungi and in many unicellular eukaryotes. In the genome of many angiosperms there are multiallelic *self-sterility loci*, sometimes with hundreds of alleles, which cause the (haploid) pollen that carries a certain allele not to develop a functional pollen tube through the (diploid) tissues of the carpel of a plant that carries the same allele. In the majority of cases studied thus far, incompatibility is monofactorial, i.e. dependent on alleles at a single locus (generally indicated by S); in the Poaceae, however, self-incompatibility is controlled at the level of two loci (S and Z), close on the chromosome.

Mechanical and genetic–physiological barriers to self-pollination can coexist. Morphological differences (with two or three alternative phenotypes) exist sometimes between the flowers that respectively produce compatible ovules and pollen, while there is a physiological barrier to pollination among flowers with the same phenotype.

It is estimated that 39% of the flowering plants possess some mechanism of self-incompatibility. The taxonomic distribution of these mechanisms suggests that these evolved several times independently. In the case of the self-sterility allele, for example, different groups have evolved this mechanism at non-homologous loci (different S genes).

Self-incompatibility is of either gametophytic or sporophytic type. In the first case, which is the most common, known for example in many plants of the Papaveraceae, Solanaceae, Rosaceae, Fabaceae, Campanulaceae and Poaceae, rejection of pollen is determined by a factor present in the haploid genome of the male gametophyte and manifests itself during the growth – soon interrupted – of the pollen tube along its path through the style. In self-incompatibility of the sporophytic type, rejection is determined by the diploid genome of the mother sporophyte; this mechanism, reported for members of the Brassicaceae, Asteraceae, Convolvulaceae, Betulaceae and Caryophyllaceae, is less frequent. The diploid cells that line the inner walls of the anthers, where pollen is produced, deposit on the wall of the latter the expression products of both alleles of the autosterility locus. Once the pollen grain reaches the stigma, if incompatible, it fails to hydrate and therefore to germinate and ends up being exhausted without producing the pollen tube.

A self-incompatibility system controlled by two loci, with a mechanism similar to that of the Poaceae, but obviously with the involvement of different molecules, has been described in an animal, the ascidian *Ciona intestinalis* (Harada *et al.* 2008).

3.5.6 Mating Systems

Especially in long-lived animals with a rich social life, such as mammals and birds, the relationship between individuals of the two sexes occurs in a variety of social contexts called *mating systems*.

A **mating system** is the way in which a population or group of individuals is structured in relation to sexual behaviour: which males and females mate, with how many partners, and under what circumstances. It may also include **mate choice**, generally by the female, an important component of the evolutionary process of sexual selection (Section 3.3.1.2; Westneat and Foz 2010).

Two main systems are recognized: **monogamy**, where an individual has only one exclusive partner, and **polygamy**, where the same individual can have multiple partners.

Monogamy can lead to the formation of lifetime stable pairs or, as in most cases, of pairs that last one breeding season (*serial monogamy*, as in brown bears). Monogamy is very common among birds, much less so in mammals; examples of monogamous species are eagles, swans and gannets among the former, badger, fox and jackal among the latter.

Polygamy is articulated in the forms of exclusively male polygamy or **polygyny** (a male with more females), exclusively female polygamy or **polyandry** (a female with more males) and polygamy of both sexes, or **promiscuity in the strict sense**, in which each individual can mate with any other individual of the opposite sex in the population. All mating systems other than monogamy can be called **promiscuous in a broad sense**. Occasional episodes of promiscuity in monogamous species (not uncommon in birds) are usually called **extra-pair matings**.

Polygyny is the most common promiscuous mating system among vertebrates, but is also known in many insects. It leads to a social organization characterized by more or less stable groups, often limited to the breeding season, formed by one male and several females, sometimes called a *harem* (Figure 3.38). Polyandry is common among insects and fishes, but is also known in some frogs, turtles and birds. In mammals it is found in the naked mole-rat (*Heterocephalus glaber*, Figure 3.39), a burrowing eusocial rodent, and has also been recorded in some mustelids, including the European polecat (*Mustela putorius*), in cetaceans and primates (e.g. the marmosets and tamarins of the genus *Callithrix*; Section 5.2.5.3). Under polyandry, some important processes of sexual selection can be established, such as *sperm competition* and *cryptic female choice*.

A variant of these promiscuous systems is **polygynandry**, where two or more males have an exclusive relationship with two or more females, forming

Figure 3.38 In the southern elephant seal (*Mirounga leonina*) a male can control a harem of up to 90 females, a mating system called polygyny.

Figure 3.39 The naked mole-rat (*Heterocephalus glaber*) is an African eusocial rodent with a polyandrous mating system, one female (the queen) mating with several males.

a stable group where the number of males is not necessarily the same as the number of females. An example is the North American passerine Bicknell's thrush (*Catharus bicknelli*).

In some species, an individual can adopt different **mating strategies** depending on circumstances, which leads to a **multiple mating system**, as in the case of the dunnock (*Prunella modularis*), a European passerine whose mating system includes monogamy, polyandry, polygyny and polygynandry. Moreover, different individuals of the same species can adopt different mating strategies according to age, the social dominant or subordinate rank they occupy within the group to which they belong or their specific phenotype (typical of species with male polymorphism or polyphenism; Figure 3.13), or according to other criteria.

Mating systems have great relevance for the reproductive biology of the species that adopt them and for their evolution, especially in relation to sexual selection (Section 3.3.1.2). For a more detailed treatment of the subject, of the mating strategies of specific individuals in the context of the system generally adopted in their population and of their significance in the evolutionary dynamics, please refer to specialized treatments (e.g. Oliveira *et al.* 2008; Westneat and Foz 2010) and textbooks on ethology and evolutionary biology (e.g. Alcock 2013; Futuyma and Kirkpatrick 2018). Here we note that these different interindividual contexts can have important consequences for the subsequent parental care that one or the other of the two partners, or both, will dedicate to the offspring, an issue discussed in Chapter 4.

3.6 Uniparental Sexual Reproduction

In the last part of this chapter we discuss those forms of sexual reproduction that involve only one parent. Depending on the specific mode, this may or may not entail a certain degree of genetic reassortment in the production of offspring.

The term **metasexuality** was introduced by Scali *et al.* (2003) to indicate all the modes of reproduction (parthenogenesis, gynogenesis, hybridogenesis) that derive from sexual reproduction, make use of egg cells, but deviate from amphigony because meiosis or the fusion of gametic nuclei is suppressed, or because of other peculiarities as described in the following pages. In spite of these deviations or simplifications, metasexual reproduction mechanisms very often still benefit from some recombination and thus are not strictly clonal. For the cytogenetic aspects of the various forms of metasexuality, see Section 5.2.3.

3.6.1 Self-Fertilization

In reproduction by **self-fertilization** (or **selfing**) there is fusion of gametes or gametic nuclei produced by the same individual (Figure 3.28b). In diplontic organisms, the gametes that merge are the products of distinct meiotic processes undergone by the same number of germ cells in the same individual, and this distinguishes self-fertilization from some forms of meiotic partheno-genesis (see Section 3.6.2) where there is the fusion of two of the four nuclei deriving from the same meiosis. Self-fertilization in metazoans with sufficient simultaneous hermaphroditism has been considered in Section 3.3.2.2. In haplontic organisms all haploid germ cells (as well as all somatic ones) are produced by mitosis from a single parent cell resulting from a single meiosis. In these organisms, self-fertilization is therefore genetically equivalent to *meiotic parthenogenesis with random fusion* (Section 5.2.3.3).

The term self-fertilization is also applied to the reproduction of the sporophyte of haplodiplontic organisms, although in the latter there cannot be true self-fertilization, because the male and female gametes are in fact produced by two distinct individuals (gametophytes), even if borne on the same (monoecious) sporophyte. From the point of view of transmission genetics, this mechanism is however equivalent to the self-fertilization of a diploid organism: it consists in the fusion of products of two distinct meioses (actually, indirect products, because each meiosis has been followed by some mitoses) of the same sporophytic individual. Of those two meiotic events, one had given rise to the megaspore, the other to the microspore. In seed plants, self-fertilization is therefore a result of *self-pollination*.

In many monoecious plants, self-fertilization is rare or absent; this may be due to several causes: (i) male and female sexual organs of the same individual mature at different times; (ii) male and female sexual organs of the same individual occupy positions such as to make self-fertilization unlikely; (iii) alleles that cause self-sterility may be present, for example ones that prevent development of the pollen tube on flowers of the same plant (Section 3.5.5).

In *cleistogamous* plants (Section 3.3.2.1), self-fertilization is obligate and occurs within the flower bud, thus in a condition mechanically preventing cross-fertilization, but in any case the gametes that merge result from distinct meioses. This situation is common in *Viola* and *Impatiens*. Almost all the monoecious plants that regularly practise self-fertilization, including androdioecious, gynodioecious and trioecious species, seem to at least occasionally practise cross-fertilization, in what is called a *mixed mating system* (Section 3.3.2.2).

3.6.2 Parthenogenesis

Parthenogenesis is the form of reproduction in which the new individual develops from an unfertilized egg (Figure 3.28c). In many cases the offspring are all females genetically identical to the mother, and therefore some authors consider parthenogenesis as a form of asexual reproduction. However, as explained in Section 1.2, here we consider it as a case (or rather, as we shall see, as a set of different cases) of sexual reproduction, because parthenogenesis uses, often in a peculiar way, developmental mechanisms of gametogenesis typical of sexual reproduction, irrespective of its clonal or non-clonal outcome. To use Boyden's (1950) clear words, 'there is no such thing as an "asexual egg", regardless of whether it was produced by meiosis or mitosis.'

Parthenogenesis occurs in both plants and animals; however, the mechanisms and the associated terminology differ quite extensively between the two groups. What follows is mainly about the more diverse phenomenology in animals. Plant parthenogenesis is treated in Section 3.6.2.9.

Before exploring parthenogenesis in more detail, let's point out that this form of reproduction does not in itself exclude the presence of males. What all the different forms of parthenogenesis have in common is that females, in order to give rise to their descendants or at least to some of them, do not need the genetic contribution of male gametes, neither through karyogamy, nor through cell-contact egg activation. Although much less common than amphigonic reproduction, parthenogenesis is a reproductive mode very widespread among multicellular eukaryotes. In animals, for example, parthenogenesis is known in all major and in some minor phyla: Rhabditophora, Gastrotricha, Annelida, Mollusca, Kinorhyncha, Nematoda, Arthropoda, Echinodermata, Chordata. Parthenogenesis is found, for instance, in about 40% of the aphids and 60% of the ostracods. Different parthenogenetic lines may have evolved from the same amphigonic ancestor. Parthenogenesis would have evolved on at least three separate occasions in the aphid *Rhopalosiphum padi*, at least four times in the ostracod crustacean *Eucypris virens*, and at least five times in the cyprinodontid fish *Poeciliopsis monacha-lucida*.

Within the vertebrates, it is striking that no bird or mammal species practises parthenogenesis regularly. In the case of mammals, it is possible that they cannot abandon amphigonic reproduction because of *genomic imprinting*: in the course of gametogenesis some genes are modified in such a way that they will be able to function only if transmitted paternally, others only if transmitted maternally (Georgiades *et al.* 2001; Morison *et al.* 2005), and therefore normal development is possible only if the individual possesses both paternal and maternal genes.

Parthenogenesis occurs in many forms, which can be classified according to different criteria.

Based on the cytogenetic mechanism through which parthenogenesis is implemented, we distinguish **haploid parthenogenesis**, where haploid male offspring are generated through unfertilized *reduced eggs* (i.e. haploid, produced by ordinary meiosis) and **diploid parthenogenesis**, where diploid offspring are generated through unfertilized *unreduced eggs* (i.e. diploid). Within diploid parthenogenesis we further distinguish *ameiotic* and *meiotic parthenogenesis*, a key feature in determining the potential to produce genetic variation (see Section 5.2.3.3). In **ameiotic parthenogenesis** (or **apomictic parthenogenesis**, or **apomixis**), which is known in rotifers and in various groups of arthropods, meiosis is suppressed, so that the offspring are genetically identical to the mother. In **meiotic parthenogenesis** (or **automictic parthenogenesis**, or **automixis**), instead, meiosis occurs and the diploid condition of the egg is obtained by premeiotic chromosomal duplication or by fusion of two nuclei resulting from the same meiotic event.

Premeiotic doubling is the most common mechanism among the parthenogenetic forms of freshwater planarians (free-living flatworms) and terrestrial oligochaetes, as well as many insects and mites, and is the only mechanism in vertebrates. The marine free-living flatworm *Bothrioplana semperi* produces cocoons, each containing two primary oocytes that simultaneously undergo two peculiar maturation divisions preceded by premeiotic doubling of chromosomes. The eight nuclei thus obtained are not fertilized, but behave like free blastomeres, uniting to form the first primordium of the future embryo (Reisinger *et al.* 1974; Martín-Durán and Egger 2012).

Species that reproduce by parthenogenesis through fusion of haploid nuclei deriving from the same meiosis are found among the nematodes, enchytraeid oligochaetes, isopods and tardigrades, and in various insect groups. Details of the cytogenetic mechanisms of parthenogenesis can be found in Section 5.2.3.3.

On the basis of the sex of the individuals generated, we distinguish between **thelytokous parthenogenesis** (or **thelytoky**), which produces only females, **arrhenotokous parthenogenesis** (or **arrhenotoky**), which results in exclusively male offspring, and **amphitokous** or **deuterotokous parthenogenesis** (or **amphitoky**, or **deuterotoky**), which generates offspring of both sexes. **Thelytoky** is said to be **complete** if this is the only reproductive mode of the species (or population), while it is called **cyclical thelytoky** or **heterogony** if parthenogenesis alternates more or less regularly with amphigonic reproduction (Rieger *et al.* 1976; Figure 3.40; Section 2.3). At the species or population level, but also sometimes referred to the single individual, parthenogenesis can be **accidental** (also called **occasional parthenogenesis** or **tychoparthenogenesis**), **facultative**, or **obligate**. Finally, parthenogenesis is **spontaneous** in many populations or species, but in others it occurs only if artificially **induced**.

3.6.2.1 THELYTOKOUS PARTHENOGENESIS

Thelytokous parthenogenesis (Section 5.2.3.3) can be performed by **automixis** (**automictic** or **meiotic thelytokous parthenogenesis**) or by **apomixis** (**apomictic** or **ameiotic thelytokous parthenogenesis**). In some cases, both mechanisms occur in related forms; in others, thelytokous parthenogenesis coexists with amphigony in a group of populations that traditional taxonomy tends to describe as conspecific.

Among the Oligochaeta, where hermaphroditism is the primitive condition, numerous cases of thelytoky are known among the Lumbricidae and

Figure 3.40 Many aphids (here, *Aphis nerii*) have heterogonic cycles in which several thelytokous parthenogenetic generations (females generating only females) are followed by an amphitokous parthenogenetic generation (females generating both males and females) and finally by an amphigonic one (with biparental sexual reproduction).

Enchytraeidae, in some cases however with production of sperm and gynogenesis (Section 3.6.3). Again in the oligochaetes, various combinations of reproductive mode and genetic system occur: *diploid amphigony, polyploid amphigony, polyploid apomixis, polyploid automixis*, plus a doubtful example of *diploid thelytoky*. Automictic forms are much more common than apomictic ones. Chromosome number is restored by premeiotic doubling. Under the name *Dendrobaena octaedra* are grouped apomictic thelytokous forms, more often hexaploid, but sometimes aneuploid. Male organs may be present or missing, and sometimes there are even non-functional supernumerary testicles. Some earthworms have abnormal spermatogenesis.

In the crustacean genus *Artemia* there are amphigonic, automictic and apomictic populations; males can occasionally show up in parthenogenetic populations.

Among the dipterans, examples of ameiotic thelytoky are known in *Pseudosmittia* (Chironomidae), with diploid and triploid lines. Automictic thelytoky with fusion of meiotic products occurs in *Lonchoptera dubia* (Lonchopteridae) and in *Drosophila mangabeirai* (Drosophilidae), both diploid.

In the chrysomelid beetle *Bromius obscurus* (often cited in the past as *Adoxus obscurus*), the North American populations are diploid and amphigonic, the European ones triploid and apomictic. Among the weevils (Curculionidae) there are many cases of polyploidy (3n, 4n, 5n and 6n, and also examples of aneuploidy) associated with thelytokous parthenogenesis. In *Trialeurodes vaporariorum* (Homoptera Aleurodidae) both an arrhenotokous race with optional fertilization and a thelytokous race are known.

Characteristic of the lepidopterans is the chromosomal system of sex determination with female heterogamety (ZW or Z0 system; Section 6.1.1), and therefore thelytokous parthenogenesis entails the production of karyotypes with different sex chromosomes in the two sets. In the psychid moth *Solenobia triquetrella* s.l. this is achieved through a form of meiotic parthenogenesis with central fusion, i.e. with the fusion of two nuclei separated at the first meiotic division (White 1973; Section 5.2.3.3). In *Apterona helix* (another psychid moth), meiosis is anomalous: the spindles of the second meiotic division are close and parallel, giving rise to a double metaphase plate with 2n chromosomes. Two diploid nuclei are obtained, from which the formation of a single embryo begins. This will then be a chimera, though in a form close to genetic mosaicism (White 1973; see Section 1.4.2).

3.6.2.2 ARRHENOTOKOUS AND PSEUDOARRHENOTOKOUS PARTHENOGENESIS

In reproduction by **arrhenotokous parthenogenesis**, offspring are all male. However, populations that practise arrhenotoky are not generally made up of males only, because other forms of reproduction (usually, amphigony) accompany parthenogenesis. This reproductive mode involves a peculiar system of sex determination, the *haplodiploid system*, based on a differential level of ploidy between the two sexes: females are diploid, males haploid (Section 6.1.3). Thus, unfertilized eggs develop into males, whereas fertilized eggs develop into females. Arrhenotokous parthenogenesis, which is unknown among the plants, is limited to a few animal taxa: (i) six groups of insects – hymenopterans (Figure 3.41), thysanopterans, whiteflies (Aleurodidae) and scale insects of the Iceryini tribe among the homopterans, a clade of bark beetles (Scolytinae) and *Micromalthus debilis* among the coleopterans; (ii) some arachnids; (iii) the Oxyurida among the nematodes (Adamson 1989); and (iv) the monogonont rotifers (in the context of the heterogonic cycle described in Section 2.3).

In hymenopterans, arrhenotokous parthenogenesis is the rule, with few exceptions. Within the symphytans, some species reproduce by thelytokous parthenogenesis. Either mode can be found in closely related species, for example *Cimbex lutea* is arrhenotokous, while *C. connata* is thelytokous;

Figure 3.41 In arrhenotokous parthenogenesis, only males are born from unfertilized eggs. This is the case of haploid parthenogenesis in the hymenopterans (here, *Vespula germanica*).

Eriocampa umbratica is arrhenotokous, *E. ovata* thelytokous; *Pristiphora conjugata* is arrhenotokous, *P. pallipes* thelytokous. Some cases of amphitoky are also known. Thus in *Pteronidea ribesii*, among the progeny deriving from unfertilized eggs a few females occasionally appear, while some males appear amidst a large majority of females in the progeny of mainly thelytokous species such as *Nematus erichsoni* and *Pristiphora fulvipes*. In *Thrinax macula* arrhenotokous females, which however occasionally also generate some females, exist alongside the thelytokous females. In the parasitic hymenopterans (once called the terebrants), arrhenotokous parthenogenesis is possibly the rule, but there are many cases of thelytokous parthenogenesis. Vandel's (1931) classic monograph on parthenogenesis cited examples from the Chalcidoidea, Scelionidae, Braconidae, Ichneumonidae, and also from the Dryinidae (*Gonatopus, Haplogonatopus*) among the Aculeata. The minuscule American wasp *Trichogramma pretiosum* reproduces by arrhenotokous parthenogenesis, but a European form morphologically identical to it generates by parthenogenesis individuals of both sexes, or females only. *T. evanescens* is arrhenotokous, but a very closely related species, *T. cacoeciae*, is thelytokous. Among the ichneumonids, *Hemiteles aerator* (Europe) is arrhenotokous, but *H. tenellus* (America), practically identical to it, is thelytokous. *Lysiphlebus tritici* (Braconidae) is amphitokous. Almost all of the Aculeata reproduce by arrhenotokous parthenogenesis. In the ants of the genus *Lasius*, unfertilized workers can generate other (female) workers.

Among the Cynipidae, the cynipines that produce showy galls on oaks have a heterogonic life cycle with alternation between a parthenogenetic generation that occurs between autumn and spring and a spring–summer amphigonic generation. Individuals of the two generations are dissimilar. In particular, the ovipositors of the females differ, as do the plant organs (buds, leaves, roots) where eggs are laid and, frequently, the shapes of the galls themselves. Generational dimorphism is particularly remarkable in *Biorhiza aptera*. In this species, the parthenogenetic generation is made up of wingless females, while the amphigonic generation includes males with wings of normal length and females with wings of variable length.

In the bark beetles, arrhenotokous parthenogenesis involves a clade of about 1400 species, which includes among others the genera *Xyleborus*, *Coccotrypes* and *Ozopemon* (Jordal *et al.* 2000, 2002).

To produce sperm, haploid males evidently do not require a reduction of germ-cell ploidy. In the males of arrhenotokous mites, aleurodid homopterans, iceryine scale insects and *Micromalthus* beetles, meiosis is replaced by a single mitotic division, whereby each spermatocyte I produces only two spermatozoa. In Hymenoptera the stages of spermatogenesis are diverse, but in general the process seems to be a modified meiosis, with a first division that separates an anucleate mass from a residual nucleus that undergoes the second division, which is similar to a mitosis. But in bees this final division is in turn asymmetric, and only one spermatozoon is eventually obtained.

From true arrhenotoky we must distinguish **pseudoarrhenotoky**, a genetic mechanism of sex determination based on the elimination or inactivation of the entire paternal chromosome set (*paternal genome loss, PGL*) in the fertilized eggs that will end up developing into males (Section 6.1.3). The male-developing egg is in effect functionally unfertilized. This phenomenon is known in the bark beetle *Hypothenemus hampei*, in scale insects (many diaspidid and lecanoid Coccidae), in three genera of midges (Cecidomyiidae) and in two families of mites, including the Phytoseiidae. Partial loss of the paternal genome during spermatogenesis is known in the dark-winged fungus gnats (Sciaridae).

3.6.2.3 GEOGRAPHICAL PARTHENOGENESIS

According to the traditional description of this phenomenon, some plant and animal species include both amphigonic and parthenogenetic populations, the latter often polyploid (Vandel 1928; Horne and Martens 1999; Adolfsson *et al.* 2010; Hörandl 2006). If we adopt the biological species concept, we should however use a different set of terms. It is logically impossible, in fact, to apply the biological species concept to organisms with uniparental reproduction, so it is perhaps better to say that some amphigonic species are accompanied by populations derived from them, which practise only thelytokous parthenogenesis. The most relevant aspect of this phenomenon is the geographical and ecological distribution of populations with different reproductive modes, and this is why it is called **geographical parthenogenesis**

(Table 3.9). Parthenogenetic populations usually occupy marginal areas of the species' range, subject to difficult or even extreme environmental conditions, and are unusually quick to colonize new areas.

A typical example is the presence of parthenogenetic populations, almost always polyploid, of weevils (Curculionidae) of the genera *Peritelus*, *Polydrusus*, *Barynotus* and above all *Otiorhynchus* (Figure 1.20) in Alpine areas and in the northernmost regions of Europe, which have been free of ice for only a few thousand years, or even less. The greater capacity for colonization demonstrated by these populations can be partly attributed to the short-term advantage of uniparental reproduction (i.e. a single individual can found a new population); in part, however, it seems to be due to their polyploid condition (Lundmark and Saura 2006).

In plants as well, patterns of geographical parthenogenesis (in the form of *agamospermy*, i.e. uniparental reproduction via seeds; see Section 3.6.2.9) are probably caused by a combination of factors that would explain why parthenogenetic taxa often have larger distributional ranges or reach higher latitudes and altitudes than their amphigonic relatives (Hörandl 2006).

However, there are cases in which the geographic or ecological distribution of the parthenogenetic and amphigonic populations does not follow the general pattern. For example, in the centipede *Lamyctes emarginatus* males are known only from insular populations of the Canaries and Azores; the lizard *Lepidophyma flavimaculatum* (Xantusiidae) is parthenogenetic only in the less disturbed areas of humid tropical forest; in two Psocoptera (*Reuterella helvimacula* and *Cerobasis guestfalicus*) the amphigonic populations are found at higher latitudes than the parthenogenetic ones; and *Drosophila mangabeirai*, the only *Drosophila* with obligate parthenogenesis, lives in hot regions of Central and South America.

3.6.2.4 ACCIDENTAL PARTHENOGENESIS

Parthenogenesis occurs accidentally in many species which are normally amphigonic, for example in some birds such as the zebra finch (*Taeniopygia guttata*) (Schut *et al.* 2008), chicken (*Gallus domesticus*), pigeon (*Columba livia*) and especially turkey (*Meleagris gallopavo*), where the phenomenon can affect one egg in five. Parthenogenesis is also known in the Indian rock python (*Python molurus*), the Komodo dragon (*Varanus komodoensis*, Figure 2.16), the bonnethead shark (*Sphyrna tiburo*), some species of *Drosophila* and a few stick insects (*Menexenus semiarmatus*, *Clitumnus extradentatus*).

3.6.2.5 PAEDOGENESIS

Advancement of reproduction to a juvenile stage is called **paedogenesis**. In the most common cases, paedogenesis is achieved through optional larval or pupal parthenogenesis. In insects, where it has evolved at least six times

Table 3.9 Arthropod species with geographical parthenogenesis and more than one ploidy level (data after Lundmark and Saura 2006)		Amphigonic	Parthenogenetic		
		2n	2n	3n	4n
CRUSTACEA					
Anostraca	*Artemia tunisiana* (amph.)/ *parthenogenetica* (parth.)	+	+	+	+
Ostracoda	*Eucypris virens*	+	+	+	−
Isopoda	*Trichoniscus pusillus*	+	−	+	−
INSECTA					
Blattodea	*Phyllodromica subaptera*	+	+	−	−
	Pycnoscelus indicus (amph.)/*surinamensis* (parth.)	+	+	+	−
Phasmatodea	*Bacillus grandis* (amph.)/ *atticus* (parth.)	+	+	+	−
Orthoptera	*Saga pedo*	+	−	−	+
	Warramaba virgo	+	+	−	−
Coleoptera	*Otiorhynchus scaber*	+	+	+	+
	Otiorhynchus singularis	+	−	+	−
	Simo hirticornis	+	−	+	+
Lepidoptera	*Solenobia fumosella* (amph.)/*lichenella* (parth.)	+	−	−	+
	Solenobia triquetrella	+	+	−	+
DIPLOPODA					
Julida	*Nemasoma varicorne*	+	+	−	−

independently, the phenomenon is particularly widespread in the gall midges (Diptera Cecidomyiidae), where it occurs as either **larval parthenogenesis** (*Miastor, Heteropeza, Mycophila*) or **pupal parthenogenesis** (*Tecomyia populi, Henria psalliotae*).

As mentioned in Section 2.8, in *Heteropeza pygmaea* (formerly known as *Oligarces paradoxus*) three different reproductive modes are known: amphigony, parthenogenesis by adult females and paedogenesis. The last mode of reproduction features *gynogenic* (or *thelygenic*) larvae, which produce other larvae similar to themselves; *androgenic* (or *arrhenogenic*) larvae, which produce larvae destined to develop into adult males; *amphigenic* larvae, which produce both larvae like themselves and male larvae; and finally *'super' gynogenic* larvae, which produce larvae destined to develop into adult females. Parthenogenesis, in both adult and larval stages, is ameiotic, and the final arrangement of sex chromosomes that determines the sex of the offspring (X0 male, XX female) depends on the course of mitosis, with or without loss of an X chromosome, respectively. This, in turn, is influenced by a factor produced in the mother's brain and circulating in its hemolymph in response to environmental or physiological conditions, the nutritional status especially.

Reproduction by paedogenesis has also been reported in a group of tiny marine invertebrates (Loricifera), the first species of which was described as recently as 1983. In particular, *Urnaloricus gadi* (Figure 3.42) is viviparous and paedogenic; according to Heiner and Kristensen's (2008) reconstruction, the *pre-megalarva* stage, which contains a large ovary with few oocytes, moults into a *cystigenous megalarva* in which other oocytes are formed. This larva moults in turn, presumably passing through a post-larval stage and eventually becoming a *ghost larva*. The latter is surrounded by the cuticle of the two previous stages.

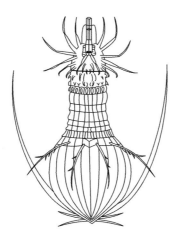

Figure 3.42 A larva of the loriciferan *Urnaloricus gadi*, a small marine animal that reproduces by larval paedogenesis.

In the ghost larva the oocytes eventually mature into eggs that will not be fertilized, but generate embryos that will develop into larvae, reabsorbing all the ghost larva's tissues.

3.6.2.6 CYCLICAL PARTHENOGENESIS

In at least 15,000 species belonging to several metazoan clades, thelytokous parthenogenesis alternates with amphigony in a heterogonic life cycle (Section 2.3). This happens in most representatives of the monogonont rotifers, digeneans, cladocerans (water fleas) and cynipid hymenopterans (two tribes, Cynipini and Pediaspidini), as well as in many aphids (Adelgidae and Aphididae) and the beetle *Micromalthus debilis*. In some of these groups, females are diploid and males are haploid.

3.6.2.7 HYBRIDS AND PARTHENOGENESIS

In interspecific hybrids, in which meiosis is generally difficult or impossible, natural selection favours cytological mechanisms that allow the production of diploid eggs able to develop anyway, either following fertilization (which however leads to triploidy, an inconvenient (odd) chromosomal complement in itself) or, better, without fertilization, that is, by parthenogenesis. Perhaps all parthenogenetic vertebrates are of hybrid origin; further examples are found in gastropods, crustaceans, curculionid beetles, stick insects and orthopterans.

Among the squamate reptiles there are about 30 parthenogenetic species (Table 3.10). Some of these include both amphigonic and diploid or triploid parthenogenetic populations. The North American lizard genus *Cnemidophorus* includes stabilized hybrid forms with different degrees of ploidy. Some of these derive from hybridization between the diploid species *C. tigris* (T) and *C. inornatus* (I), of which they conserve the genotypes, in an arrangement that can be diploid (*C. neomexicanus*, with TI genotype) or triploid (*C. perplexus*, with TII genotypes). Under the name of *C. tesselatus* are described several hybrid biotypes, two of which, with 3n+1 karyotype, are perhaps trihybrids to which *C. tigris*, *C. septemvittatus* and *C. sexlineatus* have contributed, while other hybrids, all diploids, seem to share only genetic material of *C. tigris* and *C. septemvittatus*.

3.6.2.8 INFECTIOUS PARTHENOGENESIS

Among arthropods, parthenogenesis can be induced by some endosymbiotic bacteria, especially those of the genus *Wolbachia* (Proteobacteria; Figure 3.43), with examples in the hymenopterans (about 50 species belonging to different families – Pteromalidae, Aphelinidae, Platygastridae, Encyrtidae, Scelionidae, Trichogrammatidae, Figitidae, Cynipidae) and thysanopterans (*Franklinothrips vespiformis*, *Heliothrips haemorrhoidalis*, *Hercinothrips femoralis*), and also in

Table 3.10 Parthenogenetic reptiles, all of hybrid origin (based on data in Avise 2008, modified and integrated). Some of these species (indicated by *p* in the second column) reproduce exclusively by parthenogenesis; in the others (indicated by *p/a*) amphigonic populations are also known

LACERTIDAE		
Darevskia	*p/a*	Caucasus and neighbouring areas
D. armeniaca	*p*	Southern Turkey, Northern Armenia, Azerbaijan
GEKKONIDAE		
Heteronotia binoei	*p/a*	Australia
Lepidodactylus lugubris	*p/a*	Islands of Indian and Pacific Oceans
Hemidactylus garnotii	*p*	Southeast Asia, Indonesia; introduced in Florida, Puerto Rico, Hawaii
Nactus pelagicus	*p/a*	Queensland, New Guinea etc.
TEIIDAE		
Aspidoscelis uniparens	*p*	Arizona to Northern Mexico
Cnemidophorus spp.	*p*	America
Kentropyx borckiana	*p*	South America
GYMNOPHTHALMIDAE		
Gymnophthalmus underwoodi	*p*	South America
Leposoma percarinatum	*p*	South America
XANTUSIIDAE		
Lepidophyma flavimaculatum	*p/a*	Central America

continues

Table 3.10 (cont)		
SCINCIDAE		
Menetia greyi	*p/a*	Australia
TYPHLOPIDAE		
Ramphotyphlops braminus	*p*	Southeast Asia

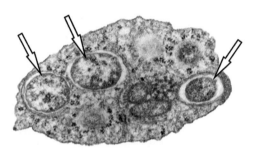

Figure 3.43 Bacterial cells (*Wolbachia*, indicated by arrows) inside an insect cell.

mites (*Brevipalpus phoenicis* and some species of the genus *Bryobia*). All these arthropods have a haplodiploid sex-determination system (Section 6.1.3), in which unfertilized eggs, otherwise expected to develop into males, give birth to females (Normark 2003). Microbial parasites other than *Wolbachia* cause a similar effect in a number of hymenopterans belonging to different families (Tenthredinidae, Aphelinidae, Signiforidae, Encyrtidae, Trichogrammatidae, Eulophidae, Cynipidae).

3.6.2.9 PARTHENOGENESIS IN PLANTS
The terms used to describe phenomena of parthenogenesis in plants, particularly in seed plants, are very different from those in use for other eukaryotes: as a mode of uniparental reproduction derived from amphigony, plant parthenogenesis reflects the peculiarities of sexual reproduction in this group (Section 2.1).

In ferns, forms of parthenogenesis comparable with the diplospory of flowering plants (see below) are very widespread. Two mechanisms are distinguished, respectively called **Döpp–Manton type agamospory** and **Braithwaite type agamospory** (Mogie 1992): in the first case, diploid spores are produced because of a premeiotic endomitosis; in the second case, the first meiotic division is followed by the fusion of its products, which creates a transitory tetraploid condition that is reduced back to diploidy with the second meiotic division.

In gymnosperms, parthenogenesis is purely accidental, and no species adopts this reproductive mode routinely.

In flowering plants (Table 3.11), processes related to parthenogenesis are commonly referred to as **apomixis** (or **agamospermy**, or **apogamy**), while, in the terminology used here, *apomixis* is only one of the two main mechanisms of parthenogenesis, i.e. ameiotic parthenogenesis, contrasted to *automixis*, or meiotic parthenogenesis. Angiosperm apomixis is a form of asexual reproduction through seed, although according to some authors the term apomixis would encompass all modes of vegetative reproduction in plants, and would thus be a feature common to most perennial plants. However, the latter use is strongly discouraged by other authors (e.g. van Dijk 2009), considering that (i) vegetative reproduction does not necessarily pass through a single-cell stage (propagules, in fact, can be multicellular), (ii) daughter plants produced by vegetative propagation generally settle in the vicinity of the mother plant (while seeds very often allow dispersal over greater distances), and (iii) seeds can enter a phase of dormancy, thus resisting adverse seasons or conditions.

In angiosperms, apomixis generally includes three main processes: (i) modification or avoidance of meiosis, or *apomeiosis*, (ii) *parthenogenesis* (here, strictly, development from an unfertilized egg cell), and (iii) modified endosperm development (*pseudogamous* or *autogamous*, i.e. with or without fertilization of the central cell, respectively). Thus parthenogenesis is an element of apomixis for botanists, whereas apomixis is a form of parthenogenesis for zoologists. The cytogenetic details of apomixis are very diverse, but botanists operate a primary distinction between *gametophytic apomixis* and *sporophytic apomixis* (Figure 3.44; van Dijk 2009).

In **gametophytic apomixis** (or **agamogenesis**), the female gametophyte phase is conserved, but the megaspore that gives rise to the megagametophyte is unreduced. This can be obtained by *diplospory* or by *apospory*.

In **diplospory** meiosis can be replaced by a non-reductional division, equivalent to a mitosis (**mitotic diplospory**, e.g. in *Hieracium*; Figure 3.45), or maintained in association with a first division restitution, i.e. a functional suppression of the first meiotic division (**meiotic diplospory**, e.g. in *Taraxacum*), which in any case leads to an unreduced egg. As yet another alternative, a complete meiosis is preceded by endomitosis (replication of the chromosomes not followed by division of the nucleus) of the mother cell of the megaspore (e.g. in *Allium*). In all cases, in diplospory the embryo develops from an unreduced (2n) egg cell. Through the most common cytogenetic

Table 3.11 Number of genera of flowering plants in which some form of apomixis is known (data after de Meeûs *et al.* 2007, with modifications)

	Gametophytic apomixis: diplospory	Gametophytic apomixis: apospory	Sporophytic apomixis
Adoxaceae		1	
Alliaceae	1		
Amaranthaceae	2	1	
Amaryllidaceae	2		1
Araceae		1	
Asteraceae	15	27	1
Balanophoraceae	1		
Betulaceae	1		
Boraginaceae		2	1
Brassicaceae	1	1	
Burmanniaceae	1		
Cactaceae			1
Casuarinaceae	1		
Cyrillaceae		1	
Cucurbitaceae	3	2	1
Globulariaceae		1	
Hyacinthaceae		1	
Hypericaceae		1	
Limoniaceae	1		
Malpighiaceae		1	
Melastomataceae			1
Myrtaceae		1	
Nymphaeaceae	1		

Table 3.11 (cont)	Gametophytic apomixis: diplospory	Gametophytic apomixis: apospory	Sporophytic apomixis
Ochnaceae		1	
Onagraceae			1
Orchidaceae		2	4
Poaceae	9	31	1
Polygonaceae		1	
Rhamnaceae	1		
Rosaceae	5	12	1
Rutaceae		4	
Saururaceae	1		
Solanaceae			1
Taccaceae		1	
Thymelaeaceae	1		1
Trilliaceae			1
Urticaceae	3	1	1

mechanisms with which this can be implemented, diplospory generates off-spring genetically identical to the mother plant.

In addition to the usual haploid megagametophytes, in the case of **apospory** a second unreduced (2n) megagametophyte is produced within the ovule, starting from a somatic cell of the sporophyte. Within the single ovule a competition starts between the two gametophytes, at the end of which usually the *aposporous* gametophyte (the one not derived from a spore) produces a diploid egg cell that will develop, without being fertilized, into an embryo genetically identical to the mother plant (Hojsgaard *et al.* 2013).

In the other main type of plant apomixis – **sporophytic apomixis** (also called **adventitious embryony**) – embryos of somatic origin are formed by diploid cells of the ovule, thus of sporophytic origin, that surround the mega-gametophyte (e.g. in many orchids and *Citrus*; Figure 3.46). However, unlike in

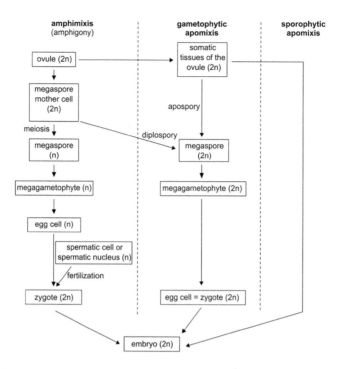

Figure 3.44 The main types of parthenogenesis in angiosperms.

Figure 3.45 Gametophytic apomixis by mitotic diplospory is observed in *Hieracium* (here, *H. pilosella*).

apospory, these cells do not give rise to a gametophytic generation, which is in fact skipped; the life cycle, originally with alternation of generations, is thus reduced to a monogenerational cycle (Section 1.3.2). This type of apomixis is also called *nucellar embryony*, from the name of the ovule tissues from which the embryos originate, or *adventitious embryony*, since the somatic embryos are produced in addition (rather than as an alternative) to the embryo of gametophytic origin, or even *polyembryony*, because one seed often starts developing two or more seedlings (but see Section 3.1.2.4).

Figure 3.46 Sporophytic apomixis is described for the genus *Citrus* (here, the orange tree, *C. sinensis*).

In most apomictic angiosperms, the endosperm develops only after fertilization of the central cell (2n), in a process called *pseudogamy* or *pseudogamic apomixis*, which is a modification of double fertilization and in many ways recalls gynogenesis in animals (Section 5.2.3.4). Failure to develop endosperm leads to seed abortion. However, some apomictic species of the Asteraceae have developed a form of apomixis independent of pollen, in which endosperm is produced without the fertilization of the central cell. Therefore, apomictic plants generally maintain the male function, thus being able to function as pollen donors in crossings with amphigonic lines or lines with non-obligate parthenogenesis.

Diplospory and apospory are sometimes known to occur together in one genus, or in one species, and even in one ovule. Apomixis is sometimes facultative and controlled by environmental factors. For example, in *Hieracium flagellare* there can be both amphigonic and agamospermous flowers in the same flower head.

Some cases are known of an amphigonic species originated by crossing between two apomictic forms. This has been recorded in *Hieracium* for two lines introduced in New Zealand (Chapman *et al.* 2003); another case has been obtained under experimental conditions in *Ranunculus auricomus* (Nogler 1984).

3.6.3 Gynogenesis

In **gynogenesis** the egg is not fertilized, but must be activated by a sperm, although this does not contribute genetically to the formation of the new individual that develops from the egg (Figure 3.28d). Together with androgenesis (Section 3.6.4), gynogenesis is a form of *pseudogamy* (Section 5.2.3.4). Gonochoric gynogenic species are composed exclusively of females; therefore,

Figure 3.47 Reproduction by gynogenesis is known for several planarian species of the genus *Dugesia*.

egg activation necessarily depends on sperm produced by males of related species.

Gynogenesis is known, for example, in various freshwater planarians (Figure 3.47), including a triploid and a tetraploid biotype of *Dugesia benazzii*. Pseudogamy occurs in other planarians too, such as *Polycelis nigra* and some biotypes of *Schmidtea lugubris* and *Polycelis polychroa*. Some polyploid strains of *Schmidtea* and *Polycelis* reproduce (often or always) asexually. Loss of amphigony is sometimes incomplete. Non-exclusive gynogenesis is known for example in *Schmidtea mediterranea*, a simultaneous hermaphrodite planarian (Benazzi and Benazzi Lentati 1992; Beukeboom *et al.* 1996; Vila-Farré and Rink 2018).

Among the insects, gynogenesis is known only for a few species, among which are some populations of the geometrid moth *Alsophila pometaria* and the triploid beetle *Ptinus mobilis*. In the latter, the maturation division of oocyte I is delayed until the penetration (not followed by karyogamy) of a spermatozoon deriving from a mating with the related diploid species *P. clavipes*.

Some populations of goldfish (*Carassius auratus*), a normally amphigonic species, are gynogenic: their eggs are activated by spermatozoa of other cyprinids. Also gynogenic are some North American freshwater fishes of the genus *Phoxinus*, similar to the European common minnow (*P. phoxinus*; Devlin and Nagahama 2002). In another fish genus, *Poeciliopsis*, six metasexual forms are derived from crossings between *P. monacha* and four other amphigonic forms. Of these six hybrids, three are triploid and gynogenic, the other three are diploid and hybridogenic (Section 3.6.5): in all cases, metasexual forms require insemination by a related amphigonic form. In a related genus, *Poecilia formosa* is the result of natural hybridization between *P. mexicana* and *P. latipinna*. Its progeny, all female, develop from diploid eggs resulting from oogenesis in which the first meiotic division is suppressed. Eggs are activated by spermatozoa produced by males of amphigonic species of *Poecilia* occurring in the same waters (Devlin and Nagahama 2002).

The salamanders of the genus *Ambystoma* are a classic example of gynogenesis, and are believed to have been reproducing this way for over a million years. It is probable, however, that in these animals the fertilization of an egg, capable of introducing new alleles into the gene pool, is a rare but not impossible event (see *paternal leakage* in Section 5.2.3.4).

3.6.4 Androgenesis

Androgenesis is a form of pseudogamy in which the penetration of the sperm cell into the egg is not followed by karyogamy, because the chromosomes of the oocyte are absent or inactivated; therefore, the zygote from which the new individual eventually develops contains only paternal chromosomes (Figure 3.28e). Depending on the specific cytogenetic mechanism of androgenesis and the mechanism of sex determination, androgenic species or lines are bisexual or male-only (Section 5.2.3.7; Schwander and Oldroyd 2016).

Androgenesis has evolved several times in the freshwater bivalves of the genus *Corbicula* (Figure 3.48; Hedtke *et al.* 2008). In cross-breeding between closely related species of these molluscs, recombination of the nuclear genome is rare, but the contribution of both mitochondrial genomes is frequent (in bivalves, cases of double, i.e. both maternal and paternal, inheritance of mitochondria are not rare; see Section 5.2.3.1). Cases of spontaneous androgenesis have been reported in laboratory strains of *Drosophila melanogaster* (Komma and Endow 1995), in hybrid strains of some plants (Chen and Henen 1989) and in complexes of hybrid forms of stick insects of the genus *Bacillus*

Figure 3.48 Reproduction by androgenesis is observed in several evolutionary lineages in the bivalve genus *Corbicula*.

(Mantovani and Scali 1992). Reproduction by exclusive androgenesis has been reported in some eukaryotes belonging to distantly related evolutionary lineages: in the Saharan cypress (*Cupressus dupreziana*) (Pichot *et al.* 2001), in four species of *Corbicula* and in some salmonid and cyprinid fishes (Devlin and Nagahama 2002).

A notable case is provided by the little fire ant *Wasmannia auropunctata*, where haploid males are generated by androgenesis, while the queens generate other queens by parthenogenesis. Normal amphigonic reproduction brings together male and female genomes only in the workers, but these are sterile. This creates a singular situation for a sexually reproducing species, where males and females reproduce without exchanging genetic material, as if they were two distinct species. Recognizing them as such would result in a unique case of a species composed exclusively of males (Queller 2005).

3.6.5 Hybridogenesis

Hybridogenesis is a reproductive mechanism midway between amphigonic and uniparental sexual reproduction (Figure 3.28f). In its simplest form, the egg of a hybridogenic female is fertilized and the paternal genome is expressed in the offspring (F_1); however, only the maternal genome is transmitted to the next generation (F_2). The female gametes are therefore partial clones (*hemiclones*) of the maternal genome; accordingly, this form of reproduction has been dubbed *hemiclonal* (Section 5.2.3.5).

Hybridogenesis is known in some fishes of the genus *Poeciliopsis*, in the green frogs of the genus *Pelophylax* and in the stick insects of the *Bacillus rossius-grandii* complex.

One of the most thoroughly investigated hybridogenic systems consists of the European green frogs currently assigned to the genus *Pelophylax* (Figure 3.49). The overall picture is very complex and involves several species, so we present here only the most frequent condition in the system traditionally described as a complex of three species, *P. lessonae*, *P. ridibundus* and *P. esculentus*. The latter is actually a hybrid between the other two species and is the prototype of reproduction by hybridogenesis. If we indicate with L/L and R/R the diploid genome of *P. lessonae* and *P. ridibundus*, respectively, the genome of an F_1 hybrid between the two species will be L/R. However, the same chromosomal complement is also found in subsequent generations, because of the peculiar way the hybrid transmits its genome. In the hybrid, in fact, meiosis involves the elimination of the L genome, whereby the females of *P. esculentus* can only produce eggs with the R genome. The continuity of the hybrid through the generations is therefore dependent on the availability of partners that can provide an L genome, i.e. male individuals of the parental

Figure 3.49 *Pelophylax esculentus* is a hybrid species, belonging to a hybridogenic complex of green frogs.

species *P. lessonae*. Consequently, *P. esculentus* can survive only under syntopy (local coexistence) with *P. lessonae*. Although less frequent, there are also hybrid populations that transmit only the L genome and therefore can only reproduce with the contribution of an R genome produced by males of *P. ridibundus*.

A similar reproductive mechanism is seen in the diploid hybrid *Poeciliopsis monacha-lucida*, a natural hybrid between females of *P. monacha* and males of *P. lucida*, which is represented by females only and reproduces by mating with males of *P. lucida* or another related species. Other hybrids between these two species are instead gynogenic (Section 3.6.3).

Chapter 4: Parental Investment in Sexual Reproduction

In the course of their lives, organisms spend time and energy on a number of activities and functions, of which reproduction is only one – think of growth, defence against predators and pests, and others. The study of how many and which energetic resources are used for reproduction, how much time is devoted to it and how this time is distributed over the course of life is an extremely varied chapter of biology, generally dispersed through specific topics of different disciplines, from gametogenesis in developmental biology to parental care in behavioural ecology.

Rather than paying attention to these disciplinary divisions, we analyse what is usually termed *parental investment*, trying to draw up a balance sheet for sexual reproduction, taking into consideration the different costs (energy, metabolism, survival risk) an animal or a plant faces in relation to reproduction. In particular we will discuss the different strategies that a parent can use to contribute to the survival of the offspring, from those involved in the production of gametes or spores, to those implemented before, during or after the moment at which the products of reproduction (eggs, embryos, larvae or juveniles) are released. We focus here exclusively on sexual reproduction in the most canonical form, namely amphigonic (biparental) reproduction, although the concept of parental investment extends also to the different modes of asexual reproduction and uniparental sexual reproduction (e.g. parthenogenesis). A comparative analysis of *sexual and asexual parental investment* is particularly interesting in those organisms that at certain times in their lives can opt for one or the other reproductive mode (this applies to many plants and many marine invertebrates). However, with the exception of modelling studies on possible trade-offs between the two types of investment (Roff 2002), this kind of study deals with very specific aspects of reproductive biology and the life history of a group. For this reason we refer the reader to the specialist literature on the subject (e.g. Flatt and Heyland 2011).

As indicated above, the topics discussed in this chapter extend far into the subject areas of other disciplines: developmental biology, physiology, ecology, ethology, but also sexual selection and the evolution of life-history traits among the themes of evolutionary biology. We therefore limit the following discussion to an outline of the subject, illustrated by some significant examples.

4.1 A First Look at Parental Investment Costs

Reproduction is expensive. The expenditure entries in the budget are of different natures and are not all easy to quantify. These costs vary enormously from group to group, depending on their life cycle and reproductive mode(s).

4.1.1 Reproduction Costs in Animals

In animals, the first entry in the expenditure column to be considered is the cost of the production of gametes, especially the eggs, which, although less numerous (often, by several orders of magnitude) than the sperm cells produced by the same species, usually represent a much more conspicuous cost, because of the amount of yolk stored in them (Section 3.4.1.3). However, when the yolk resources supplied to each egg are very small or even absent, the egg can be tiny, with a diameter in the order of one-tenth of a millimetre, as in many marine invertebrates and in parasitic worms such as *Fasciola hepatica*. The eggs of parasitoid insects, laid in the egg or larva of another insect, are often remarkably small (Figure 4.1); the smallest so far measured are perhaps those of the tachinid fly *Clemelis pullata* (previously known as *Zenillia pullata*), which measure just 0.027×0.020 mm.

We should not, however, underestimate the cost represented in some animals by the production of spermatozoa: in the tiny nematode *Caenorhabditis elegans*, in which there are both male and hermaphrodite individuals, reproductive activity significantly reduces the life span of males, but not that of hermaphrodites; in this case, it seems that the limiting factor affecting the duration of life is spermatogenic activity (van Voorhies 1992). The production of masses of sperm or spermatophores (Section 3.4.1.3), delivered to females that use them for trophic purposes, as we will soon explain, can also be very expensive.

Let's go back to the costs of reproduction borne by the mother. Provisioning the offspring with food often extends well beyond the stage of production of mature eggs into the various forms of matrotrophy described in Section 4.4.4, each of which has its own costs, often very high. To these direct costs, corresponding to the production of gametes and the supply of nourishment to the embryo or the young, must be added indirect costs, especially those – very widespread but not universal – related to exploratory behaviour in search of a partner, to nest building, to excavating burrows, to producing cases or other devices for the protection of the eggs. A special cost has been identified in a small bird, the zebra finch (*Taeniopygia guttata*), in which the parents' immune system has been found to be weakened in proportion to the number of nestlings they have to feed. Another cost arises from the lack of food intake,

197

Figure 4.1 The eggs of some species of the ichneumonid genus *Rhorus*, parasitoids of the larvae and eggs of other hymenopterans, are less than 0.15 mm long.

even for prolonged periods, that must be endured by the parent that takes charge of incubating the eggs. In this regard, one case that stands out is that of the male emperor penguin (*Aptenodytes forsteri*), which fasts for the entire duration of incubation (62–67 days) in the extreme environmental conditions of the southern winter on the margins of the Antarctic continent. However, equally remarkable is the case of the scolopendromorph and geophilomorph centipedes (Figure 4.2), where the mother remains coiled around the eggs, fasting as well for several weeks.

The costs faced by the male are generally much more modest than those borne by the female, but the case of the emperor penguin is not unique. Sperm production can be a significant expense in the case of promiscuous mating systems (Section 3.5.5), so that the strategy adopted for distributing the gametes between different and frequent episodes of insemination (*sperm allocation*) can be an important aspect of the breeding strategy, as in some fishes (Birkhead *et al.* 2008). Furthermore, the production of male gametes is usually accompanied by the secretion of a **seminal fluid** that helps or allows sperm to reach the eggs, but can also have other functions, for example nourishment for the female, or defence of the sperm carried with it against

Figure 4.2 A female of the geophilomorph centipede *Strigamia maritima* coiled around the eggs.

the sperm of competitor males. The production of gametes and seminal fluid, which together make up the **semen** (or **ejaculate**), can therefore be loaded with additional functions, and thus costs.

As mentioned above, some insects produce large amounts of sperm which, to a large extent, end up nourishing the female after being absorbed into specialized structures, such as the *organ of Ribaga* in the bed bug (*Cimex lectularius*). In many orthopterans, a substantial nutritional contribution is provided to the female by the trophic spermatophores they get from the males. The spermatophore of the Mormon cricket (*Anabrus simplex*, tettigoniids) is enormous, weighing a quarter of the male's entire weight. Less expensive are the bridal gifts, very common among birds, but also found among insects, such as the prey enclosed in a silk wrapper that the male of many empidid flies offers to his partner. The ultimate price is paid by a male that is eaten by the female during mating, as sometimes happens in some Mantodea (mantises) and Araneae (spiders).

4.1.2 Reproduction Costs in Seed Plants

For the sporophyte of the seed plants (Spermatophyta), the dominant generation in their haplodiplontic cycle, we must take into account both the costs related to pollination and those connected with the production of seeds and fruits, including dissemination (the costs sustained by the gametophyte will be mentioned in Section 4.6.2).

For plants that rely on anemophilous pollination, whose efficiency is low, the greatest cost is represented by the production of large amounts of pollen, of which only a small fraction will reach a female (or bisexual) flower of the same species. For plants with entomophilous or ornithophilous pollination, the production of bulky structures (large and colourful corollas, or showy

Figure 4.3 The inflorescence of *Calla palustris*, with the showy white spathe that helps to attract pollinator insects and holds them for a while in contact with the tiny flowers that line the long spadix.

involucral bracts such as the inflorescence in the arum family; Figure 4.3), nectar, perfumes, as well as an extra amount of pollen consumed by pollinators, represents a significant cost.

We will discuss in Section 4.6.3 the costs associated with seed dispersal, including specialized fruit parts such as the abundant fleshy pulp of drupes and berries and the structures that facilitate dispersal, such as the wing of the samara (the dry fruit of maples, *Acer* species) or the seed head of dandelions and thistles.

4.2 Fecundity

The term 'fecundity' has two main meanings, one qualitative, the other quantitative. The first refers simply to the ability of an individual, or a part thereof, to reproduce. For many organisms, including our species, this capacity is also called *fertility*.

In the second sense, discussed here, **fecundity** is instead a measure of the numerical abundance of descendants, and can be applied to individuals but also to species or higher-rank taxa. Fecundity can be referred to a single reproductive episode, or to a single reproductive season, or to the whole life of an individual (*lifetime fecundity*).

The distinction between fecundity in the single reproductive season and fecundity over the whole life of the organism is clear for many animals and plants, where multiple reproductive seasons occur in an individual's life. In other cases, however, such a distinction becomes meaningless because the reproductive phase can be prolonged for a very long time with uninterrupted production of offspring. Exemplary, in this regard, is the serranid fish *Diplectrum formosum*, a simultaneous hermaphrodite which releases gametes every two days during an adult life span extending up to eight years (Bubley and Pashuk 2010).

4.2.1 Measuring Fecundity

Although the notion of fecundity applies without difficulty to a large part of the living world, the way of measuring it effectively is not universal. This depends on the reproductive modes of the organism of interest, its life cycle and the aspects of its biology (physiology, ecology, evolution) we want to investigate. In asexual reproduction, fecundity is generally measured as the number of propagules (unicellular or multicellular, Section 3.1.2) produced by a reproductive event, in a reproductive phase, or within a certain time interval.

In amphigonic metazoans, the usual numerical disproportion in favour of male gametes suggests that we should measure fecundity based on the number of female gametes, which are the numerically limiting resource. In the case of an Australian passerine bird, the splendid fairywren (*Malurus splendens*), the female lays six eggs at a time, while the male always has 8 billion sperm cells at his disposal, potentially suitable to fertilize as many eggs. In the coho salmon (*Oncorhynchus kisutch*), the corresponding comparison is 3500 eggs to 100 billion sperm cells. Fecundity estimates in animals sometimes refer to the number of eggs produced irrespective of whether they are fertilized, an acceptable choice in cases where it is actually possible that a very large fraction of eggs produced by an individual are indeed fertilized. The case is different for animals in which the maximum number of offspring generated by a female is much lower than the total number of eggs produced, as in mammals, for which it seems reasonable to measure fertility based on the number of offspring born (Table 4.1).

In the case of spermatophytes, the most obvious indicator of fecundity is the number of seeds produced, while for terrestrial plants with gametophytes not dependent on sporophytes (bryophytes and pteridophytes), fecundity is expressed as the number of spores produced.

4.2.2 r/k Reproductive Strategies

In the different species of plants and animals, the fraction of the incoming energy (nutrition) directed towards reproduction, rather than used for maintenance and growth of the individual, is very diverse. This disparity is not limited to total parental investment, but also (or mostly) relates to the way in which this investment is used.

Table 4.1 Litter size per parturition event in some mammal species (data after Bourlière 1964)	
Common hedgehog (*Erinaceus europaeus*)	4–6
European mole (*Talpa europaea*)	1–7
Bats	1(2)
Nine-banded armadillo (*Dasypus novemcinctus*)	4–5
European hare (*Lepus europaeus*)	1–4
House mouse (*Mus musculus*)	4–7
North American beaver (*Castor canadensis*)	1–6
Japanese macaque (*Macaca fuscata*)	1
Common chimpanzee (*Pan troglodytes*)	1
Cetacea, all species	1
Bighorn (*Ovis canadensis*)	1
Giraffe (*Girafa camelopardalis*)	1–2
Elk, moose (*Alces alces*)	2
Wild boar (*Sus scropha*)	3–12
Mountain zebra (*Equus zebra*)	1
Asian elephant (*Elephas maximus*)	1(2)
Lion (*Panthera leo*)	2–6

In ecology, we label as **r** and **k** two extreme reproductive strategies corresponding to two opposite ways in which the expected reproductive success can be maximized:

- *r-strategy*: many small eggs (or many seeds) and no parental care
- *k-strategy*: fewer offspring, sizeable nutritional contribution and prolonged parental care

The reproductive strategies actually observed in nature are found in a continuum between these two extremes.

The *r* and *k* letters with which we usually designate these opposite reproductive strategies refer, respectively, to the population's intrinsic growth rate

Figure 4.4 The giant grouper (*Epinephelus lanceolatus*) has an *r*-type reproductive strategy, producing thousands of relatively small eggs.

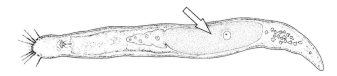

Figure 4.5 The microscopic gnathostomulid *Austrognathia microconulifera* (0.7 mm), which lives in the sediments on the sea floor, has a *k*-type reproductive strategy, with the production of a few relatively large eggs. An egg close to maturation is indicated by the arrow.

(*r*) and the carrying capacity of the environment (*k*), that is, respectively, to the exponent value according to which the population would increase in the absence of limiting density-dependent factors, and to the asymptotic value towards which population size tends, due to limited environmental resources.

The *r–k* axis cuts off the extreme economy of producing very few eggs with little yolk (or very few seeds with little endosperm) and abandoning them to their fate (a losing strategy due to the very low probability of survival of the offspring) and the extreme burden of producing many eggs or many large seeds and/or providing long, demanding parental care to a very numerous offspring (simply not a practicable strategy).

Here are examples of fecundity in plants and animals with different reproductive strategies along the *r–k* continuum. Among the animals, a very large number of eggs is produced by most sessile marine invertebrates such as sponges and corals, which release their gametes (female as well as male) into the water, where fertilization takes place. The fecundity of many fish species is also very high. For example, a large grouper (Serranidae) (Figure 4.4) can produce in its lifetime many tens of millions of eggs. By contrast, in some small marine invertebrates, including many interstitial species (Figure 4.5), a *k*-type strategy prevails, with few large eggs.

Large numbers of eggs per clutch are produced by brachyuran decapods (Vogt 2016). Examples are *Metacarcinus anthonyi* and *Scylla tranquebarica*, with

3.8 and 5 million eggs per clutch, respectively. In some *Cancer* species, the total number of eggs produced by a female during a lifetime was estimated to be more than 20 million, but in *Callinectes sapidus*, which has up to 18 broods, this value may be exceeded.

Among the insects, most species produce egg batches ranging from a few tens of eggs to a few hundred eggs; however, egg number is greatly reduced in the case of pupiparous dipterans and the other matrotrophic species mentioned below (Section 4.4.4). In other cases, very high fecundity values are reached. The queen of the African driver ant *Dorylus wilverthi* lays up to 3–4 million eggs every 25 days (Raigner and van Bovan 1955); among the non-social insects, the highest fecundity has been found in the Australian hepialid moth *Trictena atripalpis*: in the ovaries of a female that had already deposited 29,100 eggs were found another 15,000 fully developed eggs (Tindale 1932).

In most birds the individual clutch rarely exceeds 4–5 eggs, except for some groups (Jetz *et al.* 2008), especially among the galliforms, in which about 20 species (such as the grey partridge (*Perdix perdix*), the red-legged partridge (*Alectoris rufa*), the quail (*Coturnix coturnix*) and the wild turkey (*Meleagris gallopavo*)) lay on average more than 10 eggs, up to 19 in the Daurian partridge (*Perdix dauurica*) and 20 in the wattled brushturkey (*Aepypodius arfakianus*). Slightly more than 10 eggs on average are also produced by a dozen species of anseriforms (10 species of Anatidae and 3 of Dendrocygnidae), by several passerines (e.g. tits, Paridae) and by three species of rails among the gruiforms. On the other hand, many small birds, e.g. the hummingbirds (Trochilidae), generally lay only two eggs at a time; others, for example many seabirds, many columbids, some swifts (*Apus*) and several nocturnal birds of prey, lay just one egg. The weight of bird eggs ranges between 0.2 g in some hummingbirds and about 1.5 kg in the ostrich (*Struthio camelus*) (see also Section 3.4.1.3). Fecundity data for some species of mammals are shown in Table 4.1.

Investment of time is also a cost. In the case of birds, for example, this includes both incubation and fledging periods. The length of incubation may be as low as 10 days in some small passerines, and 13–14 days in other passerines such as the European skylark (*Alauda arvensis*) and the nightingale (*Luscinia megarhynchos*), but it is up to three weeks in the chicken and 40–45 days in the golden eagle (*Aquila chrysaetus*).

In plants, the number of seeds produced in a season (Greene and Johnson 1994) depends both on the number of fertilized flowers (female and/or hermaphrodite), and on the average number of seeds produced per flower. The latter number is particularly high in those plants, such as orchids, with very small seeds and no, or very little, endosperm. In orchids, the nourishment of the embryo in the heterotrophic phase depends on symbiotic interaction with a fungus (Smith and Read 2008).

A diversity of reproductive strategies along the *r–k* continuum is also found in the bryophytes: a mature sporophyte of *Archidium alternifolium* releases only 16 spores, but there are 50 million in a mature sporophyte of *Dawsonia lativaginata* (Kreulen 1972).

4.3 Temporal Distribution of Reproductive Effort

Another characteristic aspect of parental investment is the distribution of reproductive effort during an individual's life (see also Section 2.9). In many species, reproductive activity is restricted to one short period in an animal's life and is followed very soon by its death, especially in males. In other cases, the breeding season extends without appreciable interruptions over a period of weeks or months; in still other cases, the individual has multiple reproductive seasons, more or less circumscribed in time and separated by long periods of reproductive rest. On a large scale, these different reproductive strategies have a great influence in determining the resources destined for a single reproductive episode.

However, the distribution of reproductive activity over time and space may also reflect more specific adaptive strategies. In insects, for example, distributed oviposition can be an effective strategy for preventing parasitoid attack. Most lepidopterans and many other insects lay their eggs in batches, but females of some *Heliconius* butterflies lay only one egg at a time; in this way, the impact of possible attack by a parasitoid wasp ready to lay its eggs in them is reduced, since each of the butterfly's eggs must be individually identified and reached before being parasitized.

Similarly, in plants, the distribution of reproductive activity over time may correspond to specific adaptive strategies. Many tree species that are dominant in some forest communities and produce seeds and/or fruits particularly favoured by local primary consumers alternate, more or less regularly, between several years of ordinary seed production and single years of particularly abundant production. For example, at temperate latitudes this occurs in the beech (*Fagus sylvatica*) every 4–5 years, with even more exceptional fruiting every 10–15 years. During the years of abundant fruiting, the production of seeds and fruits saturates the herbivores' demand, thus increasing the chances of survival of the new plant generation. At the same time, the irregular occurrence of these exceptional years does not allow herbivores to evolve specific adaptations to exploit them fully.

The same principle of saturation of herbivorous demand is also possibly at the origin of the extraordinary synchronism in the flowering of all the individuals in many bamboo species (Poaceae Bambuseae). Depending on the species and environmental conditions, in these large – often huge – members of the grass family (more than 1400 species are known), different flowering modes and calendars may occur. In the herbaceous species and in some groups

Figure 4.6 In the bamboo *Phyllostachys bambusoides* all plants bloom at the same time, every 120 years. Flowering of the current generation is expected around year 2090.

with woody stems, such as *Schizostachyum*, flowering is continuous, or at least repeated over the years. In other species the flowering season may appear continuous at the population level, but the individual plants bloom for a short time, different from that of other plants of the same species. In most woody bamboos, however, all the plants of a particular species bloom at the same time, regardless of where they grow and independent of the climatic conditions to which they are exposed. This mass flowering is repeated at rather regular intervals, different in the different species, ranging from a few years in several species up to 60 in *Phyllostachys nigra* and over 120 in *Ph. bambusoides* (Figure 4.6). These synchronously flowering bamboos are semelparous, i.e. they die at the end of the flowering period (Veller *et al.* 2015).

In addition to the number of breeding seasons in an individual's life (Section 2.9), another factor affecting the temporal distribution of reproductive effort is the **age at the onset of reproductive maturity**. This feature, which is a fundamental trait in the theory of life-history evolution (Stearns 1992), is defined as the age at which an individual begins to reproduce sexually. As discussed in Section 2.9, this moment does not necessarily coincide with the time at which typical morphological and/or physiological characteristics of the adult are reached, although it may be strictly dependent on these. In animals, the age at which the first sexual reproduction takes place in what is conventionally described as the adult stage varies greatly from species to species (Figure 4.7), and may also differ significantly between related species. This is true even if we

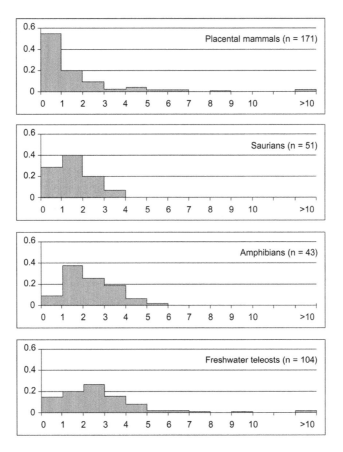

Figure 4.7 Frequency distribution of age (years) at reproductive maturity in samples of four vertebrate taxa. (Data from Bell 1980)

disregard the cases in which the individual reproduces in a somatic condition that we can define as larval or juvenile, as in paedogenesis (Section 3.6.2.5) and the first of the two reproductive phases of dissogony (Section 2.9).

The age at the onset of reproductive maturity can vary greatly even within a species and even within a population, as an effect of genetic polymorphism and/or environmental causes. In a study of some Hawaiian populations of *Gambusia affinis*, it was observed that the most precocious individuals of this small freshwater fish reached maturity within 35–40 days, while the latest ones did not reproduce before an age of 140–160 days (Stearns 1992).

There is also a direct relationship, even within higher taxa such as mammals and birds, between the age at the onset of reproductive maturity and the average size of the adult animal. However, other aspects of the biology of an organism may have influenced the evolution of this feature, as can be seen for example in mammals (Box 4.1). Where there is marked sexual dimorphism,

---

Box 4.1 Pregnancy, Lactation and Sexual Maturity in Mammals

Pregnancy is very short in marsupials, where the newborn, which often might be considered a larva, is hosted for a longer period in the mother's marsupial pouch: in the Virginia opossum, for example, the intrauterine life is only 13 days. The pregnancies of small placentals are also short, however – for example 16 days in the golden hamster, 21–24 days in the house mouse, and just 42 days even in an animal as large as the European hare. There are significant differences between the different orders, even between animals of comparable size. Carnivores, in particular, have rather short intrauterine development (about 2 months in cat and dog, 3½ months in tiger and lion); the pregnancy of domestic cattle is much longer (approximately 9 months, as in humans), and it is longer still in camels (about 400 days), giraffes (465 days), rhinos (573 days) and elephants (660 days). In spite of its size, the pregnancy of the blue whale is shorter, estimated at about 11–12 months. At birth, a kangaroo weighs just 30 millionths of the weight of the mother, while in some bats the newborn is up to 40% of the maternal weight.

Lactation is also quite short in carnivores (4–6 weeks in the cat, 6 weeks in the dog, 10 weeks in the lion) compared, for example, to primates (over a year in the chimpanzee) and ungulates (up to 2 years in rhinos).

Sexual maturity is reached very early in many rodents (19 days in the Norway lemming, 55–70 days in the guinea pig). By contrast, the beaver, which is mature at an age of about 2 years, is not particularly precocious even taking into account that as an adult it is remarkably large for a rodent: compare the hippopotamus (maturity at 2½ years), bison (3 years), horse (3–4 years), brown bear (5 years) and especially the blue whale (4 years). Age at maturity is however considerably higher in elephants (12–15 years) and rhinos (up to 20 years).

---

especially in body size, males and females often reach reproductive maturity at different ages. In fishes, females often mature later than males, while the opposite is observed in birds and mammals (Stearns 1992). In plants with the sporophyte dominant over the gametophyte, such as tracheophytes, the age at the onset of reproductive maturity can coincide with entry into the *adult reproductive phase*, the last of the three phases into which the post-embryonic development of the sporophyte is normally divided (Poethig 2003; see Section 2.9). This is characterized by the acquisition of competence for sexual reproduction through the development of reproductive organs such as strobili or flowers.

Table 4.2 Age at which the first flower or strobilus is produced by some seed plants under natural conditions (after Clark 1983)	
Pacific silver fir (*Abies amabilis*)	30 years
Sitka spruce (*Picea sitchensis*)	20–35 years
Western white pine (*Pinus monticola*)	7–20 years
Rocky Mountain bristlecone pine (*Pinus aristata*)	20 years
Douglas fir (*Pseudotsuga menziesii*)	20 years
Western hemlock (*Tsuga heterophylla*)	20–30 years
California redwood (*Sequoia sempervirens*)	5–15 years
Giant sequoia (*Sequoiadendron giganteum*)	20 years
Western redcedar (*Thuja plicata*)	15–25 years
Grapevine (*Vitis* spp.)	1 year
Hybrid tea roses (*Rosa* sp.)	20–30 days
Cherry, plum, almond and relatives (*Prunus* spp.)	2–8 years
Pear and relatives (*Pyrus* spp.)	4–8 years
Apple and relatives (*Malus* spp.)	8–10 years
Pedunculate oak (*Quercus robur*)	25–30 years
European beech (*Fagus sylvatica*)	30–40 years
Orange, lemon and relatives (*Citrus* spp.)	5–8 years
Common ivy (*Hedera helix*)	10 years

In fast-growing herbaceous species, the stages of development that precede sexual maturity (*juvenile vegetative phase* and *adult vegetative phase*) can last as little as a few days, and there are generally few vegetative structures typical of the juvenile phase, or none at all. On the contrary, in tree species the achievement of reproductive maturity may require, in the most extreme cases, up to 30–40 years (Table 4.2), and the retained juvenile structures are often a significant part of the soma of the adult plant.

In long-lived perennials, the age at the onset of reproductive maturity may vary within one species even more than in animals, depending on the environmental conditions to which the plant is exposed during the

pre-reproductive phases of development. Because plants can continue to produce new reproductive organs from apical meristems, adverse environmental
conditions, pest attacks, damage by herbivorous animals or other damage from
external causes can strongly affect the reproductive output during the adult
phase (Taiz and Zeiger 2010).

4.4 Investment of Animals in Egg and Embryo Development

A traditional classification of animal species based on the spatial and trophic
relationships between the egg (or embryo) and the maternal body divided
them into the three classes of *oviparous*, *ovoviviparous* and *viviparous* animals.
In oviparous animals the egg is expelled early from the mother's body, sometimes even before being fertilized; in the ovoviviparous it is retained in the
genital tract, but no intimate and direct relationship is established between the
embryo and the maternal tissues; finally, in viviparous animals, the embryo,
which develops inside the mother's body, is nourished through specialized
structures, e.g. the placenta of almost all mammals. More recently, however
(since Wourms 1981), a tendency to abandon this traditional partition has
been increasingly accepted, in favour of a double classification that considers
separately (i) the state in which the offspring leave the maternal body and (ii)
the ways in which the mother provides them with nourishment.

We therefore distinguish between **oviparous** animals, which lay eggs, and
viviparous animals (among which are now also included those that were
classified as ovoviviparous), which instead retain them. Independently, we
distinguish between **lecithotrophic** animals, in which the mother's contribution to the nutrition of the offspring ends with storing yolk in the egg, and
matrotrophic animals, in which the mother transfers nourishment to the
offspring at later stages (Figure 4.8).

This adjustment to the classification allows us to frame in a more suitable
way even some apparently atypical cases, including the monotremes. These
animals – the platypus (*Ornithorhynchus anatinus*) and the echidnas
(*Tachyglossus aculeatus* and the three *Zaglossus* species) – are the only oviparous
mammals, but they are also matrotrophic, albeit in ways different from those
of marsupials and placentals. In the latter, during intrauterine life there is
passage of nutrients from the mother to the fetus through the placenta, while
in the monotremes the embryo receives a substantial nutritional contribution
through the thin shell of the egg, before this is deposited. In the platypus, the
egg initially measures just 3 mm, yolk included, but later, at the time of
deposition, it reaches 15 × 17 mm. Thus, alongside the conditions of oviparity
in the traditional sense, associated with lecithotrophy, and viviparity,

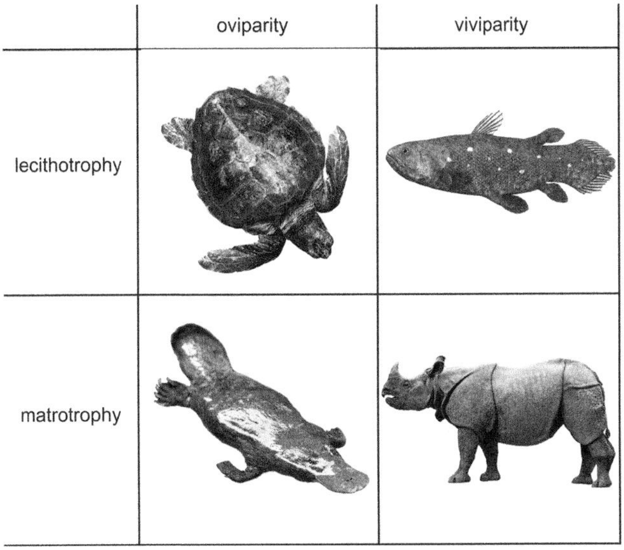

Figure 4.8 Different types of parental investment in the development of eggs and embryos in animals, with an example for each combination: the loggerhead sea turtle (*Caretta caretta*), the coelacanth (*Latimeria chalumnae*), the platypus (*Ornithorhynchus anatinus*) and the Indian rhinoceros (*Rhinoceros unicornis*).

associated with matrotrophy, we find not only a condition in which lecithotrophy is accompanied by giving birth to offspring that have already completed embryonic development (corresponding to the traditional category of ovoviviparity), but also the reverse condition, exemplified by the monotremes.

4.4.1 Oviparity

In oviparous animals, eggs are released either before or after fertilization. In the latter case, the eggs can be released following a single mating (or, at least, after the mating that has led to insemination and has not been made void by the elimination of spermatozoa during a subsequent mating) or following multiple inseminations. The eggs may be released all at once or in a number of separate laying events, in groups or individually, even over a long time span.

Eggs released into water may be covered by gelatinous coatings of various origins and consistencies, or more rarely they may be enclosed in protective cases like the parchment-like *cocoons* of some leeches (erpobdellids, which attach them to submerged objects) and some cartilaginous fishes such as skates (Rajidae; Figure 4.9) and dogfishes (*Scyliorhinus*).

In terrestrial environments, both land vertebrates and several lineages of arthropods have evolved measures to avoid rapid desiccation of the eggs. Apart from the transition to viviparity and the evolution of different forms of incubation (see next section), the problem can be solved in four main ways:

Figure 4.9 A ray egg in its protective case.

1.   First, there are animals that spend most of their lives out of the water, in particular as adults, but return to the water to release their eggs. Among the vertebrates, for example, this strategy is found in most of the amphibians, and among the insects in mayflies (Ephemeroptera), some of the dragonflies (Odonata Anisoptera), stoneflies (Plecoptera), caddisflies (Trichoptera) and several Diptera including mosquitoes (Culicidae).

2.   For many insects, instead, dehydration of the egg is prevented by laying them inside the tissues of another organism. This may be a victim (generally the egg or larva of another insect) that will provide nourishment to the larva, but in other cases protection is given by the tissues of a plant that will be abandoned by the insect at the time of hatching. This *endophytic deposition* is seen in damselflies (Zygoptera) and hawker dragonflies (Aeshnidae), and in many Orthoptera.

3.   A third solution is represented by the production of shells and/or protective cases, the latter sometimes used to cover the single egg (the *individual egg cases* of the stick insects), otherwise to protect a whole batch of eggs, as exemplified by the *oothecae* of cockroaches (Blattodea), mantises (Mantodea) and the primitive termite *Mastotermes darwiniensis* (Isoptera).

4.   A last series of adaptations involves the active behaviour of the mother (or, more rarely, of the father or of both parents) that modifies the environment by digging a den or a gallery or adapting an existing one, or building a nest to house the eggs – where the offspring may additionally benefit from parental care extending beyond oviposition.

4.4.2 Viviparity and Incubation

In **viviparous** animals the embryo develops within the reproductive system, the body cavity or the tissues of the parent until the release (birth) of the offspring in a post-embryonic condition, as a larva or juvenile.

Animals are said instead to practise **incubation** when the offspring, released as a zygote, embryo or even in a post-embryonic stage, is maintained in close contact with the surface of the mother's body or inside a fold (such as the mantle cavity of molluscs) or a pouch, sometimes but not always specialized for this function, or even in the gastric system (see also Section 4.5). Whatever the details, the relationship with the parent may consist solely in the protection of the offspring from environmental injuries, from predators or parasites, or be combined with the direct supply of nourishment (*matrotrophy*; see Section 4.4.4). In addition, the relationship can extend beyond the end of embryonic development.

In the case of viviparity, the locations in which eggs can be retained are the most diverse: the mesohyl of poriferans; the mesoglea of some hexacorals; the parenchyma of acoelomorphs, cydiophorans and many non-parasitic rhabditophorans; the general body cavity, be it a pseudocoel (e.g. in the acanthocephalans and in matriphagous nematodes; Section 4.4.4), the coelom (some species of polychaetes, asteroids, holothurians and many bryozoans) or a haemocel (two species of mites (Acari), the strepsipterans and some other insects), the ovary (a couple of medusae, most of the matrotrophic nemertines, numerous insects and several echinoderms) or the genital ducts (matrotrophic parasitic rhabditophorans, gastropods, scorpions, numerous insects, a few isopods, the matrotrophic species among the onychophorans, nematodes and some others).

In the case of incubation, the eggs released through the mother's genital opening can be retained until hatching in sacs or oothecae in contact with the mother's body, or in *brood pouches* (often called marsupia) which are morphologically external, but nevertheless capable of providing almost total protection. These behaviours, which in some cases extend beyond hatching, verge to a variable degree on viviparity. In addition to the mammals traditionally grouped in the order of marsupials (divided into up to seven orders, in current classifications), animals provided with a brood pouch include peracarid crustaceans (isopods, amphipods, etc.; Figure 4.10) and several anurans. For the taxonomic distribution of animals that practise some form of incubation associated with matrotrophy, see Table 4.3.

In insects, viviparity is almost always accompanied by lecithotrophy, with some notable exceptions (Table 4.4). Viviparity is also common in fishes, involving about half of the cartilaginous fishes (Chondrichthyes), the coelacanth (*Latimeria*) and a few hundred teleosts belonging to the following families: Sebastidae, Comephoridae, Anablepidae, Goodeidae, Jenynsidae, Poeciliidae, Clinidae, Embiotocidae, Labrisomidae, Zoarcidae, Brotulidae, Bythitidae, Parabrotulidae and Hemiramphidae. Isolated cases of viviparity also occur among the amphibians (some caecilians (Apoda), the anurans

Figure 4.10 A female of the isopod crustacean *Caecidotea communis,* with the ventral marsupium filled with eggs.

Nectophrynoides occidentalis and *Eleutherodactylus jasperi* and, among the uro-deles, some populations of *Salamandra salamandra*). The phenomenon is more widespread among the squamates, with examples in different families of saurians (Chamaeleonidae, Agamidae, Iguanidae, Gekkonidae, Scincidae, Cordylidae, Xantusiidae, Lacertidae (as in some populations of *Zootoca vivi-para*), Anguidae and Xenosauridae), in amphisbaenians and a number of snakes (examples among the Typhlopidae, Aniliidae, Boidae, Tropidophiidae, Acrocordidae, Colubridae, Elapidae and Viperidae).

4.4.3 Lecithotrophy

There is **lecithotrophy** when the mother's contribution to the nutrition of the offspring ends with supplying the egg with yolk. In almost all animals, oogenesis includes, in addition to the reduction of the chromosome number through meiosis, also filling the cytoplasm with a diversity of molecules, some of which (accumulated in the form of mRNA or already translated into pro-teins) will control the first stages of the embryo's morphogenesis, while the rest will satisfy its nutritional needs. The nutritious materials supplied to the egg are collectively known as **yolk**, and their production is called **vitellogenesis**.

Table 4.3 Distribution of viviparity and incubation associated with matrotrophy in the animal kingdom (based on Ostrovsky *et al.* 2016). The names of taxa in which all members are matrotrophic are given in bold

	Number of matrotrophic species	Viviparity (V) vs. incubation (I) (number of species in parentheses)
PORIFERA		
Demospongiae	24	V
Calcarea	6	V (transition to I in 2 spp.)
Homoscleromorpha	3	V
Hexactinellida	1	V
CNIDARIA		
Scyphozoa	2	V (1), I (1)
Hydrozoa	1	V
Anthozoa	2	V
DICYEMIDA	~110	V
ORTHONECTIDA	24	V
ACOELOMORPHA	2	V
RHABDOCOELA		
Digenea	~18000	V
Cestoda	17	V
Monogenea	~450	V
Turbellaria	8	V
ENTOPROCTA	5	I
CYCLIOFORA	2	I and V
BRYOZOA		
Cyclostomatida	626	V
Gymnolaemata	~130	I, V (5)

continues

Table 4.3 (cont)		
	Number of matrotrophic species	Viviparity (V) vs. incubation (I) (number of species in parentheses)
Phylactolaemata	87	I
NEMERTEA	14	V
SYNDERMATA		
Acanthocephala Eoacanthocephala	3	V
Acanthocephala Palaeacanthocephala	1	V
Acanthocephala Archiacanthocephala	1	V
Rotifera Monogononta	1	V
GASTROTRICHA	1	V
MOLLUSCA		
Gastropoda	23	V (18), I (5)
Bivalvia	42	I
ANNELIDA		
Polychaeta	19	V
Clitellata	3	I
NEMATODA		
Chromadorea	33	V
Enoplida	3	V
LORICIFERA	1	V
ONYCHOPHORA	86	V
ARTHROPODA		
Arachnida **Pseudoscorpiones**	3385	I

Table 4.3 (cont)		
	Number of matrotrophic species	Viviparity (V) vs. incubation (I) (number of species in parentheses)
Arachnida **Scorpiones**	1753	V
Arachnida Acari	2	V
Crustacea Isopoda	13	I, V (2)
Crustacea **Cladocera**	37	I
Crustacea Anomopoda	19	I
Crustacea Ctenopoda	1	I
Crustacea Decapoda	1	I
Insecta Blattodea	1	V
Insecta Dermaptera	13	V
Insecta Hemiptera	~5030	V
Insecta Psocoptera	4	V
Insecta Diptera	~810	V
Insecta **Strepsiptera**	~600	V
Insecta Coleoptera	15	V
ECHINODERMATA		
Holothuroidea	32	V (14), I (18)
Asteroidea	10	V (6), I (4)
Ophiuroidea	8	V (1), I (7)
Echinoidea	7	I
Crinoidea	2	V (1), I (1)
CHORDATA		
Ascidiacea	6	I, V (1)

continues

Table 4.3 (cont)		
	Number of matrotrophic species	Viviparity (V) vs. incubation (I) (number of species in parentheses)
Thaliacea **Salpida**	48	V and I
Chondrichthyes	351	V
Osteichthyes	110	V, I (4)
Amphibia	38	V, I (3)
Reptilia	uncertain	V
Mammalia	5750	V, I (5)

We speak of **primary vitellogenesis** when yolk is supplied to the egg through the blood or other body fluids or derives from intestinal contents, with no contribution from other cells. This occurs in ctenophores, annelids, bryozoans, chaetognaths and echinoderms. In many insects, yolk proteins are synthesized in the fat bodies and pass into the haemolymph as *vitellogenins*, reach the oocyte by passing through the intercellular spaces between the follicular cells, and are finally taken up by endocytosis.

In **secondary vitellogenesis**, nourishment is instead provided by abortive oocytes (often the sister cells of the oocytes that develop into eggs) or by mesodermal cells of the ovary wall. In some cases (poriferans, cnidarians, gastropods, cladocerans) the abortive oocytes can be phagocytized by the egg, in other cases (rotifers, nematodes, some insects) thin cytoplasmic bridges are formed through which storage materials are brought to the egg cell. In *Drosophila*, each oocyte is accompanied by 15 *follicular cells*; these number 47 in the honey bee (*Apis mellifera*).

In many other animals (brachiopods; polyplacophorans, solenogasters and cephalopods among the molluscs; malacostracan crustaceans, insects, arachnids; holothurians, tunicates and vertebrates) the eggs are surrounded by a *follicular epithelium* which represents a local differentiation of the ovary wall. In mammals, the oocyte is surrounded by two follicles (the internal *primary follicle* and the external *secondary follicle*), while the surrounding tissues form a double, strongly vascularized *follicular theca*. Between the follicular cells are formed spaces that unite to form the *follicular antrum*. The result is the *tertiary follicle* or *Graafian follicle*, within which the oocyte occupies a marginal position, inside a *cumulus oophorus* formed by follicular cells.

Table 4.4 Viviparity in insects	
Ephemeroptera	Viviparity in some Baetidae only
Dermaptera	Viviparity with lecithotrophy in *Marava arachidis*, with matrotrophy in two very specialized genera: *Hemimerus*, parasites of large African rodents of the genus *Cricetomys*, and *Arixenia*, parasites of Indonesian bats
Blattodea	The oviparous forms lay groups of eggs protected by a robust ootheca; in the Blaberidae, which are viviparous, the ootheca is retained in the female genital tract until the completion of embryonic development, but its presence seems to completely exclude matrotrophy; however, things may be different in *Geoscaphus*, the only genus in which no trace of an ootheca is left
Plecoptera	Viviparity in four lecithotrophic species
Psocodea	Very few viviparous species, one of which (*Archipsocopsis fernandi*) is possibly matrotrophic
Thysanoptera	Viviparity in several species of Phloeothripidae
Homoptera	Viviparity and matrotrophy widespread, especially in aphids
Heteroptera	Viviparity only in the Polyctenidae, parasites of bats
Coleoptera	Viviparity in six families only: Micromalthidae (*Micromalthus debilis*, the family's only known species, mentioned in Chapter 3), Carabidae (two species of the genus *Pseudomorpha*), Staphylinidae (many species of the aleocharine group, e.g. those of the genus *Corotoca*, which lay larvae ready to transform into pupae), Tenebrionidae (16 species of the tribe Pedinini and *Alegoria castelnaui* of the Ulomini; Dutrillaux *et al.* 2010), Chrysomelidae (about 50 species, among which *Chrysolina varians* and several species of *Oreina*) and Cerambycidae (*Borneostyrax*)
Strepsiptera	All viviparous and matrotrophic
Hymenoptera	Viviparity in the Ichneumonidae of the genus *Polyblastus*
Diptera	In this order viviparity has evolved at least ca. 60 times independently and occurs, in different forms, in 22 families. In some instances, viviparity is optional, in others obligate.

continues

Table 4.4 (cont)	
	Among the latter there are species in which only one larva at a time develops within the mother's body, with either lecithotrophy or matrotrophy. Characteristic of the so-called pupiparous dipterans (Glossinidae or tsetse flies; Hippoboscidae, blood-sucking parasites of birds and mammals; Streblidae and Nycteribiidae, all parasites of bats) is *adenotrophic viviparity*. In these insects, the larva, while inside the mother's body, feeds on secretions that allow it to complete development: born as a mature larva, it transforms immediately into a pupa, without taking any more food
Trichoptera	Viviparity in five lecithotrophic larviparous species of the genus *Triplectides*
Lepidoptera	Viviparity in a couple of moths of the genus *Monopis*

Depending on the amount of yolk they contain, mature eggs can be classified into **alecithal** (without yolk), **oligolecithal** (with modest amounts of yolk) and **macrolecithal** (with abundant yolk). On the basis of the distribution of yolk material with respect to the entire egg mass, they can instead be described as **isolecithal** (with uniformly distributed yolk, a condition typical of oligolecithal eggs), **telolecithal** (with yolk prevalently packed towards one of the egg's poles) and **centrolecithal** eggs (with yolk in the innermost part).

In many animal species, some of the eggs end up as a source of nourishment for the other eggs (or for the embryos that emerge from them: *oophagy*, see next section). These are mainly ancillary oocytes that maintain their integrity until the moment they transfer the products synthesized within them to a nearby egg or are swallowed up during oogenesis. Somatic cells such as coelomocytes, or other eggs, can also become targets of phagocytosis.

In rhabditophorans, yolk can be stored in the egg, as usual (*endolecithal egg*), or in specialized cells surrounding the gamete, forming with the latter a **compound egg** (*ectolecithal egg*). The endolecithal condition, phylogenetically primitive, is characteristic of catenulids, macrostomids and polyclads, generally referred to as *archoophores*. In the remaining free-living rhabditophorans and in the neoderms (the parasitic forms classified as Digenea, Monogenea and Cestoda; Figure 4.11), collectively referred to as the *neoophores*, the yolk is stored instead in particular cells of the gonad (*vitelline cells*), which eventually form part of the compound egg. Vitelline cells and

Figure 4.11 The fluke *Clonorchis sinensis*, a flatworm parasite of humans, produces compound eggs, i.e. with reserve substances stored in cells other than the egg cell.

germ cells are usually produced in spatially distinct segments of the ovary called the *vitellarium* and the *germarium*, respectively.

Compound eggs are not exclusive to neoophoran flatworms. A coating of somatic cells is found, for example, around the eggs of some oviparous poriferans, and in some cases at least it is known that these somatic cells are internalized by the developing embryo and most likely perform a nutritive function. Functional equivalents of the vitelline cells of the neoophoran flatworms are found in some nemertines, crinoids, polychaetes (e.g. *Pygospio elegans*) and gastropods (e.g. *Nucella lapillus*). In the latter case, within the capsule surrounding the fertile egg some non-viable eggs are also deposited, which will serve as nourishment during post-embryonic development; in *Buccinum undatum*, within each capsule a few hundred eggs are released, of which only 10–20 survive.

4.4.4 Matrotrophy

Matrotrophic animals are those in which there is a direct and continuous maternal contribution of nutrients other than the yolk supplied to the egg and the food collected in the surrounding environment for the benefit of the offspring. Matrotrophy can be associated with viviparity or incubation of eggs (Table 4.3) and can also extend beyond the end of embryonic development, as occurs in the feeding of offspring with milk by female mammals. In some cases, in spite of the apparent oxymoron, matrotrophy during post-embryonic development can also involve the male parent. Next to the best-known example of lactation in mammals (remember that the females of monotremes, which are oviparous, also feed their newborn with milk), we should mention the *dermatophagy* of some fishes and amphibians, in which the juveniles feed on specialized epithelia of the parent, and the so-called *crop milk* that is fed to young pigeons (Columbidae). The latter is a product of the physiological degeneration of the epithelial cells that line the crop of the adult bird (Figure 4.12). Father and mother collaborate in providing this nutrition, rich in fats and proteins, which for pigeon chicks represents the only food during the first three days of life. A similar food is also given to their offspring by other

Figure 4.12 In pigeons (here, the rock dove, *Columba livia*), parents feed their chicks with derivatives of their crop's epithelial cells. This is a form of matrotrophy called dermatophagy, in this case involving the male parent as well as the female.

birds, such as flamingos (Phoenicopteridae) and the emperor penguin (*Aptenodytes forsteri*).

Depending on the nature of the nourishment consumed by the embryos or by the developing juveniles, five forms of matrotrophy can be distinguished:

1. In **oophagy**, there is ingestion of other eggs (*trophic eggs*) present in the mother's genital tract or of the products of their resorption, a situation known for free-living rhabditophorans, a polychaete, several nematodes and five echinoderms belonging to different classes, excluding sea urchins. In the Alpine salamander (*Salamandra atra*), the larvae feed on unfertilized eggs present in the uterus where they are developing. In ants (Formicidae), sterile eggs produced by workers (or by the queen) are sometimes supplied as food to the larvae they care for. Trophic eggs are food provisions that do not require direct mother–offspring contact (Trumbo 2012).

2. In **adelphophagy**, nourishment is provided in the form of sibling embryos present in the female's genital tract. The few known cases are the bull shark (*Carcharias taurus*) and some populations of fire salamander (*Salamandra salamandra*; Figure 4.13), as well as a gastropod, a genus of dipterans, four isopods, several asteroids and two ophiuroids. In most angiosperms, the embryo gains nutrition from the triploid cells of the secondary endosperm resulting from a fertilization parallel to that from which the seed originates (Section 4.6.3), and this could be described as a form of adelphophagy. Even closer to adelphophagy is the way the embryo is fed in some scale insects. Here the nutrients derive from a pentaploid tissue (*mycetoma*, which also hosts unicellular symbionts) deriving from the fusion of a diploid nucleus of the embryo with the nucleus of the first polar body that has not completed the

Figure 4.13 In some populations of the fire salamander (*Salamandra salamandra*), adelphophagy is observed, a form of matrotrophy in which nourishment for an embryo consists of its sibling embryos.

second meiotic division (and is therefore diploid) and with the nucleus of the second polar body, which is haploid (Schrader 1923).

3. When nutrients are supplied directly from the liquids present in the mother's body cavity (or, in some cases, from the tissues of the mother) and are absorbed or phagocytized, matrotrophy is described as **histotrophy**, a mode found in a few mites, a decapod crustacean, a beetle and some echinoderms of the different classes, excluding crinoids. It also occurs – limited to the early stages of embryo development – in a number of poriferans, cnidarians, nematodes, molluscs, polychaetes, bryozoans and crustaceans (branchiopods and isopods). In the ovoviviparous cockroach *Thorax porcellana*, care is extended when hatching nymphs ride in a specialized compartment on the dorsum and obtain liquid nourishment by using their mandibles to pierce the mother's cuticle (Trumbo 2012).

4. **Histophagy** occurs when the offspring ingests tissue or glandular secretions, fragments of cells, but also portions of tissue, especially from the sometimes hypertrophic epithelium of the mother's genital tract, or, in rare cases, whole organs or even the whole mother (these extreme cases are referred to as *matriphagy*, as in the mite *Adactylidium*; Section 5.2.3.1). Histophagy (not always adequately documented) appears to be present in all pseudoscorpions, in many polychaetes and dipterans, and in several other animals, including some scorpions, isopods and echinoderms. A particular case of histophagy occurs in some species of caecilians (amphibians), where the newborns feed off their mother's shreds of skin (Wilkinson *et al.* 2013; Figure 4.14). Matriphagy from the outside of the mother's body also occurs in some spiders such as *Stegodyphus lineatus* (Salomon *et al.* 2015; Figure 4.15). In this extreme (suicidal) form of provision of parental care, the mother is first consumed in regurgitating food and products of the dissolution of her internal organs to

Figure 4.14 In the limbless amphibian *Microcaecilia dermatophaga* the young feed on pieces of the mother's skin, which they actively remove thanks to specialized juvenile teeth; this form of matrotrophy is called histophagy.

Figure 4.15 In the spider *Stegodyphus lineatus*, matrotrophy takes the extreme form of matriphagy, with the newborns devouring the mother's body.

feed the newborn spiderlings, then, once deceased because of this, is completely devoured by her offspring.

5. **Placentotrophy** requires close contact or even fusion between some specialized tissues of the mother, forming the *placenta*, and those of the fetus, allowing continuous exchange of fluids between them. Characteristic of placental mammals, in which there is considerable structural and functional diversity, this form of matrotrophy is also found in some representatives of different groups of invertebrates (poriferans, cnidarians, acoelomorphs, gastrotrichs, rotifers, entoprocts, cycliophorans, platyhelminths, nemerteans, annelids, molluscs, bryozoans, arthropods, onychophorans and nematodes).

Although nutrition qualifies as a form of parental care (see next section), we do not classify as matrotrophy the widespread cases in which the offspring receive food collected in the surrounding environment, sometimes reworked mechanically or with the help of hydrolytic enzymes. This food may be stored in a suitable place, where it remains available to the larvae or juveniles that hatch from the eggs laid in these clusters of food or in the immediate vicinity, as a rule in burrows or tunnels dug by the parent; or it may be administered directly to the offspring by the parent or by a helper (the most significant examples of which are the sterile workers among the eusocial insects), often in nests prepared by the parent, alone or with the help of the partner and/or one or more related individuals.

4.5 Parental Care in Animals

Parental care is a specific behaviour of the parents, or at least one of them, that allows the development of the offspring in suitable conditions or at least increases their chances of survival. This care can be provided at different stages of the offspring's life, including both prenatal behaviours such as egg guarding, and postnatal care such as active protection from predators and parasites, and feeding the young.

In some animal groups there is extreme diversity in the relationships between adults and offspring, sometimes even between closely related species. Here we report some particularly significant examples of the behaviours that have evolved among the anurans, while we refer to Table 4.5 for a summary of the strategies adopted by fishes and to Box 4.1 for some aspects of matrotrophy (pregnancy and lactation) in mammals.

Among the terrestrial vertebrates, anurans are the group with the greatest diversity of incubation patterns, in which one of the parents carries around eggs or tadpoles on itself or within itself. In the obstetric toad (*Alytes obstetricans*; Figure 4.16), the male wraps strings of eggs laid by one or more females around his hind legs, and carries them around until hatching. In the Suriname toad (*Pipa pipa*), instead, the female places eggs and tadpoles into deep depressions in the skin of her back and carries them until metamorphosis is completed. In the *Gastrotheca* frogs of Central and South America, the female is provided with a sort of dorsal marsupium, which in some species has specialized areas through which exchanges of gas, water and excreta occur between mother and offspring. In *Rheobatrachus silus* the female literally ingests the newly fertilized eggs, and incubation (5–6 weeks) takes place inside the maternal stomach. In Darwin's frog (*Rhinoderma darwinii*), the male hosts the newly hatched tadpoles in his vocal pocket, where they will remain, feeding on

Table 4.5 Classification of oviposition and parental care strategies in fishes (after Balon 1975, 1984, simplified)

1	**No parental care**
1.1	Eggs released in open environment
1.1.1	Eggs released in water
1.1.2	Eggs released on substrate
1.1.3	Eggs released in sand, out of water
1.2	Eggs laid in protected sites
1.2.1	Eggs released under stones
1.2.2	Eggs released in substrate slits
1.2.3	Eggs released on invertebrates
1.2.4	Eggs released on the beach
2	**Eggs guarded**
2.1	Eggs released on the substrate
2.1.1	Eggs released on rocky bottom
2.1.2	Eggs released on submerged vegetation
2.1.3	Eggs laid in the sand, out of the water
2.1.4	Eggs released into the water
2.2	Eggs laid in a nest
2.2.1	Nest of pebbles and gravel
2.2.2	Nest in the sand
2.2.3	Nest of plant materials (loose or glued together)
2.2.4	Nest of bubbles
2.2.5	Nest in a natural cavity
2.2.6	Nest of mixed material
2.2.7	Eggs laid in a sea anemone
3	**Eggs retained by the parent**
3.1	Eggs held outside the body (incubation)

Table 4.5 (cont)	
3.1.1	Eggs retained by the mother until they are transferred to an oviposition substrate
3.1.2	Incubation in the mouth
3.1.3	Incubation in the branchial chamber
3.1.4	Incubation in a pouch of the mother's body
3.2	Eggs held inside the body (viviparity in the broader sense)
3.2.1	Viviparity optional
3.2.2	Obligate viviparity without matrotrophy
3.2.3	Obligate viviparity with matrotrophy

Figure 4.16 A male obstetric toad (*Alytes obstetricans*) carrying the eggs of one or more females with which it has mated.

substances derived from the parent's tissues, until they develop into tiny froglets.

When parental care is the responsibility of only one parent, this is almost always the mother. However, in addition to Darwin's frog, the care of the offspring is entirely entrusted to the father in other animals of very different groups. Among these are the seahorses (*Hippocampus* spp.), where the male has a ventral pouch, also present in several species of needle fish such as *Syngnathus* and *Nerophis*, which belong to the same family. Paternal care is also observed in some millipedes (Diplopoda), in which the male remains coiled around the eggs laid by his partner, until hatching. In the polychaete *Neanthes arenaceodentata* the two partners form a common mucous tube inside which they release their gametes. The female dies shortly afterwards, while the male remains with the directly developing offspring, until hatching, about 10 days after fertilization, and even for a further three weeks thereafter. The incubation of the eggs is also the full responsibility of the male in some birds, among which, as mentioned in Section 4.1.1, is the emperor penguin. Even more extraordinary is the effort produced by the male Australian malleefowl (*Leipoa ocellata*), which spends five hours a day for six long months to build and maintain the gigantic mass of sand and decaying matter in which it makes room for the eggs laid by the female, one at a time; a huge amount of material is reworked each time. The heat required for incubation is released by the decomposition of the organic material of which the nest is largely composed, and the continuous rearrangements performed by the male ensure that the embryos will have the appropriate conditions for development until hatching.

Among the invertebrates, a most notable example of paternal parental care is provided by the pycnogonids, the males of which carry the bulky egg masses produced by the females on a pair of specialized appendages (ovigers; Figure 4.17). A male can carry up to a dozen groups of eggs at a time, produced by different females, with embryos at different stages of development. Paternal parental care of some large water bugs (Belostomatidae) is less demanding: in the genus *Lethocerus*, males guard the eggs glued by females onto the stems of

Figure 4.17 A male of the sea spider *Nymphon rubrum* carrying the masses of eggs produced by the female, held by a pair of specialized front appendages called ovigers.

marsh plants, just above the water's surface, and keep them wet, while the males of the related genus *Belostoma* carry them until hatching on their backs, where the females have laid them.

In some animals, in addition to the parental care provided by one or both parents, care is supplied by other individuals. These helpers are sometimes individuals specialized for this function, as in eusocial insects (worker bees, ants, termites), sometimes relatives that do not reproduce, in the current season at least. This is the case of the Florida jay (*Aphelocoma coerulescens*), in which young birds often help their parents, feeding their siblings or guarding and protecting them from predators. Another example, which actually involves non-kin, 'unaware' helpers, is seen in the parasitic behaviour of the European cuckoo (*Cuculus canorus*) and another 60 species of the same family: in these birds the female lays the eggs in the nests of other birds, entrusting them with the whole business of incubation and subsequent nutrition of the cuckoo chick.

4.6 Investment of Plants in Gametophyte and Sporophyte Development

Most plants, including all embryophytes, have a haplodiplontic life cycle. It therefore seems appropriate to distinguish the parental investment destined for the development of the gametophyte (an investment made by the sporophyte of the previous generation), from that intended for the development of the sporophyte (one made by the gametophyte of the previous generation). However, the more general issues of parental investment by the gametophyte and the sporophyte in a haplodiplontic cycle also apply to multicellular algae with haplontic and diplontic cycles, respectively.

4.6.1 Gametophyte Development

In plants where the gametophyte's life does not depend on the sporophyte that generated it (e.g. many algae, mosses and ferns), the parental investment of the latter is usually limited to the supply of reserve substances to the spore, and possibly, in land plants in particular, also of protective coating layers which increase their chances of survival and dispersal (Section 3.1.2.1). In plants with a heterosporous sporophyte (Section 3.3.2.1), microspore and megaspore are recipients of differential parental investment.

However, there are many cases in which the gametophytes, during more or less extensive phases of their development, benefit from the protection and nourishment of the sporophyte that generated them. This is observed in some taxa of algae belonging to different groups (e.g. in brown algae (Phaeophyceae) of the Fucales and Ascoseirales; Section 7.2), but is in fact typical of spermatophytes. Here the megagametophyte is retained and nourished by the parental sporophyte; the latter also, although indirectly, ends providing nutrients for some time to the sporophytes of the next generation that the megagametophyte will eventually generate (see next section). The microgametophyte, i.e. the pollen grain, will develop only thanks to the nutrients given to the microspore by the parental sporophyte. In the angiosperms, a specialized tissue of the anthers, the *tapetum*, which surrounds the microspore mother cells, produces a nutritive *locular fluid* that contributes to the development and maturation of pollen (Pacini 2010).

4.6.2 Sporophyte Development

Among the plants in which the gametophyte generation is dominant over the sporophyte generation, there are groups where the latter is not autonomous. In bryophytes the nutrition of the sporophyte depends, for a longer or shorter time, on the gametophyte, through cells or tissues specialized for the transfer of nutrients from parent to offspring, like the placenta of mammals. The same applies to the carposporophyte of the Florideae among the red algae (Rhodophyceae) (Figure 7.7), which is also dependent on the gametophyte from the nutritional point of view. In tracheophytes, where the sporophyte is dominant over the gametophyte generation, the female gametophyte is relatively small and has only limited trophic and protective functions, in any case restricted to the embryonic phase of the sporophyte's development.

In ferns and in *Equisetum* (horsetail, Figure 4.18), the embryonic sporophyte is retained by the gametophyte that nourishes it during early development, until it produces the first leaves and the first roots, thus becoming independent. In parallel, the small parent gametophyte generally fades out and eventually dies, a phenomenon analogous to the extreme matrotrophy of some animals (see Section 4.4.4).

In spermatophytes, the embryonic sporophyte is retained by the megagametophyte, but this, in turn, is held back by the sporophyte of the previous generation. Ultimately, nutrition and protection during the development of a young sporophyte are provided by the tissues of the sporophyte that produced the parent megagametophyte. However, the trophic resources for the early developmental stages of the new sporophyte are provided in various ways. In gymnosperms the egg cells are gigantic, stuffed with carbohydrates and proteins; in these plants, it is the haploid cells of the female gametophyte that continue to grow and function as a nutritive tissue, called **primary endosperm**, destined to be incorporated into the seed (see next section). In angiosperms, instead, the trophic reserves available for the development of the

Figure 4.18 During the early stages of development, the sporophyte of horsetails (here, fertile stems of *Equisetum arvense*) is nourished by the gametophyte that generated it.

new sporophyte come from a triploid tissue, called **secondary endosperm**, which derives from the parallel fertilization of another cell of the same mega-gametophyte by the same pollen granule (*double fertilization*; Section 3.5.4.2). In some respects, the secondary endosperm could be considered a twin embryo of the new sporophyte, so that this mode of parent-to-child transfer of resources, through a sibling, is somehow equivalent to adelphophagy in animals, as mentioned in Section 4.4.4.

In spermatophytes, the parental investment in the young sporophyte, which generally goes well beyond the formation of the egg cell, is modulated mainly through the development of the seed and, in the flowering plants, of the fruit. We therefore turn our attention now to the development of these two structures.

4.6.3 Seed and Fruit Development

The **seed** is a composite structure that develops from the ovule in the vast clade of plants called the spermatophytes (Section 3.4.2.1). A seed includes: (i) the *embryo*, or the new sporophyte, deriving from the fertilization of a female gametophyte cell by a male gametophyte cell; (ii) nutritive tissue, almost always represented by the *endosperm*, consisting of tissues of the parent megagametophyte (*primary endosperm* of the gymnosperms) or tissues resulting from a parallel fertilization (*secondary endosperm* of the angiosperms); and (iii) a coating (*integument*) which may be contributed to by tissues of both the megagametophyte and the sporophyte of the previous generation. Therefore, tissues of three different generations are usually found in a seed: two successive sporophytic generations and the gametophytic generation between them.

Accompanied or not by the fruit (see below), the seed is a structure suitable for dispersal. At maturity, the seed enters a phase of *quiescence* or *dormancy* during which many of its metabolic functions are suspended, to be resumed if and when the conditions for germination occur. There are huge differences between species both in the minimum time of dormancy and the duration of potential viability, or germinability (from a few weeks to several years). Generally, germinability declines rapidly after the first year, or after a few years, but individual seeds can survive much longer than the average for their species. For horticultural purposes it is considered that the germinability of onion seeds is limited to two years, that of fennel, aubergine, pea, parsley, celery and spinach to three years, that of tomato, pumpkin and watermelon to four, that of chard, melon and cucumber to five. The length of time a seed can remain viable and capable of germinating depends not only on the species, but also, and to a considerable extent, on the conditions in which it is stored. The oldest seed of which the germinability has been verified belongs to the arctic narrow-leaved campion (*Silene stenophylla*), with an age of $31,800 \pm 300$ years for a sample of seeds that were stored by a squirrel and remained protected until today under the permafrost (Yashina *et al.* 2012); this dating was carried out through ^{14}C, as in the case of some date palm seeds (*Phoenix dactylifera*), 2000 years old and also germinable, found during archaeological excavations in the palace of Herod the Great at Masada, Israel (Sallon *et al.* 2008). Equally reliable is the dating (1300 years) of germinable lotus seeds (*Nelumbo nucifera*) found in the dried bed of a lake in northeastern China (Shen-Miller *et al.* 1995). For more information on the physiology of seed development, germination and dormancy, see Bewley *et al.* (2013).

The way in which the nourishment provided by the parent is stored in the young embryo may vary from group to group. Among the angiosperms, in most eudicots the *cotyledons*, or embryonic leaves, accumulate the nutrients that will be used during and after germination. During embryonic development, the cotyledons become filled with starch, oils and proteins, while the endosperm, the source of nutrients, is reduced until it is exhausted. In monocots, by contrast, the single cotyledon generally remains thin, while the endosperm persists in the mature seed. During germination, the cotyledon, now called a *haustorium*, acts as an absorbing and digestive tissue, transferring nutrients from the endosperm to the embryo. Between these two extremes, there exists a huge variety of intermediate developmental modes, where both the cotyledons and the endosperm participate in the nourishment of the embryo during germination. A mature seed in which the endosperm is very abundant, as in wheat (*Triticum* spp.), is said to be *albuminous*, while if at maturity the endosperm is distributed or absent (many legumes (Fabaceae), including peas and beans), the seed is said to be *exalbuminous*.

The extent of growth and embryonic development that occurs before dormancy is extremely variable. Orchids and bromeliads, for example, have tiny, almost dust-like seeds, in which the embryo is only a small sphere of cells without cotyledons, radicle or vascular tissue. At the other extreme, in wheat, the embryo at the time of germination is already at a very advanced stage of development, with six small leaves. Considering that a fully developed wheat plant generally has only 10 or 12 leaves, more than half of the leaves of the new sporophyte are produced while it is enclosed in the parent gametophyte, in turn enclosed in the previous parent sporophyte. In angiosperms, seed mass varies over 10 orders of magnitude (Igea *et al.* 2017), from the minute 1 μg seeds of some orchids to the seeds of an endemic palm tree from the Seychelles, the sea coconut (*Lodoicea maldivica*), which can weigh more than 18 kg with a diameter of 50 cm. As in the common coconut (*Cocos nucifera*), most of the mass of this gigantic seed ('milk' and 'pulp') consists of the triploid endosperm.

The **fruit** is a plant structure, typical of the angiosperms, responsible for the protection and dispersal of seeds. The fruit does not contribute directly to the nourishment of the embryo, but it plays a major role in determining the chances of survival of the latter. For this reason, the processes of development of the mother plant that lead to its production, and the investment of energy in it, can be compared with parental care in animals.

From the point of view of development, the fruit derives from ovarian tissues (Section 3.4.2.1) and is therefore, strictly speaking, a structure exclusive to the flowering plants (for an in-depth study of the physiology of fruit production and maturation see Nath *et al.* 2014). However, in some gymnosperms (for example in *Ginkgo*, *Taxus* and *Juniperus*) a fleshy envelope develops around the seed – and although this cannot strictly be called a fruit, given its different developmental origin, from the functional point of view it assumes a role similar to that of the fruits of the flowering plants.

In the angiosperms, the fruit may derive exclusively from ovarian tissues (**true fruit**, for example in the cherry (*Prunus avium*); Figure 4.19a), or its formation may also involve tissues of the receptacle, stamens, sepals or petals. In this case it is called an **accessory fruit** (or **false fruit**, as in the pear (*Pyrus communis*), where a large part of the fruit develops from the receptacle; Figure 4.19b).

a) b)

c) d)

Figure 4.19 (a) The simple true fruit of the cherry tree (*Prunus avium*). (b) The simple false fruit (accessory fruit) of the pear tree (*Pyrus* sp.). (c) The aggregate fruit (compound fruit) of the bramble (*Rubus ulmifolius*). (d) The composite fruit (multiple fruit) of the pineapple (*Ananas comosus*).

Table 4.6 Seed-dispersal agents		
Agent	Descriptive term	Examples
Animal	Zoochory	
Attached to the animal	Epizoochory	Beggarticks or bur-marigolds (*Bidens* spp.): fruits (achenes) have two long toothed appendages that stick to the hair of mammals
Eaten by the animal	Endozoochory	Dog rose (*Rosa canina*): showy false fruit (rosehip) eaten by birds that disperse the still germinable seeds with their excrement
Birds	Ornithochory	Dog rose (see above) Swiss pine (*Pinus cembra*): seeds stored in hidden places by the spotted nutcracker (*Nucifraga caryocatactes*), which will feed on them in the winter, but some of the seeds are not retrieved by the bird and produce new plants

Table 4.6 (cont)		
Agent	Descriptive term	Examples
Mammals	Mammalochory	Oaks (*Quercus* spp.): seeds dispersed by rodents
Bats	Chiropterochory	Banana (*Musa* spp.) and mango (*Mangifera indica*): seeds dispersed by frugivorous bats (*Cynopterus* spp. and others)
Ants	Myrmecochory	Myrtle (*Myrtus communis*) and Italian buckthorn (*Rhamnus alaternus*): seeds dispersed by ants that feed on an appendix of the seed (elaiosome) without damaging it
Wind	Anemochory	Poplar (*Populus* spp.), maple (*Acer* spp.)
Water	Hydrochory	Sea rocket (*Cakile maritima*), European white water lily (*Nymphaea alba*)
The plant itself	Autochory	Squirting cucumber (*Ecballium elaterium*): by explosion of the fruit

The way carpels fuse in the flower has important consequences for the kind of fruit that is eventually produced. A fruit derived from a single carpel or from several carpels fused together is a **simple fruit** (as in the bean), while if a single fruit derives from several separate carpels we have an **aggregate fruit** (or **compound fruit**, like a blackberry (*Rubus* spp.); Figure 4.19c). Finally, if a single accessory fruit develops from an inflorescence, this is a **composite fruit** (or **multiple fruit**, as in the pineapple (*Ananas comosus*) and the fig (*Ficus carica*); Figure 4.19d).

Fruits are also classified into dry and fleshy. Some dry fruits open at maturity, allowing the dispersal of seeds (**dehiscent dry fruits**), while others remain closed (**indehiscent dry fruits**). Fleshy fruits, most of which are indehiscent, are generally destined for consumption by animals that contribute in various ways to their dissemination (Table 4.6).

Parental investment varies greatly in the number of fruits (generally high in the case of small dry fruits, low for large fleshy fruits), energy reserves for the development of each fruit (generally low for dry fruits, higher for fleshy ones), and the number of seeds released with a single fruit (from one, e.g. in the coconut (*Cocos*), to almost 4 million in the *Cynoches* orchid).

Chapter 5: Genetics and Cytogenetics of Reproduction

Of the many features that distinguish the different modes of reproduction, this chapter discusses the effects of adopting a given reproductive mode on the genetics of hereditary transmission and its consequences for the production of genetic variation through generations.

Dealing with these topics across the whole range of life forms, as in this book, necessitates a number of generalizations, which risks producing an inaccurate picture of the genetics of transmission through reproduction. Thus, before starting a detailed exploration of these topics, it will be appropriate to highlight some of these generalizations, in a list of caveats to be borne in mind while reading the chapter.

- The mode of reproduction is not the sole determinant of the quality and quantity of genetic variation that reproduction generates. It is supplemented by a large number of complementary factors, among which there are the characteristics of the **genetic system** of the organism (White 1984). This includes both the organization of the hereditary material (e.g. the total amount of DNA, the number of chromosomes and number of sets of homologous chromosomes) and the way in which the genetic material is transmitted across generations (e.g. depending on the mating system or the type of life cycle).

- Not all the DNA of a eukaryotic organism is in the nucleus, and not all the DNA of a prokaryote is found in its single chromosome. We will take into consideration all the genetic elements of the organisms, especially because many aspects of the transmission genetics of one of the genomes of a given organism are not generally shared with the other genomes of the same organism.

- Following the tradition of classical genetics, in the following discussion we adopt the simplification of thinking of chromosomes as if they were made of genes separated by non-coding sequences. However, the way hereditary information is organized along a chromosome is more complex than a simple concatenation of discrete coding units (Pearson 2006). Some of these aspects might have substantial effects on the genetics of transmission and the evolutionary processes, but their study is still at an early stage.

- The transmission genetics of hereditary characters does not cover all the effects of reproduction on evolutionary processes. This view derives from a concept of evolution that considers the terms 'genetic' and 'heritable' as synonyms. However, there are other forms of inheritance, the importance of which we are only beginning to understand. Many traits of the phenotype, morphological, physiological and behavioural, can be transmitted from one generation to the next independent of the transmission of specific DNA sequences. The different alternative modes of hereditary transmission, collectively called *epigenetic inheritance systems* (Jablonka and Lamb 2005), are currently the subject of intense studies, especially in an evolutionary framework.

- Finally, the tag 'genetic' applied to a phenotypic character should not be understood as a mark of a rigidly determined phenotype, since several distinct phenotypes may correspond to the same genotype. Examples of this phenomenon, known as *phenotypic plasticity*, are the *seasonal polyphenism* of some butterflies, the *caste polyphenism* of social insects and the *environmental sex determination* of some reptiles (Fusco and Minelli 2010; Section 6.2.1).

5.1 Asexual Reproduction

What is there to say about the genetics of asexual reproduction? There are of course several different mechanisms by which new individuals are generated, or segregate from already existing individuals (Chapter 3, Figure 5.1), but surely the very definition of asexual reproduction (Chapter 1) tells us that it should mean the production of offspring that are perfect (or almost perfect) genetic copies of the parent. The genetics of asexual reproduction would therefore seem to coincide with the genetics of DNA replication. But things are not exactly as they seem.

Firstly, the genetics of asexual reproduction only partially coincides with the genetics of DNA replication, because some forms of genetic reassortment are also regularly found in hereditary mechanisms associated with this mode of reproduction. Secondly, DNA replication is not a chemical process that runs independently of the constitution and the dynamic state of the cell or the body as a whole. The genome is not that secure repository of genetic information, necessary for the development and the survival of the organism, that a too simple view of the phenomena of inheritance would suggest. It should rather be seen as a very dynamic cellular component, continually and actively engaged in the managing of mutational events. Mutations result from the intrinsic limited

Figure 5.1 *Hydra*, a freshwater hydrozoan reproducing asexually by budding.

stability of chemical bonds, from the non-absolute precision of DNA replication mechanisms, and from the movements along the chromosomes or between chromosomes of genetic elements of different type and origin (Pearson 2006). The effects of these different sources of instability on the copying fidelity through asexual reproduction depend in turn on the organization of the hereditary material in the different genomes of an organism, either in the nucleus and the cytoplasmic organelles of eukaryotic cells, or in the chromosome and the plasmids of prokaryotic cells. DNA molecules in the different genomes vary greatly in number, size and constitution. They may or may not be associated with structural proteins, such as histones, or with enzymatic complexes, such as a DNA repair system. Moreover, DNA replication can be variously coordinated with either the cell cycle or the life cycle, if the former does not coincide with the latter. All these aspects can significantly affect the introduction of novel genetic variants through asexual reproduction.

In summary, although from the point of view of the genetics of transmission, asexual reproduction is a form of **clonal reproduction** – i.e. a form of reproduction that, in principle, does not involve changes in the genetic information across generations, thus producing **clones** (Box 5.1) – actually, for the reasons explained above, this is true only to a limited extent. A large number of processes and events, more or less strictly related to reproduction, make it possible that a certain amount of genetic (and epigenetic) variation is generated *within* a clone.

Box 5.1 Clones

The term **clone** is used with two meanings which, although similar, are in fact distinct. It may be used to indicate (i) a biological entity (e.g. a DNA segment, a gene, a genome, a cell, a multicellular organism) that is genetically identical to another (e.g. 'A is a clone of B'), or (ii) the set of genetically identical entities that are derived from the replication of a single biological entity, their common ancestor (e.g. 'A, B and C belong to the same clone'). In the first case, 'clone' simply indicates genetic identity between two or more entities. In the second case, the term takes on a genealogical meaning, indicating a set of genetically identical lineages, whose members are called *clonemates*.

Owing to a number of genetic events, including mutations, 100% pure clones are very improbable, especially if the whole genome is considered. For instance, it is well established that the somatic cells of a multicellular organism actually have some level of *intraclonal genetic variation* (strictly speaking, an oxymoron). For this reason, when we refer to the genetic identity of the members of a clone, it is implied that we are neglecting variation due to mutations of recent origin. Since the level of genetic variation is in general not known outside the studied loci in a clone, it has been suggested that a clone should be more correctly called a **clonotype**, in analogy with the concept of genotype, which can be defined on a limited number of loci (Lushai and Loxdale 2002).

Beyond the actual level of copy fidelity, the distinctive feature of every form of clonal reproduction is the replication and transmission of genetic material in the absence of processes that combine or recombine DNA molecules from distinct sources, as occurs with sex and recombination in the broad sense. The concept of clonality is generally referred to the replicative processes of the cell, either in the reproduction of unicellular individuals or in cell proliferation within one multicellular individual. However, it also applies to genetic processes operating within the cell, at the level of DNA sequences, single DNA molecules (chromosomes) or whole genomes (Avise 2008).

DNA sequences and DNA molecules. DNA replication that precedes cell division usually involves all the different genomes of the cell in their entirety. However, it is possible that at different times of the cell cycle single DNA molecules (chromosomes), or segments of different length of a DNA molecule, can replicate (clonally) independent of the rest of the genome of which they are part. Examples are found in the chromosomes of the cytoplasmic organelles of the eukaryotic cell, or in the plasmids of the prokaryotic cell, which can replicate in phases of the cell cycle other than those of the division of the cell and independent of the division of other molecules of the same type in

continues

Box 5.1 (cont)

the same cell. Other examples are provided by the proliferation of sequences due to replicative transposition (*copy-and-paste* type) of mobile genetic elements (**transposons**).

Genomes. Clonal multiplication of the eukaryotic nuclear genome coincides with mitosis, but this does not necessarily include a division of the nucleus and the cell. Nuclear divisions that are not followed by cytokinesis can produce a multinucleated condition called a *plasmodium*, whereas a whole nuclear genome duplication not followed by the division of the nucleus, a process known as **endomitosis**, results in an increase in the ploidy level of the nucleus, i.e. in the number of homologous sets of chromosomes it contains.

The genomes of cytoplasmic organelles, mitochondria and plastids, may have a transmission system that differs from species to species. However, organelles are frequently transmitted through only one of the two parents. For instance, at fertilization in eukaryotes with oogamety, the male gamete almost exclusively supplies its haploid nuclear genome, so that the mitochondrial DNA in the zygote (and the offspring) is predominantly, or exclusively, of maternal origin. Even in some species with isogamy, the cytoplasmic organelles can be transmitted by only one of the two parents, characterized by a specific mating type. Moreover, even in the case of an equal contribution of organelles by both parents, the genomes of these rarely recombine. In other words, these DNA molecules are transmitted clonally not only during their multiplication within the same organelle (which generally contains several copies of them), in the multiplication of the organelles within a cell, and in the cellular proliferation of a multicellular individual, but even from one generation to the next through sexual reproduction.

Sex chromosomes of the eukaryotic cell. Some chromosomes may exhibit clonal behaviour even in sexual reproduction. In species with chromosomal sex determination (like the XY system of mammals, see Section 6.1.1), the two sex chromosomes can be very different and present very limited homologous regions, so that they do not recombine in meiosis. In practice, the sex chromosome that is found only in the heterogametic sex (in mammals, the Y chromosome) is transmitted clonally (unaltered). Paternal transmission occurs if the heterogametic sex is the male, as in mammals, but there is maternal transmission if the heterogametic sex is the female, as in birds. The other sex chromosome (in mammals, the X chromosome) has instead a type of transmission similar to that of autosomes, because in the homogametic sex (in mammals, the female) it is found in duplicate and can recombine.

Perfect genetic similarity probably does not exist in nature, so that any reference to *genetic identity* among the members of a clone implicitly ignores genetic variation of more recent origin (Lushai and Loxdale 2002). Asexual reproduction is a form of clonal reproduction, but not all forms of clonal reproduction fall into the category of asexual reproduction. We defined asexual reproduction (Section 1.2) as a process by which a portion of the body of an individual (parent) becomes a new independent individual (offspring) without the involvement of sexual processes or their derivatives, such as the production of gametes. Accordingly, cases of uniparental reproduction that do not fulfil these criteria, even if they have a clonal outcome (such as in ameiotic parthenogenesis), are not to be considered asexual. Such phenomena will be treated here in the context of sexual reproduction (Section 5.2).

5.1.1 Genetic Variation Resulting from New Mutations

Genetic or chromosomal mutations may cause the members of a clone not to be perfectly identical to each other. For the sake of clarity, we distinguish between mutations associated with the reproductive processes in the strict sense and mutations that may arise in other phases of the cell cycle or the life cycle. However, with regard to the introduction of new genetic variation, their effects are not generally distinguishable.

5.1.1.1 MUTATIONS RESULTING FROM ERRORS IN DNA REPLICATION

DNA replication is an extremely precise process, but it is not perfect. The frequency of error is very low (the mutation rate per base pair per replication is in the order of 10^{-9} to 10^{-10}; Lynch 2010), but not zero. In the reproduction of unicellular organisms, mutations can make the daughter cells genetically different from one another and from the parent cell. In the same way, modified cell lines may arise during the cell proliferation that characterizes the development of multicellular organisms. The amount of tissue affected by the mutation depends on its location in the cell lineage of the developing organism. In general, the earlier the mutation, the larger the quantity of tissue exhibiting it. If a multicellular organism reproduces asexually from cells belonging to one of these modified cell lines, the offspring will not have a genotype perfectly identical to the (original) parental one.

5.1.1.2 MUTATIONS RESULTING FROM OTHER CAUSES

During the proliferation of a clone of unicellular organisms, or during the lifetime of a multicellular organism, the genome of some cells may mutate due to chemical or physical factors of various kinds, the most effective of which are known as *mutagens*. Similarly to the case of mutations produced during DNA replication, we can thus generate modified clones (in the case of

Figure 5.2 Variegation in plants (here, on the leaves of *Tradescantia fluminensis*) is often due to a form of genetic mosaicism in the shoot apical meristem. This can involve nuclear or plastid mutations affecting either flavonoid pigments or chlorophyll in leaves or flowers. (Bossinger & Spokevicius 2011)

unicellular organisms) or modified cell lines within the same organism (in the case of multicellular organisms). Some cells can also genetically mutate because of the horizontal transfer of genes between different regions of the same genome (*gene transposition*), or between the genomes of different genetic elements of the same organism, for example between the nuclear genome and the genome of a cytoplasmic organelle. In any case, asexual reproduction starting from modified cells will generate progeny with a genotype that is no longer identical to the parental one.

5.1.1.3 MUTATIONS AND MOSAICISM

Whatever the origin of the mutations, these can establish a condition where different genomes, which originated from the genome of the same cell, are present and simultaneously expressed in the same individual, a condition known as **genetic mosaicism** (not to be confused with *genetic chimerism*, discussed in Section 5.2.5.3). Mosaicism (Figure 5.2) can occur during the (putatively clonal) cellular proliferation that characterizes the development of a multicellular organism, both in species that reproduce asexually and in sexually reproducing ones. In sexually reproducing multicellular organisms there is a distinction between *somatic mosaicism*, which has no effect on the genetic constitution of the offspring, and *germline mosaicism* (also called, somewhat inaccurately, *gonadal mosaicism*), which, by affecting cells destined to produce gametes, can have effects on the genome of the progeny. Obviously, this distinction does not apply in the case of asexual reproduction. In metazoans, the mutation rate of the cells of the somatic compartment is higher than that in the germline, where the mutation rate by cell division is more similar to that of unicellular eukaryotes. It is estimated that in a middle-aged human being ($6 \cdot 10^9$ bp per diploid nuclear genome, with ~10^{13} somatic cells) point mutations alone can amount to ~10^{16}, most of which, for purely statistical reasons, are in regions of non-coding DNA (Lynch 2010).

Mosaicism is not a phenomenon exclusive to multicellular eukaryotes, since genetic variation can originate by mutation also in the genome of cytoplasmic organelles. The presence in the same individual of organelles with different genomes is called **heteroplasmy**. We distinguish *intercellular heteroplasmy*, the variation among organelles of different cells, from *intracellular heteroplasmy*, when variation is among organelles of the same cell. Heteroplasmy may have causes other than mutation. For instance, it may arise through syngamy (producing intracellular heteroplasmy) or as a consequence of the particular mechanism of replication and segregation of organelles at cell division (producing intercellular heteroplasmy). These aspects of the genetics of the organelles are addressed in Sections 5.1.3, 5.2.2.4 and 5.2.3.1.

A form of mosaicism can also affect the plasmids of prokaryotic cells. This may originate either from mutations or as a result of sexual exchanges (Section 5.2.1). The random distribution of plasmids between daughter cells after cell division may create genetic differences between members of a prokaryotic clone.

Finally, it should be noted that the condition of a genetic mosaic can be transmitted, completely or partially, to the progeny. In addition to intracellular heteroplasmy, which can easily be transmitted through both sexual and asexual generations of unicellular and multicellular eukaryotes alike, other forms of genetic mosaicism in multicellular eukaryotes can pass from one generation to the next through asexual reproduction, when this involves multicellular, potentially genetically heterogeneous, propagules (Section 3.1.2).

5.1.2 Genetic Variation Resulting from Recombination

Similar to mutation, some mechanisms of genetic reassortment, more precisely of **genetic recombination**, such as crossing over and gene conversion (explained in Section 5.2.2.3), can undermine the perfect genetic identity otherwise shared by the members of a clone. Likewise, recombination can transform a multicellular individual into a genetic mosaic, in which cell populations with distinct genomes coexist. In any case, in asexual reproduction starting from cells with a modified genome, the offspring will not have a genotype that is perfectly identical to the parent's original one.

In the nucleus of diploid eukaryotic cells, **mitotic recombination** is a relatively rare event compared to the recombination associated with meiosis (Section 5.2.2.3), in terms of frequency per cell division (LaFave and Sekelsky 2009), but it does still occur. Similarly to meiosis, the two main forms of homologous recombination in mitosis are crossing over and gene conversion, and in mitosis the two can also take place together during the same recombination event.

While meiotic crossing over actually occurs during meiosis, the **mitotic crossing over** takes place more frequently during the interphase of the cell cycle preceding mitosis, rather than during mitosis itself. Following the phase of DNA synthesis, at the end of which each chromosome is made up of two sister chromatids, two homologous chromosomes can pair and exchange parts. At the subsequent mitosis, one of the two equiprobable configurations for the segregation of the sister chromatids (no longer identical after the crossing over) of this pair of homologous chromosomes leads to the production of homozygous genotypes at all the loci located downstream from the point of crossing over along the chromosome arm. The probability of such an event, which can result in the phenotypic expression of recessive alleles, is therefore 1/2 (Figure 5.3).

Similar to mitotic crossing over, **mitotic gene conversion** occurs more frequently during the phases of the cell cycle that precede mitosis. It consists in the unidirectional (non-reciprocal) transfer of genetic information between two sequences with a high degree of similarity, such as those that can be found along the arms of homologous chromosomes in mitosis. In this way, a gene or a part of a gene acquires the sequence of another allele at the same locus. The effects on the homogenization of the sequences depend on whether the conversion occurs before or after the phase of DNA synthesis (Figure 5.4).

In the first case, all possible products of mitosis will show homozygous genotypes at the locus (rarely more than one) affected by the conversion. In the second case only 50% will be affected. The length of the DNA segment affected by a typical gene conversion event is generally not very extensive, usually limited to only a portion of a gene.

Among prokaryotes, homologous recombination is not necessarily associated with the incorporation of exogenous DNA that characterizes sexual processes in these organisms (Section 5.2.1). In fact, the DNA of different plasmids of the same individual can recombine in different ways, although 'hybrid' plasmids seem to be produced more frequently through gene transposition.

5.1.3 Genetic Variation Resulting from Stochastic Segregation

Another source of genetic variation can be found in those mechanisms of multiplication and/or distribution of hereditary material among the daughter cells during cell division that are not rigidly controlled. These mechanisms easily lead to the onset of disparities in the genetic constitution of the offspring, and their iteration through cellular cycles tends to amplify these

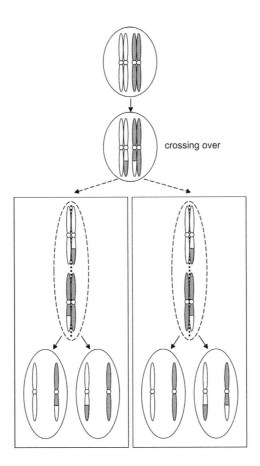

Figure 5.3 Mitotic crossing over. After DNA synthesis, two homologous chromosomes can pair and exchange parts. One of the two possible outcomes of mitosis, which therefore has a probability of 1/2, is the production of genotypes homozygous at all the loci downstream of the crossing over. The figure shows only one pair of homologous chromosomes in a diploid nucleus (oval). White and grey indicate chromosomes (or parts thereof) from different gametes (or gametic nuclei). The dotted line indicates the metaphase plate. The two dashed arrows leading to the two panels refer to the two alternative courses of the same process, which have the same probability of occurring. Continuous arrows connect the phases of mitosis.

differences. This random component in the hereditary genetics of cell division is found in several systems.

5.1.3.1 RANDOM SEGREGATION OF PLASMIDS IN THE PROKARYOTIC CELL

The plasmids of a prokaryotic cell (Box 5.2) that are present in a low number of copies (1–5) rely on a specific cellular apparatus to segregate equally, at cell division, in the two daughter cells (Salje *et al.* 2010), while plasmids present in a higher number of copies are divided at random between them. For those plasmids that are present in an intermediate number of copies (say, between 6 and 12), there is a non-negligible probability that one of the two daughter

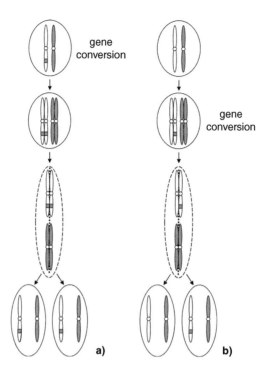

Figure 5.4 Gene conversion at mitosis. Two homologous chromosomes can pair and a DNA segment of one of them can acquire the sequence of the other. (a) Conversion occurs before DNA replication. Both products of mitosis have homozygous genotypes in the segment affected by the conversion. (b) Conversion occurs after DNA replication. Only one of the two mitotic products has homozygous genotypes in the segment affected by the conversion. The figure shows only one pair of homologous chromosomes in a diploid nucleus (oval). White and grey indicate chromosomes (or parts thereof) from different gametes (or gametic nuclei). The dotted line indicates the metaphase plate. Continuous arrows connect the phases of mitosis.

cells does not receive any copy (Krebs *et al.* 2011). Therefore, some genetic variation at the level of extrachromosomic DNA (but in the case of integration of the plasmid as an episome, even at the level of the bacterial chromosome) can be introduced in a clone through the stochastic process of distribution of genetic material associated with cell division.

5.1.3.2 MITOTIC SEGREGATION OF CYTOPLASMIC ORGANELLES

Allelic variation in the DNA of both mitochondria and plastids of a single cell (intracellular heteroplasmy), resulting from allelic differences between the organelles inherited from the two parents or from recent mutations, can be lost over generations. This can happen owing to the stochastic transmission of the organelles' genomes.

At cell division, the random distribution of the organelles between the daughter cells can result in an unequal distribution of the different genomic variants between them. Subsequent mitotic divisions can further increase

---

Box 5.2 Genetic Elements of the Prokaryotic Cell

Prokaryotic cells may contain several distinct genetic elements, each capable of replicating independently (Krebs *et al.* 2011).

The **bacterial chromosome** is a double-stranded, generally circular, DNA molecule. Its size varies from 0.6 Mbp, with about 470 genes, for the human parasite *Mycoplasma genitalium*, to nearly 8 Mbp, with more than 7500 genes, for some nitrogen-fixing bacteria. All free-living bacteria, however, have a genome larger than 1.5 Mbp (with about 1500 genes). The chromosome of the archaea has a size generally between 1.5 and 3 Mbp, but *Nanoarchaeum equitans* has a genome of only 0.5 Mbp (Waters *et al.* 2003): the smallest genome known for a cellular living being (i.e. excluding viruses). Generally, 85–90% of the prokaryote DNA encodes polypeptides, so the number of genes is closely related to the size of the genome. In the cytoplasm of the prokaryotic cell there are also numerous **plasmids**: small molecules of double-stranded DNA, generally (but not always) circular, able to replicate independently, which are maintained in a stable number of copies. Some types of plasmid are found exclusively as independent molecules, while others, called **episomes**, can be reversibly incorporated into the bacterial chromosome.

Most plasmids are rather small (less than 10 kbp), but are usually present in a high number of copies per cell (15–20, in some cases thousands). Other plasmids are much larger (up to 100 kbp) and are present in few copies or even in a single copy. Some bacteria also have *megaplasmids*, with a size up to one-third of the bacterial chromosome. At each cell division, the plasmids that are present in a low number of copies (from 1 to 5) usually rely on a specific cellular apparatus to segregate equally in the two daughter cells, while the plasmids present in a larger number of copies divide between them at random.

Other extrachromosomal genomes that are frequently found in the prokaryotic cell are those of **bacteriophages** (or **phages**), viruses that have infected the cell. These DNA molecules can be found free in the cytoplasm, or integrated into the bacterial chromosome as **prophages** (or **temperate phages**).

---

these disparities, to the point of obtaining cells with only one of the alternative allelic variants. At this point, intracellular heteroplasmy will be irreversibly lost, at least until mutation or syngamy occurs. In its place, in a multicellular organism, a form of intercellular heteroplasmy will be established. This stochastic process is called **mitotic segregation**.

A similar process can occur between the different chromosomes of the same organelle, which, we recall, may be up to a few tens per organelle (Box 5.3).

Box 5.3 Genetic Elements of the Eukaryotic Cell

Eukaryotic cells contain distinct genetic elements, each capable of independent replication (Strasburger *et al.* 2002; Krebs *et al.* 2011; Alberts *et al.* 2015).

The **nuclear genome** consists of one or more homologous sets of distinct double-stranded DNA molecules (nDNA), which correspond to the chromosomes. The number of chromosomes for each set varies from species to species, from a minimum of one in the ant *Myrmecia pilosula* and the nematode *Parascaris univalens*, to a maximum of about 1600 in the radiolarian *Aulacantha*, while the number of homologous sets (or ploidy level) can be one (haploid), two (diploid), or more (polyploid). The total amount of DNA for each set (*haploid genome size*) spans more than four orders of magnitude, from 12 Mbp in the yeast *Schizosaccharomyces pombe* to 133 Gbp in the lungfish *Protopterus aethiopicus*. Generally, in eukaryotes a conspicuous percentage of nuclear DNA does not encode polypeptides, and the number of genes encoding polypeptides is not strictly related to genome size. The number of genes ranges from a minimum of 4300 in the yeast *Saccharomyces cerevisiae* to about 60,000 in the unicellular human parasite *Trichomonas vaginalis*. In some unicellular eukaryotes and in some tissues, or in specific developmental phases of different groups of multicellular eukaryotes, a cell can have multiple nuclei (*multinucleate condition*). It is referred to as a *plasmodium* if its condition originates by multiplication of the nucleus within a single cell, while it is called a *syncytium* if it originates by fusion of previously distinct uninucleate cells.

The **mitochondrial genome** consists of a variable number of mitochondrial chromosomes, each formed by a double-stranded DNA molecule (mtDNA), generally circular (but linear in some taxa, e.g. in medusozoans and in some unicellular eukaryotes), between 6 and 2400 kbp in size and containing 5–100 genes encoding proteins or RNAs. There can be many chromosomes within one organelle (up to some tens) and many mitochondria within one cell (up to tens of millions in some egg cells). The chromosomes of plant mitochondria are generally much larger than those of animals (200–2000 kbp vs. 16–17 kbp), owing to the significant presence of non-coding sequences. Some mitochondrial genes of plants and fungi contain introns, but those of animals do not. Some eukaryotes, such as the vertebrate parasites of the genus *Giardia* (diplomonad protists) and some loriciferan species living in sediments of the hypersaline anoxic basins of the Mediterranean (Danovaro *et al.* 2010), lack mitochondria by secondary loss, because they descend from eukaryotic ancestors with mitochondria (Embley 2006).

Box 5.3 (cont)

In some species (e.g. in the ciliate *Nyctotherus*), elements of the mitochondrial DNA can also be found in some cytoplasmic organelles called *hydrogenosomes*, which are believed to have evolved by degeneration from mitochondria.

The **plastid genome** consists of a variable number of plastid chromosomes, each formed by a circular double-stranded DNA molecule (ctDNA, or cpDNA), with size between 70 and 400 kbp. Chloroplast chromosomes of the embryophytes are about 140 kbp long and contain 120–130 genes, of which 90 encode polypeptides. There can be many chromosomes within one organelle (usually 20–40, but also up to a few hundred) and many plastids within one cell (even more than 100). Plant ctDNA genes have introns. Plastids are found in many eukaryotic clades, including green plants, red algae, brown algae and several unicellular eukaryotic clades, which have acquired them independently, through the evolution of symbiotic relationships with photosynthetic organisms (Baurain *et al.* 2010). Some lines of these clades have subsequently lost the plastids, returning to a form of heterotrophic nutrition. In some clades in which the plastid has been acquired by endosymbiosis with a photosynthetic eukaryote (*secondary*, or higher-level, *endosymbiosis*), vestiges of the nuclear genome of the latter remain, in the form of a *nucleomorph* (e.g. in cryptomonads).

Organelles are generally very dynamic structures, able to change shape, divide and even fuse. Furthermore, cell-to-cell transfer of both mitochondria and plastids is documented. Growth and multiplication of the organelles within the cell are part of normal cellular metabolism and are not limited to the S phase of the cell cycle, when nuclear DNA is replicated. Replication of the organelles can be part of the normal turnover of these subcellular structures, which can degenerate over time. However, variation in their number is sometimes an aspect of the physiological response to different types of environmental or (in the case of a multicellular organism) systemic stimuli. In the case of intracellular heteroplasmy, organelle fusion creates the conditions for genetic recombination. However, this seems to be a rare event, demonstrated for yeast mitochondria, but less certain for other organelles and organisms (Krebs *et al.* 2011). Furthermore, the dynamics of change of mitochondrial and plastid DNA from one generation to the next are quite different from those of the nuclear DNA, and there can be considerable variations between groups of taxa. For instance, in metazoans mitochondrial DNA accumulates mutations faster than the nuclear DNA, while in green plants, mitochondrial and plastid DNA evolves at lower rates than nuclear DNA (Lynch *et al.* 2006).

Since the synthesis of DNA that precedes the division of an organelle lacks any strict control over which particular chromosomes are replicated, some of these can be replicated more times than others, and some may not be replicated at all. If these different DNA molecules of a dividing organelle differ from each other, due to mutation or a previous fusion between two organelles with different genomes, it is possible that the different chromosomal variants are not equally represented in the genomes of the deriving organelles. As in the process of segregation of organelles in mitosis, subsequent divisions of the organelles may further increase these disparities, until there are organelles presenting only one of the alternative allelic variants.

The hereditary transmission of the genome of the cytoplasmic organelles is a type of **non-Mendelian inheritance**, because the lineage does not present the distribution of characters expected by segregation according to Mendel's laws. This is also called *cytoplasmic inheritance*, to stress the fact that it affects the transmission of extranuclear genomes. However, the use of this term is strongly discouraged, to avoid confusion between the inheritance of the organelles' genomes, which depends on specific nucleotide sequences, and forms of epigenetic inheritance associated with physicochemical or structural characteristics of the cytoplasm (Section 5.1.4).

5.1.3.3 AMITOSIS IN CILIATES

In the asexual reproduction by binary fission of ciliates, the division of the hyperpolyploid macronucleus (generally one per cell; Section 5.2.5.1; Figure 5.5) follows a process that is not strictly clonal, profoundly different

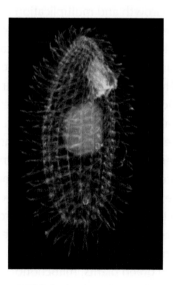

Figure 5.5 The ciliate *Tetrahymena thermophila*. The macronucleus, a cell nucleus replicating by amitosis, can be seen in the centre of the cell.

from the conventional mitosis that runs in parallel in the diploid micronucleus (generally also one per cell). The division of the macronucleus, which is called **amitosis**, consists in the division of this subcellular compartment into two compartments of approximately equal size, which contain roughly the same number of chromosomes. The process begins with a simple approximately equatorial constriction, which progresses until the two daughter macronuclei are completely separated, dividing the numerous chromosomes without condensation of the latter and without the production of any apparatus similar to a microtubule spindle.

The statistics of hereditary transmission through the successive divisions of a macronucleus is similar to that of a sampling problem without reintroduction, described in probability theory by a hypergeometric distribution (Bell 1988). Without going into the mathematical details of this stochastic process, let's consider a chromosome whose N copies in a given individual possess two alternative allelic variants (a 'heterozygous' condition) with the same frequency ($p = q = 0.5$). Across generations, the frequency of each variant will fluctuate randomly in the offspring, with a tendency towards increasingly extreme frequency values, until a homozygous condition is achieved, where one of the two variants will be present in an individual with frequency $p = 1$ or $p = 0$ (i.e. all the chromosomes with the alternative allele, $q = 1$). Homozygous conditions act as *absorbing boundaries*, i.e. extreme values of a distribution that can only increase their frequency, in this case because from homozygous individuals no heterozygotes can derive (other than by mutation). Thus, the frequency of heterozygous individuals in the clone will inexorably decrease and finally disappear. The speed of this process depends on the number of copies of the chromosome that segregate at each division.

5.1.4 Epigenetic Variation

In addition to the intraclonal variation that may derive from structural changes in the genome, rapid changes and differentiation in DNA *expression* can occur in clonal lines because of epigenetic effects. The suite of processes that, without modifying the DNA sequences, can produce heritable differences among the members of a clone (cells or individuals) is known as **epigenetic drift**. Accordingly, the heritable phenotypic variation in cells and organisms that does not depend on variation in DNA sequences is called **epigenetic variation**. Among the best-known mechanisms of epigenetic change is *chromatin marking*, which consists of chemical modifications of DNA bases (e.g. cytosine methylation), or proteins that are closely associated with it (e.g. histone acetylation). Although these modifications do not alter the primary sequence of the protein product synthesized by a marked gene, they can nonetheless have important effects on its expression, so as to be a constitutive part of the gene-expression control system. DNA methylation is also involved in the regulation of gene expression in prokaryotes. Many types of epigenetic marking can persist across DNA replication, thus transmitting the marking to the daughter cells.

In many species of eukaryotes with separate sexes, chromatin marking is at the origin of the phenomenon of *genomic imprinting*, for which some genes of an individual are expressed only in the version (allele) found on one or the other parent's chromosome. In the cells of an individual's germline, these markings, which are responsible for the *monoallelic* expression of some loci, are first erased and then replaced by the marking consistent with the individual's own sex.

Other types of chromatin marking can survive meiosis, and are therefore transmitted through sexual generations. The stability of these epigenetic modifications and their transmissibility through the generations are the subject of many recent studies and heated debate, especially with regards to the role they may have in the evolutionary processes. This is a very complex and controversial area of studies that cannot be properly developed here. A first introduction to these issues can be found in Jablonka and Lamb (2005) and Moore (2017).

5.1.5 Unconventional Mechanisms of Cell Division in Eukaryotes

In asexual reproduction, eukaryotic cells do not always divide by simple mitosis, and mitosis itself can progress through alternative mechanisms. At the level of the genetics of hereditary transmission, these particular mechanisms of cell division may have sizeable effects on the production of genetic variation, which differ from case to case.

5.1.5.1 DIVISION OF MULTINUCLEATE CELLS

We have already seen (Section 5.1.3) that, owing to the particular organization of the genetic elements in the ciliate cell, cell division in these protists includes a mitosis of the micronucleus accompanied by the amitosis of the macronucleus. In other organisms with unusual organization of the hereditary material, for example the *binucleate* diplomonads, or the heterokaryotic hyphae of ascomycetes and basidiomycetes, cell division requires specific mechanisms for correct distribution of the nuclear genome among the daughter cells. In addition, there are multinucleate unicellular eukaryotes (e.g. *Pelomyxa* among the amoebozoans and *Opalina* among the heterokonts; Figure 5.6) that divide by a process called **plasmotomy**, which consists in the division of the cell of a multinucleate parent into two, also multinucleate, daughter cells.

Figure 5.6 The multinucleate unicellular eukaryote *Opalina* divides by plasmotomy, a process that distributes the nuclei between two daughter cells, also multinucleate.

5.1.5.2 VARIANTS OF MITOSIS

Mitosis is the fundamental process of cell division, and thus of clonal reproduction in eukaryotes. A detailed description of this important biological process can be found in general biology and cell biology textbooks (e.g. Alberts *et al.* 2015). But a little-known fact, which deserves to be reported here, is that mitosis, especially in a number of unicellular eukaryotes, can take place in ways that differ from the conventional one.

Several variants of mitosis can be distinguished, mainly on the basis of the persistence of the nuclear envelope and the location and symmetry of the mitotic spindle (Raikov 1994). Mitosis is said to be *open*, *semiopen*, or *closed*, depending on whether the nuclear envelope disappears completely (as in conventional mitosis), or remains substantially intact but with some fenestrations that will be traversed by the spindle fibres, or remains intact, respectively. In addition, we distinguish between *orthomitosis*, when the spindle is bipolar and symmetrical (as in conventional mitosis), and *pleuromitosis*, when the spindle is asymmetric, not forming from opposite cell poles. Finally, in closed mitosis, the process is said to be *intranuclear*, when the spindle forms within the nucleus, or *extranuclear*, when the spindle forms outside the nucleus and makes contact with the chromosomes through the persisting nuclear envelope. Different variants of mitosis arise from the combination of these characteristics (Figure 5.7), resulting, for example, in *closed intranuclear orthomitosis* or *semi-open pleuromitosis*. Out of the eight possible combinations, cases of *closed extranuclear orthomitosis* and *open pleuromitosis* are not known in nature.

5.1.6 Intermittent Episodes of Sexual Reproduction

Many forms of asexual reproduction can alternate, in a more or less regular way, with sex events. Even if occasional or rare, these episodes of **sexual**

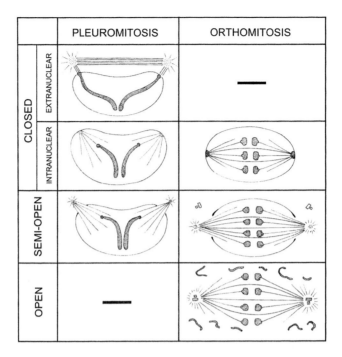

Figure 5.7 Variants of mitosis. Mostly among the protists, mitosis can run in different ways that can be distinguished on the basis of the persistence of the nuclear envelope and the location and symmetry of the mitotic spindle. The result is eight classes, two of which are virtual: (i) *extranuclear closed pleuromitosis* (e.g. *Trichomonas*); (ii) *intranuclear closed pleuromitosis* (e.g. microsporidia and some forams); (iii) *semi-open pleuromitosis* (typical of the Apicomplexa); (iv) *open pleuromitosis* (no known case); (v) *extranuclear closed orthomitosis* (no known case); (vi) *intranuclear closed orthomitosis* (e.g. *Trypanosoma*); (vii) *semi-open orthomitosis* (e.g. *Chlamydomonas*); (viii) *open orthomitosis* (the usual condition in multicellular eukaryotes, also common in different groups of protists).

leakage can have considerable effects on the genetic structure of clonal lines. We treat this subject more extensively when dealing with uniparental sexual reproduction (Section 5.2.4).

5.2 Sexual Reproduction and Sex

In Chapter 1 we defined **sexual processes** as the biological processes through which new combinations of hereditary material are created from different sources. These therefore include the phenomena of genetic reassortment, such as syngamy and genetic recombination in a broad sense (Section 5.2.2.1). Sexual processes are an important source of genetic variation in populations, able to amplify variation that originates by mutation, through a combinatorial elaboration of those mutations.

For the sake of completeness, in this chapter we also consider sexual processes that are not associated with reproduction, such as bacterial conjugation

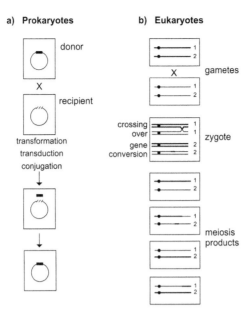

Figure 5.8 Schematic representation of the differences between sexual processes in prokaryotes and eukaryotes. (a) In prokaryotes, a fragment of the genome of an individual donor (thick line) is acquired by a recipient individual through transformation, transduction or conjugation. In the case shown here, the exogenous fragment integrates into the recipient's genome through homologous recombination, replacing the homologous segment in the original genome of the recipient (striped line). (b) In eukaryotes, two haploid gametes merge into a diploid zygote. The two chromosomes of the haploid nuclear genome are indicated by numbers 1 and 2. The chromosomes of the two gametes are denoted by lines of different thickness. Recombination between the genomes of the two gametes occurs by gene conversion, crossing over and independent assortment of chromosomes at meiosis.

(Section 5.2.1), in which an individual's genome is transformed by the contribution of genetic material from a donor individual, and the conjugation of ciliates, where two individuals mutually exchange genetic material and become genetically identical (Section 5.2.5.1).

Although there are many similarities, both genetic and biochemical, between sexual processes in prokaryotes and eukaryotes, from a cytogenetic point of view sex occurs in such different modes in these two groups (Figure 5.8), that we prefer to treat them separately.

5.2.1 Sex in Prokaryotes

Genetic exchanges in prokaryotes are not associated with reproduction, which generally occurs by simple binary or multiple fission. For this reason, these are frequently designated as **pseudosexual** or **parasexual** processes (not to be confused with *parasexuality* in fungi, discussed in Section 5.2.5.2). In addition to not being associated with reproduction, these phenomena differ substantially from those we will describe for eukaryotic organisms, owing to the different organization of the prokaryotic cell and the peculiar characteristics of its genetic system (Box 5.2).

Sexual processes in prokaryotes have an 'extremely promiscuous' nature. Comparative genome analyses have revealed the very dynamic character of acquisition, loss and transfer of genes among individuals of the same species, as well as among individuals of different species. While species of eukaryotes that diverge in their genome by more than 2% are generally not able to exchange DNA, prokaryotes can easily acquire and incorporate DNA from organisms whose genomes differ by more than 25% from their own (Cohan 1999). It has been argued (e.g. Fraser *et al.* 2007) that, through an appropriate series of intermediate steps and vectors, it is probably always possible to transfer genes between any two species of bacteria. Not surprisingly, some authors have advanced the idea that the whole world of prokaryotes might be considered as a single 'superspecies' or even as a single 'global superorganism', characterized by a single network of interconnected gene pools that has been called the *bacterial pan-genome* (Lapierre and Gogarten 2009).

5.2.1.1 MECHANISMS OF GENETIC EXCHANGE IN PROKARYOTES

In bacteria and archaea, the cytogenetic mechanisms through which DNA sequences of distinct individuals end up in the same cytoplasm, so as to possibly produce some kind of genetic reassortment, are the (pseudo)sexual processes of *transformation, transduction* and *bacterial conjugation* (Figure 5.9).

As with many other aspects of the biology of prokaryotes, sexual processes are better known in bacteria than in archaea (Cavicchioli 2007), and therefore the general treatment that follows refers mainly to the former group.

Transformation

The **transformation** of a prokaryotic cell is the alteration of its genome that follows the acquisition of DNA from the surrounding environment (exogenous DNA). This can happen simply by chance, but many bacterial species have surface proteins, specialized for absorbing exogenous DNA, which specifically recognize and transfer only DNA from closely related bacterial species. The DNA thus incorporated can be integrated into the bacterial chromosome by recombination.

Transduction

In **transduction**, gene transfer between bacterial cells is mediated by the action of a phage (or bacteriophage), i.e. a virus that infects bacteria. There are two types of transduction, each resulting from the alteration of one of the two most common types of replication cycle of phages.

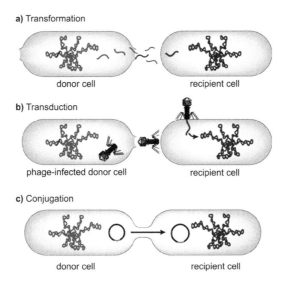

Figure 5.9 Schematic representation of sexual processes in prokaryotes. In *transformation*, DNA (in the figure, the DNA released by a lysed cell) is acquired from the environment; it can be integrated into the bacterial chromosome by recombination. In *transduction*, gene transfer between bacterial cells is mediated by a virus that infects bacteria (a phage). The new bacterial DNA can recombine, in homologous regions, with the DNA of the infected cell. In *conjugation*, two prokaryotic cells join temporarily through a cytoplasmic bridge. The transfer of DNA (in this example, a plasmid) takes place in only one direction, from a donor cell to a recipient cell.

In **generalized transduction**, at the end of the *lytic cycle* of the virus, which leads to the death of the bacterial host cell, a phage can accidentally incorporate a fragment of the bacterial chromosome of the lysed host cell, in place of the viral DNA. This defective phage cannot reproduce further, but it can infect a new bacterial cell, injecting the DNA from the previously lysed bacterium. This newly introduced DNA fragment can replace the homologous region of the bacterial chromosome by recombination. In generalized transduction, the bacterial genes that are transferred are a random subset of the bacterial genome, corresponding to the bacterial DNA segment accidentally incorporated by the phage.

In contrast, **specialized transduction** originates at the end of the *lysogenic cycle* of a phage, during which the virus DNA is integrated, as a *prophage*, into the chromosome of the host cell, generally at a specific site. When the prophage separates from the bacterial chromosome to start the lytic cycle, it may carry small segments of the bacterial DNA flanking the chromosomal site where it was inserted. When the viruses derived from the subsequent lytic cycle infect a new cell, the DNA of bacterial origin is injected together with the viral one. Through specialized transduction, only some genes are transferred, i.e. those located near the chromosomal site where the prophage is inserted.

The new bacterial DNA can recombine, in homologous regions, with the DNA of the infected cell. During a lysogenic cycle, some of the genes of the prophage may be expressed by the host, altering its phenotype.

Conjugation

Conjugation is the direct transfer of genetic material between two prokaryotic cells temporarily joined through a cytoplasmic bridge. DNA transfer is unidirectional, from a donor cell to a recipient cell. During conjugation, a single-strand cut in a DNA molecule (chromosome or plasmid) is produced in the donor cell. One end of the cut strand initiates the transfer to the recipient, while the double-stranded DNA structure is restored through the synthesis of complementary strands, both in the donor and in the recipient. The length of the transferred segment depends on how long the cytoplasmic bridge persists, which may also depend on contingent environmental conditions. The partial 'diploid condition' of the recipient, as a result of the DNA transferred by the donor, is soon lost by recombination, while the DNA segments that have not recombined are degraded. The genome of the recipient bacterium will thus become a recombinant one. Conjugation is generally mediated by special **conjugative plasmids**, which may or may not be integrated into the bacterial chromosome as *episomes*.

The best-known example is the **F plasmid** of *Escherichia coli*. Cells that act as donors have a special DNA sequence, called the **F factor** (F for fertility). In *E. coli*, the F factor contains about 25 genes, many of which are necessary for the construction of *sexual pili*, thin tubular structures that originate from the plasma membrane, through which the donor makes contact with the recipient and approaches it. During the conjugation between a cell that possesses the plasmid (F^+) and a cell without it (F^-), a copy of the F plasmid is transferred to the F^- cell, converting it to F^+. In this case, the sex process only produces a modification of the 'fertility' condition of the recipient. Conversely, when the F factor is integrated into the chromosome as an episome, the donor cell, which is referred to as *Hfr* (for *high-frequency recombination*), can transfer other chromosomal genes through conjugation. The replication of the bacterial chromosome of the Hfr cell begins at a site close to the F episome, but may proceed even further, producing a copy of a segment of variable length of the donor's DNA. This copy will be transferred to the recipient cell together with the F factor. Following recombination, the sequences that had not been integrated into the chromosome, like the F factor itself, are degraded. The recombinant recipient cell will remain 'sexually' F^-.

There are further modes of conjugation in bacteria. In *Enterococcus faecalis*, the future recipient cell produces and releases specific pheromones that attract donor cells (which possess a conjugative plasmid), causing them to produce an aggregation substance that mediates cytoplasmic bridge formation and DNA transfer.

5.2.1.2 QUANTITATIVE ASPECTS OF GENETIC EXCHANGE IN PROKARYOTES

In prokaryotes genetic exchange involves a small fraction of the genome (generally less than a few thousand base pairs) and the transfer is unidirectional and relatively rare, when compared to the cell replication rate (Cohan 1999). This is true in general, but the existence of considerable diversity among bacterial species in the frequency of genetic exchanges should not be overlooked. There are species with very stable clones, such as *Salmonella enterica*, while others, e.g. *Helicobacter pylori*, tend to genetically diverge so rapidly that they actually form *clonal complexes*, rather than clones (Spratt 2004).

In addition to homologous recombination, prokaryotes can easily acquire new genetic loci from other organisms. Strains that are almost identical in the sequences of genes they share can diverge by as much as 15% in those sequences that are not homologous (Cohan 1999). Furthermore, in addition to the integration of new genes and new alleles in their chromosome, bacteria can also accept plasmids of extremely divergent species and express their genes.

Despite the highly promiscuous nature of sex in prokaryotes, there are several factors that actually limit genetic exchange, so that a bacterium cannot directly exchange genes with any other bacterium with equal ease. For instance, the recombination that depends on specific vectors (phages or plasmids) is limited by the spatial distribution of potential vectors in the environment. Furthermore, specific enzymes of the prokaryotic cell, known as *restriction endonucleases*, greatly reduce the recombination rate in transduction, conjugation and, to a lesser extent, transformation. Finally, homologous recombination is strongly conditioned by the degree of homology between the sequences: recombination frequency tends to decrease exponentially as the degree of divergence increases.

5.2.2 General Aspects of Sexual Reproduction in Eukaryotes

According to the definition of sexual reproduction given in Chapter 1, its qualifying character resides in the association and reassortment of genetic information from different sources. However, in the sexual reproduction of eukaryotic organisms, whether unicellular or multicellular, the genetic exchange associated with the specific mechanisms through which it may occur does not necessarily involve all the genomes or all parts of one genome of the organism (Box 5.3). For instance, we have already seen (Box 5.1) that the genome of cytoplasmic organelles and some chromosomes of the nuclear genome may have clonal transmission, even through sexual reproduction. Unless otherwise specified, the following discussion is about sexual processes at the level of the nuclear genome.

In this section we first analyse the processes of reassortment of the hereditary material in sexual reproduction, and then explore how these processes interact in the context of the different reproductive modes described in Chapter 3.

5.2.2.1 SYNOPSIS OF THE SOURCES OF GENETIC VARIATION

Two main sex processes, characterized by their significantly random nature, can contribute to the generation of a new, original, nuclear genome. In the sequence typical of a diplontic organism, these are:

1. **Genetic recombination in the broad sense**, which occurs during gametogenesis and produces gametes that are different from the parent gametes, i.e. from those that fused to form the genome of the organism in which gametogenesis is observed. This includes three distinct events:

    1a. **Independent assortment of homologous chromosomes** at the first meiotic division (metaphase I), which results in the subsequent **independent segregation of homologous chromosomes** in the two daughter nuclei.

    1b. **Genetic recombination in the strict sense**, i.e. crossing over and gene conversion (Figure 5.10) between homologous chromosomes at the first meiotic division (prophase I, pachytene).

    1c. **Independent assortment of non-identical sister chromatids** (not identical, because of recombination 1b) at the second meiotic division (metaphase II), which results in the subsequent **independent segregation of the non-identical sister chromatids** in the two daughter nuclei of each of the two nuclei derived from the first meiotic division.

2. **Syngamy**, i.e. the merging of the haploid genomes of two gametes into the diploid genome of the zygote.

The same two processes are found in the sexual reproduction of a haplontic organism – but in reverse order, since in this case syngamy precedes meiosis. Conversely, in a haplodiplontic organism the two processes are distributed between the two phases (and generations): recombination occurs in the diploid phase (producing spores), while syngamy involves the gametes produced in the haploid phase (Section 2.1). Let's now look at these processes of sexual reproduction in more detail.

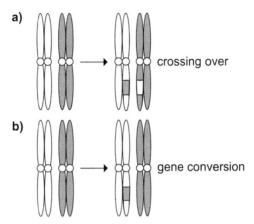

Figure 5.10 Schematic representation of the difference between crossing over and gene conversion at meiosis in a pair of homologous chromosomes. (a) In crossing over, two chromatids of two paired homologous chromosomes exchange chromosome segments reciprocally. (b) In gene conversion, a chromatid segment of one of the two chromosomes acquires the nucleotide sequence of the homologous segment of the other chromosome. Both processes can run in different ways. White and grey indicate chromosomes deriving from different gametes (or gametic nuclei).

5.2.2.2 INDEPENDENT ASSORTMENT OF CHROMOSOMES AND CHROMATIDS

During the metaphase of the first meiotic division, the pairs of homologous chromosomes, each consisting of the chromosomes deriving from two distinct gametes (or gametic nuclei), are arranged at the equatorial plate. The orientation of the chromosomes of each pair with respect to the opposite cell poles is random and independent of the orientation of other pairs. The number of possible pairing configurations (assortments) is therefore 2^{n-1}, where n is the number of homologous chromosome pairs (Figure 5.11). These alternative configurations can therefore give rise to 2^n distinct gametes, two for each possible configuration. Only two of these show the genotypes of the two parental gametes, while all the others possess different genotypes, as an effect of the random reassortment of the chromosomes.

Actually, a much larger number of different gametes can be produced in meiosis, since the independent assortment of chromosomes (*interchromosomal recombination*) adds to the recombination in the strict sense, i.e. that due to crossing over and gene conversion (*intrachromosomal recombination*, see next section).

In the metaphase of the second meiotic division, the sister chromatids of the chromosomes that recombined during prophase I are no longer identical. Also in this case, the orientation of the two chromatids of each chromosome on the metaphase plate with respect to the two cell poles is random and independent of the orientation of the other chromatid pairs. For each of the two nuclei produced with the first meiotic division, the number of possible configurations (assortments) is therefore 2^{m-1}, where m is the number of

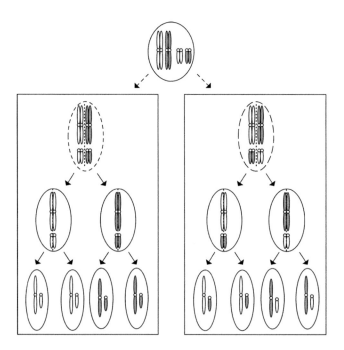

Figure 5.11 Independent assortment of chromosomes at meiosis in the absence of recombination for $n = 2$ pairs of homologous chromosomes. Chromosomes can segregate into $2^{n-1} = 2$ distinct modes, producing $2^{n-1} = 2$ types of gamete for each option. The maximum number of different gametes that can be produced is $2^n = 4$. White and grey indicate chromosomes deriving from different gametes (or gametic nuclei) of the previous generation. The two dashed arrows leading to the two panels refer to the two alternative courses of the same process, which have the same probability of occurring. Continuous arrows connect the phases of meiosis. The dotted line indicates the metaphase plate.

non-identical chromatid pairs (corresponding to the number of chromosome pairs that engaged in crossing over or gene conversion at the first meiotic division, see Section 5.2.2.3).

Combining the effects of the independent assortment of chromosomes and chromatids (Figure 5.12), for a specific recombination pattern (defined by type, number, location and extent of the recombination events) the assortment of chromosomes at meiosis can therefore be accomplished in $2 \cdot 2^{n-1} \cdot 2^{m-1}$ different ways, i.e. 2^{n+m-1}. If $n = m$, as will generally be the case in the absence of homologous chromosomes that do not recombine because of their different structures, such as the sex chromosomes in the heterogametic sex of many species (Section 6.1.1), the number of possible assortments will be 2^{2n-1}. In our species, segregation can occur in $2^{45} \approx 3.5 \cdot 10^{13}$ different ways for a primary oocyte, and in $2^{44} \approx 1.8 \cdot 10^{13}$ (half as many) for a primary spermatocyte, since the two sex chromosomes of the male (X and Y) do not recombine.

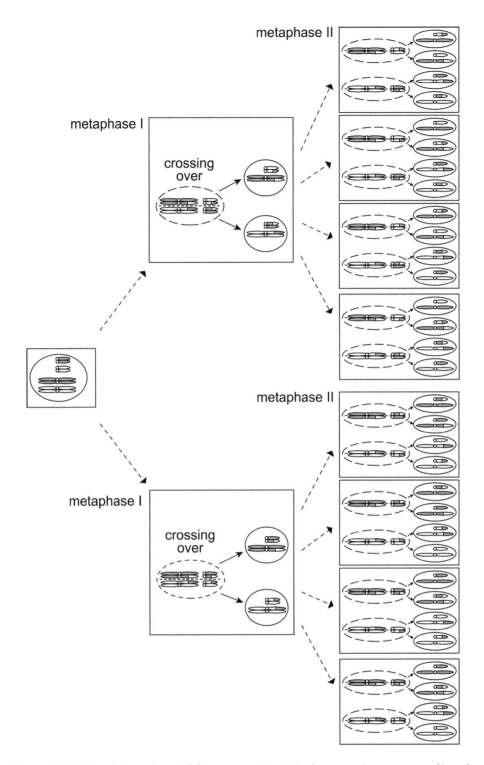

Figure 5.12 Independent assortment of chromosomes at meiosis where a crossing over occurred in each pair. The pairs of homologous chromosomes are $n = 2$, the same as the number of pairs of recombinant chromosomes $m = n = 2$. The chromosomes can segregate in $2^{n-1} = 2$ distinct modes at meiosis I, and for each of these the sister chromatids can segregate in the two nuclei in $2 \cdot 2^{m-1} = 4$ distinct ways at meiosis II, for a total of $2^{n+m-1} = 8$ segregation modes. The maximum number of different gametes that can be obtained is $2^{n+m} = 16$. White and grey indicate chromosomes (or parts of chromosomes) deriving from different gametes (or gamete nuclei) of the previous generation. Dashed arrows pointing to panels refer to alternative courses of the same process with the same probability of occurring. Continuous arrows connect the phases of meiosis. The dotted line indicates the metaphase equatorial plate.

The number of different gametes attainable from the assortment process alone is $4^m \cdot 2^{n-m} = 2^{n+m}$, where n is the number of chromosome pairs and m the number of chromosome pairs that recombined (the chromosomes of the m pairs that recombined are present in four different versions, those of the $n-m$ pairs that did not recombine are present in the two original parental versions). This number is twice the number of possible ways in which meiosis can proceed from the same set of recombination events.

5.2.2.3 RECOMBINATION IN THE STRICT SENSE

We have seen (Section 5.2.2.1) that genetic recombination in the strict sense is a process capable of producing new DNA molecules by combining nucleotide sequences of different origins. Let's examine this process more closely. From a biochemical point of view, **recombination** occurs when a nucleic acid molecule, either DNA or RNA, joins another molecule of the same type to form a new nucleotide sequence. As we have already seen, this is an important biochemical process associated with sex, but genetic recombination is not exclusive to sexual processes (Box 5.4). Particularly important for sexual processes are *homologous recombination* processes, i.e. those that take place between DNA segments characterized by high sequence similarity, such as *crossing over* and *gene conversion* during meiosis in eukaryotes (Figure 5.10).

Although from a biochemical point of view these recombination processes do not differ substantially from similar processes involved in other cellular functions, such as DNA repair, during meiosis they are regulated differently, so that they occur more frequently and are more likely to result in a genetic exchange. For instance, in meiosis recombination occurs preferably between homologous chromosomes, rather than between the sister chromatids of the same chromosome. Conversely, in the repair of double-strand breaks, it is the sister chromatids produced by recent DNA replication that more easily tend to recombine.

Crossing Over in Meiosis

The crossing over consists in the exchange of homologous portions of genetic material between two chromatids belonging to two different chromosomes of a pair of homologues (Figure 5.10a). The crossing over tends to mix the allelic content of linked loci (i.e. loci on the same chromosome) that would otherwise be transmitted as inseparable ('non-recombinant') units. The recombinant chromosomes deriving from crossing over present a new **haplotype** that is a new combination of alleles along the chromosome.

Box 5.4 Recombination

From a purely biochemical point of view, recombination is a well-defined process, but the use of this term in biology, in population genetics in particular, is not as rigorous. For some authors the term 'genetic recombination' is synonymous with 'genetic shuffling' and therefore includes all possible forms of reassortment of the hereditary material. In this case, the main sources of genetic variation, the raw material for evolutionary change under natural selection, are mutation and recombination, the latter thus becoming equivalent to what others (see below) describe as 'sex and recombination'.

For other authors, the meaning of the term 'recombination' is restricted to the formation of new associations of DNA molecules, or segments of DNA molecules; in eukaryotes, this includes the independent assortment of chromosomes (*interchromosomal recombination*) plus crossing over and gene conversion (*intrachromosomal recombination*) in meiosis. In this way, sex coincides with syngamy. This meaning of the term recombination is revealed in the phrase 'sex and recombination', which frequently occurs in the scientific literature. It also emerges in the use of the adjective 'asexual' with reference to species or populations that reproduce by parthenogenesis or other forms of metasexuality (Section 5.2.3) that do not include syngamy, or when recombination is presented as an alternative to mutation as a source of genetic variation. Throughout the book we refer to this second meaning of recombination as 'recombination in the broad sense'.

In biochemistry, the term recombination has a meaning that on the one hand is even more restrictive, because it excludes the independent assortment of chromosomes in meiosis, but on the other hand is wider, because it includes phenomena that are not necessarily connected to reproduction, either sexual or asexual, such as those involved in mechanisms of DNA repair. In biochemistry, different types of recombination are distinguished, of which the main ones are briefly described here.

Homologous recombination (also known as *general recombination* or *generalized recombination*) is the exchange of genetic information that occurs between two sequences that have a high degree of similarity (or sequence homology). Homologous recombination performs different functions in the cell. The one with the broader scope, found in prokaryotes and eukaryotes alike, is the accurate repair of double-strand breaks in DNA molecules. Ruptures can be caused by chemical or physical agents, but more often they are the result of accidents during the DNA synthesis phase of the cell cycle (e.g. caused by the block or the break of the replication fork). Repair of a

continues

> **Box 5.4** (cont)
>
> double-strand rupture without using a template sequence (i.e. without recombination) generally produces a mutation at the site of repair. Homologous recombination is also involved in the repair processes of other types of DNA damage. Moreover, in eukaryotic cells, homologous recombination has an important mechanical role in the pairing of homologous chromosomes in the synapsis phase of the first meiotic division. During prophase I, programmed double-strand breaks occur, followed by recombination, which tightly binds the chromatids of the homologous chromosomes. Some of these recombination events (in the order of 10%) will result in a crossing over. Examples of general recombination related to sex are the crossing over and gene conversion between homologous chromosomes during meiosis in eukaryotes (see below in this section) and the recombination of bacterial DNA after conjugation, translation or transposition (Section 5.2.1).
>
> **Site-specific recombination** is the exchange between DNA segments that have little or no homology, but which are located at precisely defined sites of the genome. Both DNA segments possess specific recognition sequences that make the exchange possible. An example of site-specific recombination is the integration of a phage into the bacterial chromosome during the lysogenic cycle of the virus (Section 5.2.1).
>
> **Transpositional recombination** can take place between non-homologous sequences as well, but it is achieved through the transfer of different types of transposable genetic elements (**transposons**). These are DNA sequences that can change their position within the genome, with either a copy-and-paste or a cut-and-paste mechanism. In transposition, unlike site-specific recombination, only the transposable element possesses specific sequences for its mobilization, and therefore its subsequent insertion can generally occur at any site of the genome.

Frequency of crossing over. The number of crossing overs that occurs on a pair of homologous chromosomes in meiosis is highly variable and depends on the size of the chromosomes themselves. For instance, in our species, for each pair of homologous chromosomes, an average of 2–3 crossing overs take place at each meiosis.

For the production of genetic variation, a very important parameter is the frequency of crossing over between two loci of interest on the same chromosome (linked loci), because recombination tends continuously to eliminate the association between specific alleles present at distinct loci, a condition known as *linkage disequilibrium*. The greater the distance between two loci on a chromosome, the larger the probability that at least one crossing over occurs during meiosis. For this reason, the frequency of recombination between two loci, which can be measured by different genetic tests, is an indicator of the physical distance between them (although modern sequencing techniques produce more precise maps). Actually, several factors cause the crossing-over frequency not to correspond precisely to real physical distances (quantifiable, for instance, as number of nucleotides). Among these are the facts that the probability of crossing over is not uniform over the whole length of a chromosome, that crossing over at a point tends to inhibit crossing over in adjacent regions, and that during a single meiosis, on the same chromosomal arm, more than one crossing over can take place between different chromatids, reciprocally masking the effects of recombination at certain loci.

The frequency of crossing over between pairs of linked loci varies from values close to 0%, for immediately adjacent genes, to a maximum value of 50% for very distant loci on particularly long chromosomes, the same value that is recorded for loci on different chromosomes, which are assorted thanks to the independent segregation of homologous chromosomes at the first meiotic division. It must be considered that, while an odd number of crossing overs between two loci generates recombinant haplotypes with respect to these loci, an even number of crossing overs restores the original (parental) haplotypes in the involved chromatids (Figure 5.13).

Effects of shuffling. The subdivision of the nuclear genome into multiple DNA molecules (the chromosomes) allows recombination by independent assortment, to which recombination between DNA sequences of homologous chromosomes is added. Combining the effects of interchromosomal and intrachromosomal recombination, the amount of genetic variation that can be produced at each generation is enormous.

In humans there are about 23,000 protein-encoding genes, 3000 (13%) of which are probably polymorphic. In an individual, the fraction of heterozygous loci is about 7% on average, which gives $23{,}000{\cdot}0.07 \approx 1600$ heterozygous loci. Assuming that each locus can segregate independently, an average individual could virtually produce $2^{1600} \approx 4.4{\cdot}10^{481}$ different gametes. Even limiting the number of alleles per polymorphic locus to two, which means only three possible genotypes per locus, the number of possible genotypes that could be observed in a human population would be $3^{3000} \approx 2.3{\cdot}10^{1431}$.

Calculations of this kind, which sometimes occur in the popular-science literature, are based on the possibility of having an unlimited number of crossing overs per chromosome, something that does not correspond to

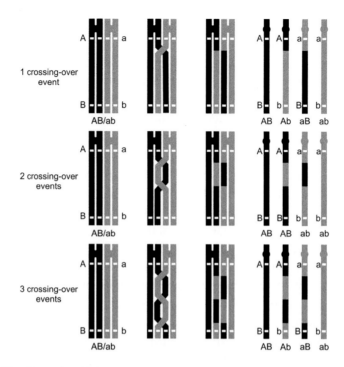

Figure 5.13 The effects of multiple crossing-over events between two loci (*A* and *B*) on the same chromosome arm. Among the four products of meiosis, an odd number of crossing overs between the two loci produces two recombinant haplotypes (*Ab* and *aB*), while an even number of crossing overs produces only parental haplotypes (*AB* and *ab*).

reality. An exact calculation would require precise statistics of the distribution of crossing overs. However, we can at least draft a better approximation than the previous one by using combinatorics. Assuming, for simplicity, a uniform distribution of the heterozygous loci in the 23 pairs of chromosomes, there would be about 70 heterozygous loci per pair of homologous chromosomes.

Assuming then a conservative number of two crossing overs per chromosome pair (for a male, only in the 22 pairs of autosomes), and counting the number of ways in which zero, one or two effective crossing overs can occur in the 69 different positions in between the 70 polymorphic loci, the number of different haplotypes per pair of homologues would be $2 \cdot (1+69+(69\cdot68)/2) = 4832$, so a woman could produce virtually $4832^{23} \approx 5.4\cdot10^{84}$ different gametes and a man $2 \cdot 4832^{22} \approx 2.2 \cdot 10^{81}$. Although these figures are much lower than the first estimate of 10^{481}, they are still huge numbers, higher than the estimated number of atoms in the universe (in the order of 10^{80}). Given that on average a man produces throughout his life a number of spermatozoa in the order of one trillion (10^{12}), if the localizations of the crossing overs are all equally likely, the probability for an individual to produce two spermatozoa with the same recombinant haplotype would be less than 10^{-57}.

Other effects. The main effect of crossing over is the creation of new allelic associations, different from the parental ones, but in some cases it can also produce new alleles or new genes, with an original nucleotide sequence. These mutational events (gene or chromosomal mutations) are addressed here because they can be more easily associated with reproductive processes, particularly with recombination during meiosis, but they can also occur in mitotic recombination (Section 5.1.2). Recombination can cause change in the nucleotide sequence of a gene in several ways.

In *intragenic recombination*, crossing over or gene conversion (see below) occurs within the boundaries of a given locus. The exchange between two homologous sequences that differ for a certain number of base pairs (different alleles of a gene or, in general, different haplotypes of any genetic marker) can generate new DNA sequences for a given locus. Recent DNA sequencing studies are revealing many cases of haplotypes and allelic variants that appear to have originated by intragenic recombination.

In *unequal crossing over*, the non-perfectly reciprocal exchange between homologous chromosomes that are not exactly aligned can result in a tandem duplication (i.e. a duplication whose products remain adjacent) of a nucleotide sequence on one of the two products of the recombination and in the corresponding deletion on the other. Unequal crossing over occurs more easily between sequences that already include tandem repeats, because out-of-register alignments can occur more easily there. The unequal crossing over, which can also occur between sister chromatids in mitosis, is probably one of the processes responsible for the high percentage of non-coding DNA sequences that are produced through chromosomal mutations, as well as for *gene duplication*, a very important phenomenon in the evolution of eukaryotes (Futuyma and Kirkpatrick 2018).

Gene Conversion in Meiosis

Gene conversion (or *recombinational loss of heterozygosity*) is a homologous recombination process (Box 5.4) through which the information of a nucleotide sequence is transferred unidirectionally from a DNA molecule (which remains unchanged) to another one, whose sequence is thus modified (Figure 5.10b). Generally, gene conversion involves a relatively short DNA segment compared to what happens in crossing over. In a typical conversion event, a gene, or a part of a gene, acquires the sequence of another allele at the same locus (*intralocus conversion*, or *intrallelic conversion*) or, as an alternative, acquires the sequence of a different locus, usually paralogous, i.e. similar to the former because the two derive from an event of gene duplication (*interlocus conversion*, or *interallelic conversion*). Gene conversion occurs at high frequency during meiosis, but, as we have seen (Section 5.1.2), it can also occur in somatic cells.

By homogenizing DNA sequences, gene conversion tends to reduce **heterozygosity**, which is measured as the fraction of heterozygous individuals at a given locus in a population. Over time, the effects of conversion homogenization can accumulate, becoming apparent both among the different allelic variants of a gene and among the different genes of a gene family. Members of a gene family that originated by duplication of an ancestral gene generally maintain high levels of sequence identity, so that they are more frequently subject to conversion (Futuyma and Kirkpatrick 2018).

5.2.2.4 SYNGAMY

Syngamy (or **fertilization**) is the process that brings together in one genome the hereditary material from two distinct gametes or gametic nuclei. This is a process that takes place at the cellular level and should not be confused with *insemination*, a process that brings the male gametes close to the female ones, and which is not necessarily, or not immediately, followed by fertilization (Section 3.5.1).

Syngamy consists of two distinct processes, which often follow each other within a short time. These are **plasmogamy**, i.e. the fusion of the cytoplasms of the two gametes into a single zygotic cell, and **karyogamy**, i.e. the fusion of the two gametic nuclei into a single nucleus called the **synkaryon**. In the interval between plasmogamy and karyogamy, the nuclei of the gametes that will merge into the synkaryon are called **pronuclei**.

In some cases, karyogamy can be deferred with respect to plasmogamy. This happens, for instance, in sexual reproduction in ascomycetes and basidiomycetes. Two monokaryotic hyphae (with one haploid (n) nucleus per cell) belonging to different mating types can merge (plasmogamy without karyogamy) producing a binucleate hypha (dikaryotic phase, n+n) that proliferates, originating a mycelium formed by dikaryotic hyphae. The dikaryotic mycelium proliferates and forms fruiting bodies that carry reproductive structures specific for each group (asci and basidia, respectively) where karyogamy occurs (diploid phase, 2n) and subsequently, by meiosis, the production of haploid spores (n) as well. The haploid spores give rise to the monokaryotic mycelium of the new generation (Figure 7.16).

Syngamy also brings about the union of the hereditary material of the cytoplasmic organelles of the two fusing gametes. In species with anisogamy, the contribution of the two parents to the set of cytoplasmic organelles of the zygote can be very unbalanced. For instance, in most animals the mitochondria are inherited from the mother, whereas in the Cupressaceae these organelles are inherited from the male parent gametophyte (Section 5.2.3.1). It should also be remembered that, while the nuclear genome is faithfully transmitted (subject to possible mutations) through the mitotic cycles that may follow the formation of the zygote (in the development of a multicellular individual, or in the asexual reproduction of a unicellular organism), the genome of organelles is subject to a process of *mitotic segregation* (Section 5.1.3) that can produce a certain degree of genetic variation in the descendants.

In a particular but prominent case, syngamy does not involve only two gametes and does not produce a zygote. In the sexual reproduction of the flowering plants, in the process known as *double fertilization* (Section 3.5.4.2), one of the two sperm nuclei of the male gametophyte merges with the two nuclei of the central cell of the female gametophyte, thus generating the triploid cell founder of the secondary endosperm, the tissue that will provide nutrients for the embryo.

5.2.3 Genetics of Hereditary Transmission Through Different Modes of Sexual Reproduction

Recombination in the broad sense and syngamy are the processes that characterize sexual reproduction. However, the diverse reproductive modes, in association with the genetic system and the mating system, entail a different relative contribution and effectiveness for the genetic reshuffling processes described in previous sections. For instance, as we shall soon see in detail, self-fertilization can severely restrict the mixing effects of syngamy, to the point of turning it into a form of clonal reproduction. In gynogenesis, there is no independent segregation of the chromosomes, and in hybridogenesis there is no crossing over. Let's thus have a closer look at the characteristics of hereditary transmission in the various modes of sexual reproduction (see Chapter 3). Where not otherwise specified, we refer to sexual reproduction in the diploid phase.

5.2.3.1 AMPHIGONY

Amphigony (or **amphimixis**, or **allogamy**) is the mode of sexual reproduction that is based on the fusion of gametes produced by two distinct individuals, to form a zygote (Section 3.5; Figure 5.14). Depending on the type of organism and its life cycle, this diploid cell can (i) correspond to a new diploid unicellular individual, (ii) be the founding cell of a diploid multicellular individual, (iii) immediately initiate meiosis, with the production of four haploid cells, in their turn new haploid unicellular individuals, or (iv) again by meiosis, produce four haploid cells, founders of four haploid multicellular individuals.

Through amphigonic reproduction, all the different types of genetic reassortment we have seen in Section 5.2.2.1 can take place in synergy, thus contributing to maximizing the production of genetic variation in each

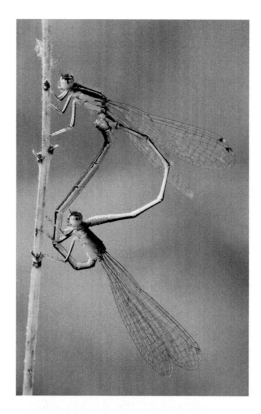

Figure 5.14 Two mating blue-tailed damselflies (*Ischnura elegans*). The male is above. The union of gametes produced by two distinct individuals characterizes amphigonic reproduction.

generation. Nevertheless, the actual result in terms of produced variation depends on many additional factors, including the genetic system of the organism, the spatial distribution of the individuals of the population, their mobility in specific phases of their life cycle or the mobility of their gametes. In addition, the distribution of sex conditions in the population, the mating system and, possibly, the social system may have significant effects as well.

Many of these factors contribute to determining the degree of **inbreeding**, i.e. the frequency of crossing between closely related individuals. To some extent, inbreeding dims the effects of genetic shuffling produced by **cross-fertilization** (or **outcrossing**, or **outbreeding**) typical of amphigonic reproduction, reducing the level of **average heterozygosity** in the population. The latter is defined as the average frequency of heterozygous individuals per locus, but it is easy to show that this value corresponds exactly to the average number of heterozygous loci per individual. The reduction in heterozygosity increases the probability for a recessive deleterious allele to be expressed in a homozygous state in the progeny. Populations with a high frequency of crossing between relatives may show a decline in the average values of some fitness components, such as survival and fertility, which is called *inbreeding depression*. However, such detrimental effects are not a necessary consequence of inbreeding, since mating systems that regularly provide for crossing between relatives are not rare. For example, the fertilized females of the mite *Adactylidium*, which feed on a single egg of a thrips (Thysanoptera), carry 6–9 eggs which hatch when they are still in their mother's body. From the eggs of a female, 5–8 females and a single male develop, which feed on the mother's tissues from the inside (*matriphagy*, Section 4.4.4). Before leaving the mother's remains, the male fertilizes his sisters and dies within a few hours. The females that are so fertilized leave what remains of their mother and look for a new thrips egg, before being devoured in their turn by their offspring. The whole cycle is completed within a few days.

In addressing asexual reproduction, we have seen (Section 5.1.3) that the genomes of cytoplasmic organelles generally show transmission modes through cell division that are different from that of the nuclear genome, and that for this reason they are sometimes labelled as *non-Mendelian genetic systems*. In sexual reproduction, cytoplasmic organelles behave in a non-Mendelian way even in syngamy. In particular, the contribution of the two parents to the organelle supply of their offspring can be very unequal. This imbalance can be established during or immediately after the formation of the zygote, because of the unequal provision of organelles by the two gametes, which in the extreme case may consist of those of only one of the two parents, or because the genome of the organelles of one of the parents is lost during development. A combination of both causes is also possible.

In the case of mitochondria, in most species of plants and animals, which typically have anisogametes, the DNA of this organelle (mtDNA) is transmitted to the offspring of both sexes predominantly, if not exclusively, through the maternal lineage (**maternal inheritance**). However, to a degree depending on the species under consideration, there is also the possibility of transmission of some paternal mtDNA. In animals, each sperm does in fact contain some mitochondria, up to a few dozen, and some of these can enter the cytoplasm of the egg during fertilization, although there are typically many thousands of copies of maternal mtDNA in the mature oocyte. In many animal species mtDNA is transmitted through the maternal lineage exclusively, as the paternal mtDNA disappears during early embryogenesis.

The limited or accidental transmission of mitochondrial DNA of paternal origin is an example of **paternal leakage**, i.e. the transmission of paternal genomic elements in a system otherwise dominated by maternal inheritance. In animals, mitochondrial paternal leakage can account for a fraction of 10^{-4} to 10^{-3} of an individual's mtDNA (Stewart *et al.* 1995). However, in recent years evidence of non-accidental transmission of paternal mtDNA has accumulated

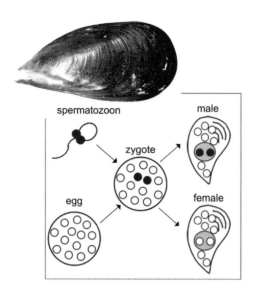

Figure 5.15 Schematic representation of the transmission of mitochondrial DNA (mtDNA) in the Mediterranean mussel (*Mytilus galloprovincialis*) through the system of *doubly uniparental inheritance*. Paternal mtDNA is represented by filled circles, maternal mtDNA by empty circles. In the adult mussels, on the right, grey ovals represent the gonads. The fertilized egg receives mtDNA from both gametes. In adults, mtDNA of paternal origin is dominant in the male gonad, and mtDNA of maternal origin in the female gonad and in the somatic tissues of individuals of both sexes.

for a number of animal and plant species. These findings question the exceptional nature of the phenomenon (Wolff *et al.* 2008). For example, a particular mtDNA transmission system, called *doubly uniparental inheritance* (DUI), has been described in several dozen species of bivalve molluscs belonging to different families. In the *Mytilus* DUI system (Figure 5.15), females inherit mtDNA only from the mother and transmit it to the offspring of both sexes. Males, on the other hand, inherit mtDNA from both parents, but only transmit the mtDNA received from the father, which will however only be maintained in their male offspring. Thus, in adult females the mtDNA of maternal origin is found both in somatic cells and in the germinal cells of the gonads, while in adult males the mtDNA of paternal origin predominates in the gonads and that of maternal origin predominates in somatic tissues (Sano *et al.* 2011).

In the case of plastids, many plants inherit these organelles from just one parent. Normally the DNA of the chloroplasts (ctDNA) is inherited maternally in the angiosperms, while it is inherited paternally in the gymnosperms. However, it has recently been realized that a characteristic of male gametes called *potential biparental plastid inheritance* (PBPI), associated with the presence of ctDNA in the latter (otherwise absent in species with exclusively maternal inheritance of the chloroplasts), occurs in a large fraction (20%) of angiosperm genera, showing that the transmission of chloroplasts by both parents is not uncommon in this group (Sodmergen 2010).

Recent studies are clearly showing that the present knowledge on transmission genetics of the organelles is very incomplete, and that some generalizations are unwarranted. Different species of a clade, even a small one, may exhibit dissimilar mechanisms of inheritance for the genomes of the cytoplasmic organelles. For example, in the Pinaceae the mitochondria are inherited through the maternal lineage, while the chloroplasts are inherited paternally. But this does not apply to all the conifers: in the Cupressaceae, the genomes of mitochondria and plastids are both inherited through the paternal lineage (Williams 2009).

Finally, amphigonic sexual reproduction can have a clonal outcome through **polyembryony** (Section 3.1.2.4). This can be considered a form of sexual reproduction with a clonal result, since it generates identical copies of the same genotype, even if different from the genotype of the parents. Alternatively, polyembryony can be considered in all respects a form of asexual reproduction at an early (embryonic) stage of development. In this second interpretation, it is good to make it explicit that the species that reproduce by obligate polyembryony (including some mammals, such as the armadillos of the genus *Dasypus*) show a form of alternation of sexual and asexual generations (metagenetic cycle; Section 2.2). The genetics of transmission is therefore different in the two generations: it has the characteristics of amphigonic reproduction in the sexual generation and those of clonal reproduction in the asexual generation.

5.2.3.2 SELF-FERTILIZATION

In **self-fertilization** (or **selfing**, or **autogamy**, or **automixis**; but see the comments at the end of this section), gametes, or gametic nuclei from the same individual (a sufficient simultaneous hermaphrodite; Section 3.3.2.2), merge in syngamy to form a zygote that thus has only one parent (Section 3.6.1). It is therefore a form of uniparental sexual reproduction. From the point of view of genetic variation, self-fertilization is an extreme form of inbreeding (who is more kin than him/herself?). In the case of a haploid organism, self-fertilization produces a zygote that is homozygous at all loci. In a haplontic cycle, from this zygote, following meiosis, haploid individuals identical to the parent and identical to each other will develop (e.g. in the green alga *Eudorina elegans*; Bell and Praiss 1986; Figure 5.16). In a haplodiplontic cycle, the completely homozygous zygote will develop into a sporophyte which generates spores all genetically identical to the gametophyte that produced it and identical to each other (*intragametophytic self-fertilization*, e.g. in the fern *Polystichum acrostichoides*; Flinn 2006).

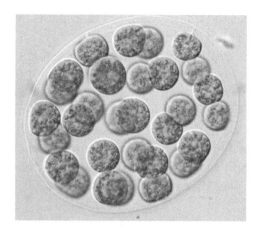

Figure 5.16 The green alga *Eudorina elegans*, a haploid organism that practises self-fertilization.

Conversely, in the case of diploid organisms, owing to the independent assortment of chromosomes and meiotic recombination, in principle the progeny could be genetically diverse and different from the parent. However, in practice, when self-fertilization is carried out regularly generation after generation, highly inbred lines emerge that tend to become clones. Moreover, at variance with clones produced by polyembryony or by parthenogenesis (Sections 5.2.3.1 and 5.2.3.3), a clone that reproduces through prolonged self-fertilization over many generations will exhibit very low levels of heterozygosity. This is because in self-fertilization heterozygosity is reduced by 50% at each generation, thus exponentially declining very rapidly toward zero (Figure 5.17). When self-fertilization alternates with cross-fertilization, even occasionally (a *mixed mating system*; see Section 3.3.2.2), the effects of the genetic reshuffling thus brought about may have sizeable, although temporary, effects on the genetic variation of the population.

These effects depend on many factors that interact in a complex way. However, it is possible to get at least an intuitive idea of the dynamics that follow a cross-fertilization event in a mixed mating system with the help of a descriptive metaphor, the so-called *fireworks model* (Avise 2008). In this metaphor, the dark night sky represents the almost complete absence of individual heterozygosity typical of highly inbred populations. Each exploding firework represents a single cross-fertilization event between two individuals from distinct clones. The sudden explosion of light depicts the high heterozygosity of the progeny, which, however, while spreading across the sky, sees its brightness gradually fade out because of self-fertilization, bringing heterozygosity back to the darkness of values close to zero. Meanwhile, another firework may explode, perhaps in another sector of the dark sky, and a new

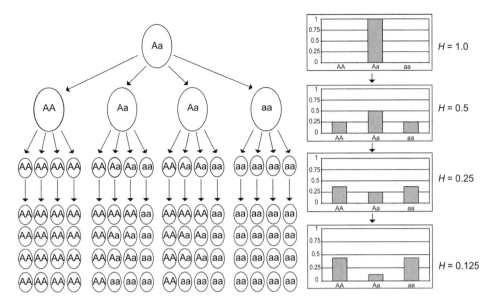

Figure 5.17 Rapid loss of average heterozygosity (*H*) across generations in a population that regularly practises self-fertilization. In each generation, on average only 50% of the offspring of individuals with heterozygous genotype (*Aa*) will be heterozygous, while all offspring of homozygous individuals will be homozygous (*AA* or *aa*). Homozygous conditions behave as *absorbing boundaries*, i.e. as extreme classes (or values) of a distribution that can only increase in frequency since individuals of such classes will never generate new individuals of different classes. The fraction of heterozygous individuals (*H*) will thus halve with each generation, according to a decaying geometric progression.

cross-fertilization event produces another explosion of genetic variation, destined also to disappear quickly.

Finally, two comments on terminology. The first concerns the term *automixis*, which is also used for other reproductive mechanisms, such as some forms of parthenogenesis (see next section). This is a consequence of the fact that different authors draw different boundaries between the different forms of uniparental reproduction. Here we have generally avoided using this term, preferring *self-fertilization* in the case of fusion of gametes or gamete nuclei produced by distinct meioses in the same individual, and *meiotic parthenogenesis* for the fusion of two (out of the four) products of the same meiosis. The differences between these two forms of syngamy are not limited solely to the mechanism at the cytological level, but obviously also involve the genetics of transmission. The second comment concerns the term *self-fertilization* when applied to haplodiplontic organisms. Strictly speaking, in these organisms there can be no true self-fertilization, even if the sporophyte is hermaphrodite, as in the case of most angiosperms, because the male and female gametes are produced by two distinct gametophytes of the generation that follows the sporophyte, a microgametophyte and a megagametophyte.

In other words, a hermaphrodite sporophyte can self-fertilize only through its offspring gametophytes. This is clearly a subtlety, since from the point of view of the genetics of transmission everything that applies to the self-fertilization of a diplontic organism is still valid.

5.2.3.3 PARTHENOGENESIS

In **parthenogenesis**, females can produce offspring without the participation of a male (Section 3.6.2). They produce 'special eggs' that can develop into a new individual without the need to be fertilized or activated by a sperm. In short, the defining feature of parthenogenesis lies in the possibility of developing from an unfertilized egg.

In **haploid parthenogenesis**, diploid females generate haploid males through *reduced eggs* (i.e. with a chromosome complement halved by meiosis) that have not been fertilized. Because of the independent assortment and crossing over, the offspring thus generated will generally have recombinant genotypes, different from the mother's parental haplotypes and different from each other. From the point of view of Mendelian inheritance, it is as if all the chromosomes were sex chromosomes in a male heterogamety system X0 (Section 6.1.1). Haploid parthenogenesis is typical of hymenopterans, some mites (*Histiostoma*) and monogonont rotifers (Section 3.6.2.2).

In the more common **diploid parthenogenesis** (actually, **polyploid parthenogenesis** in the case of polyploid species or populations), diploid (or polyploid) females produce *unreduced eggs* from which individuals with the same ploidy level as the mother and of the same sex or of both sexes (depending on the species) will develop (Section 3.6.2). What follows concerns *obligate diploid parthenogenesis*, i.e. parthenogenesis in populations that reproduce exclusively (or nearly so) through this mode. This is carried out through a large variety of alternative cytogenetic mechanisms, which in turn lead to considerable disparities in the genetics of hereditary transmission (Table 5.1). Deriving from amphigony and normal female gametogenesis, the different cellular mechanisms through which parthenogenesis is implemented can be grouped into two main cytogenetic categories. These distinguish the mechanisms that have retained meiosis from those that have lost it. For the terms used to describe parthenogenesis in plants, which differ in part from those adopted in the following paragraphs, see Section 3.6.2.9.

Meiotic Parthenogenesis

In **meiotic parthenogenesis** (or **automictic parthenogenesis**, or **automixis**; but see the comments at the end of Section 5.2.3.2) meiosis is maintained and the diploid condition of the meiotic products is obtained through different mechanisms. When this mechanism involves the fusion of haploid products of meiosis (collectively called **ootids**, irrespective of their

Table 5.1 Probability of transition to the homozygous condition for a heterozygous locus, as a function of the distance from the centromere and in different forms of parthenogenesis, and consequent reduction of the expected average heterozygosity by generation (modified from Pearcy *et al.* 2006)

Mechanism	Probability of transition to homozygosis		Reduction of average heterozygosity (*H*) per generation
	Locus close to the centromere (no crossing over)	Locus far from the centromere (crossing over frequent)	
Gametic duplication	1.00	1.00	100% (*H* reduced to zero)
Terminal fusion	1.00	0.33	>33%
Central fusion	0.00	0.33	<33%
Random fusion	0.33	0.33	33%
Fusion of the first polar nucleus with the nucleus of the secondary oocyte	0.33	0.33	33%
Premeiotic doubling	0.00	0.00	0% (*H* preserved)
Apomixis	0.00	0.00	0% (*H* preserved)

fate), these strictly derive from the same event of meiosis. This distinguishes automictic parthenogenesis from self-fertilization (Section 5.2.3.2), which involves two gametes produced by distinct meiotic processes in the same individual. Only some of the mechanisms of meiotic parthenogenesis known in animals are also observed in plants.

Production and conservation of genetic variation in a population that practises this type of parthenogenesis depends on the precise cellular mechanism with which it is carried out. In some cases, variation is maintained even though reproduction is clonal; in other cases, recombination can produce a certain degree of variation among the descendants, but with some limitations,

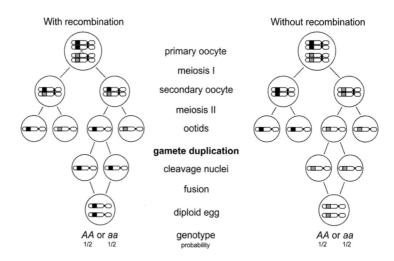

Figure 5.18 Meiotic parthenogenesis by gamete duplication. The diploid condition of the egg is obtained through duplication and subsequent fusion of one of the four products of the second meiotic division. A single pair of homologous chromosomes is shown. On the left, the case where crossing over occurred for the chromosome segment that contains a locus of interest (*A*); on the right, the case in which no effective crossing over occurred for the same locus.

since these forms of meiotic parthenogenesis, when practised regularly through many generations, tend to erase heterozygosity. The complete loss of heterozygosis is a point of no return, unless other forms of reproduction intervene, because, lacking syngamy, meiotic parthenogenesis can produce genetic variation in the offspring only if the parent is not already homozygous at all loci. The main types of cytogenetic mechanisms of meiotic parthenogenesis are described below, following Stenberg and Saura (2009), but it should be noted that this is not a complete list, and that each of them can occur through a certain number of minor variants.

Gamete duplication. The haploid egg cell divides by mitosis, producing two nuclei that subsequently merge to produce a diploid nucleus. Alternatively, only the chromosomes replicate and the products of their replication, remaining in the same nucleus (*endomitosis*), restore the diploid condition. Gamete duplication produces exclusively offspring homozygous at all loci, independent of the crossing overs that may have occurred during meiosis (Figure 5.18). Gamete duplication has been observed in the crustacean *Artemia* (Figure 5.19), in some mites and in many insects, including some *Drosophila* species.

Terminal fusion. The nucleus of the second polar body (the *sister nucleus* of the egg cell, itself an ootid deriving from the secondary oocyte) merges with the nucleus of the egg cell. In the absence of crossing over, a locus that is heterozygous in the mother will be exclusively homozygous in the offspring,

Figure 5.19 In *Artemia*, an anostracan crustacean, parthenogenesis occurs by gametic duplication. The ventral bulge contains the eggs.

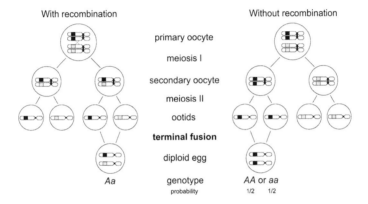

Figure 5.20 Meiotic parthenogenesis by terminal fusion. The diploid condition of the egg is obtained through the fusion of two sister products of the second meiotic division. A single pair of homologous chromosomes is shown. On the left, the case where crossing over occurred for the chromosome segment that contains a locus of interest (*A*); on the right, the case in which no effective crossing over occurred for the same locus.

whereas if there is crossing over in meiosis I its heterozygosity will be maintained in the offspring (Figure 5.20). If the locus is relatively distant from the centromere, many crossing overs may occur in between, so that the segregation of the four alleles (one for each chromatid) can be regarded as virtually independent. In this situation, the probability that a heterozygous locus (*Aa*) can be found in a homozygous state after the first meiotic division, and therefore in the progeny, approaches the probability of extracting without replacement two identical elements (*AA* or *aa*) from a set of four elements that are identical in pairs (*AAaa*), which is equal to 1/3. Thus each heterozygous locus has a probability of becoming homozygous between 1/3 (if it is relatively far from the centromere) and 1 (if it is so close to the centromere that in practice it cannot recombine). In a population that reproduces by partheno-genesis according to this mode there will be a progressive decline in average heterozygosity, at a rate that varies from locus to locus and that is greater for those closer to the centromere, but reducing anyway by not less than 1/3 in each generation. Among the species that reproduce by parthenogenesis through terminal fusion there are some nematodes, enchytraeid oligochaetes, tardigrades, oribatid mites, isopods and insects.

Central fusion. The two central ootids (by ordering the products of the second meiotic division based on their derivation from the first division), whose cell lines separated at the first meiotic division (*non-sister nuclei*, because they respectively derive from the first polar body and the secondary oocyte) merge to form the zygote. In the absence of crossing over, the offspring will be genetically identical to the mother. When the locus is relatively distant from the centromere, applying a calculation analogous to the previous case, the probability that a heterozygous locus (*Aa*) will remain heterozygous after the first meiotic division is equal to 2/3. The fusion of the two non-sister nuclei will give genotypes *AA*, *Aa* and *aa* in Mendelian proportions 1/4, 1/2 and 1/4, respectively, so that the probability of passing to a homozygous condition will be equal to 2/3·(1/4 + 1/4) = 1/3. Thus, each heterozygous locus has a prob-ability of becoming homozygous ranging from 0 (if it is so close to the centromere that practically it cannot recombine) to 1/3 (if it is relatively far from the centromere) (Figure 5.21). A population that reproduces by partheno-genesis according to this mode will progressively decline in average

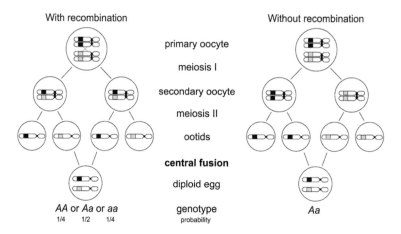

Figure 5.21 Meiotic parthenogenesis by central fusion. The diploid condition of the egg is obtained through the fusion between two non-sister products of the second meiotic division. A single pair of homologous chromosomes is shown. On the left, the case where crossing over occurred for the chromosome segment that contains a locus of interest (*A*); on the right, the case in which no effective crossing over occurred for the same locus.

heterozygosity, at a rate that varies from locus to locus and that is greater for those closer to the centromere, but reduces anyway by no more than 1/3 in each generation. Central fusion has been reported for several insects, particularly among the dipterans and hymenopterans.

Random fusion. The egg nucleus merges with any of the nuclei of the three haploid polar bodies. For a heterozygous locus, one of the two polar nuclei carries the same allele as the egg cell, while the other two carry the alternative allele. Thus, irrespective of crossing over, a heterozygous locus has a probability of 1/3 to change to homozygous after meiosis and fusion. As in the case of self-fertilization (Section 5.2.3.2), the average heterozygosity in the offspring will decline exponentially, but at a slower rate, i.e. decreasing by 1/3, rather than 1/2, in each generation, and equally for all loci. This is because in parthenogenesis the fusion is between two products of the same meiosis, and for each ootid there is only one potential partner with the same allele, against two with the alternative allele, whereas in the fusion of products of two distinct meioses, as in the case of self-fertilization, half of the potential partners (two ootids for each meiosis) carry the same allele.

Fusion of the nucleus of the first polar body with the nucleus of the secondary oocyte. The nuclei derived from the first meiotic division do not separate, or first separate and then fuse, producing a transitory tetraploid state that reduces to diploid at the second meiotic division. A mother heterozygous at locus *A* (*Aa*) will produce offspring with genotypes *AA*, *Aa* and *aa* in proportions 1/6, 4/6 and 1/6, respectively (6 = the number of combinations of 4 elements taken 2 at a time without repetition; Figure 5.22). Recombination has no limiting effect on the progressive erosion of heterozygosity, which is reduced on average by 1/3 per generation in all loci. This parthenogenetic mechanism is known in some insects and in the fluke *Fasciola hepatica*. It also corresponds to one of the modes of *meiotic diplospory* in angiosperms (Section 3.6.2.9), which is observed in *Taraxacum* and *Tripsacum*. Alternatively, if as a result of recombination and fusion of the two nuclei the chromatids of each chromosome do not separate, a mother heterozygous at the locus *A* (*Aa*) will produce offspring with genotypes *AA*, *Aa* and *aa* in proportions 1/4, 1/2 and 1/4, respectively, with a loss of heterozygosity of 1/2 in each generation. This occurs for instance in some populations of the stick insect *Bacillus atticus* (Marescalchi *et al.* 1993; Figure 5.23).

Premeiotic doubling (or *premeiotic endomitosis*). Meiosis is preceded by **endomitosis** (replication of the chromosomes not followed by the division of the nucleus) that doubles the number of chromosomes, a number that will be restored to the original value through meiosis. At the first meiotic division, all chromosomes pair with their genetically identical homologues. Thus, even if there is crossing over, the mother's genotype is passed unaltered to the

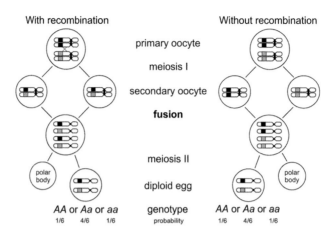

Figure 5.22 Meiotic parthenogenesis by fusion of the nucleus of the first polar body with the nucleus of the secondary oocyte. In the example shown here, the diploid condition is obtained through the reductional division of a tetraploid nucleus produced by the fusion of the products of the first meiotic division. A single pair of homologous chromosomes is shown. On the left, the case where crossing over occurred for the chromosome segment that contains a locus of interest (*A*); on the right, the case in which no effective crossing over occurred for the same locus.

Figure 5.23 Stick insects of the genus *Bacillus* (here, *B. rossius*) can reproduce through different forms of metasexuality, including parthenogenesis and androgenesis.

offspring (Figure 5.24). From the point of view of transmission genetics, this form of meiotic parthenogenesis is equivalent to *ameiotic parthenogenesis* (see below). It results in *unreduced eggs* (i.e. with the same ploidy level as the germ cells from which they derive) that are genetically identical to the somatic cells of the mother and will develop into new individuals identical to the mother and to each other. Premeiotic doubling is the most common mechanism among the parthenogenetic forms of free-living flatworms and terrestrial oligochaetes and is also known in many insects, mites, tardigrades and in all the

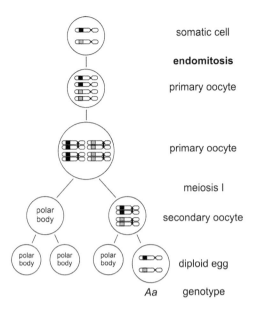

Figure 5.24 Meiotic parthenogenesis by premeiotic doubling. The diploid condition of the meiotic products is obtained by starting meiosis from a tetraploid nucleus produced by endomitosis. A single pair of homologous chromosomes is shown. In this case, the crossing over has no effect on the genotypes of the meiotic products.

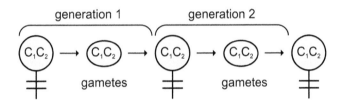

Figure 5.25 Ameiotic parthenogenesis. In this form of parthenogenesis, eggs are produced by cell divisions equivalent to mitosis. Offspring are genetically identical to the mother and to each other. Symbols C_1 and C_2 indicate the two sets of homologous chromosomes.

parthenogenetic vertebrates hitherto studied from a cytogenetic point of view. It is also found in a form of apomixis in angiosperms (e.g. in some species of *Allium*; Section 3.6.2.9).

Ameiotic Parthenogenesis

In **ameiotic parthenogenesis** (or **apomictic parthenogenesis**, or **apomixis**), meiosis is suppressed, so that the eggs are produced by a maturative cell division that, from a strictly karyological point of view, is not generally distinguishable from a mitosis. The mother's genotype is passed unaltered to the offspring (Figure 5.25). From the point of view of transmission genetics, this form of parthenogenesis can therefore be equated to asexual reproduction starting from a somatic cell. Daughters are genetically identical to the mother

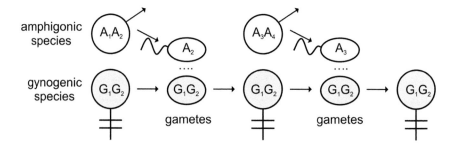

Figure 5.26 Schematic representation of the hereditary transmission of nuclear DNA in reproduction by gynogenesis. The female produces unreduced (in this example, diploid) eggs that develop only if activated by a sperm. The donor of the sperm, which in non-hermaphrodite species must belong to a different species, does not contribute any genetic material to the genome of the offspring. Symbols G_1 and G_2 indicate a pair of homologous chromosomes in the gynogenic species, while A_1–A_4 represent the homologous chromosomes in the amphigonic species (for simplicity, depicted as non-recombinant).

and to each other, subject to possible mutations occurring during gametogenesis. Extant genetic variation is preserved unaltered and is increased by the accumulation of mutations, with a tendency to complete heterozygosity. The latter may in part be opposed by **ameiotic recombination**, i.e. by *mitotic recombination* (Section 5.1.2) at the level of the germline of the apomictic species, as observed in the crustacean *Daphnia* (Omilian *et al.* 2006).

All animals with cyclical parthenogenesis (Section 3.6.2.6) reproduce by ameiotic parthenogenesis, including monogonont rotifers, cladoceran crustaceans, aphids, the beetle *Micromalthus*, cecidomyiid dipterans and cynipid hymenopterans. These are joined by numerous representatives of cnidarians, flatworms (both free-living 'turbellarians' and parasitic digeneans), nematodes, gastrotrichs, bdelloid rotifers, gastropods, oligochaetes, and arthropods of different groups.

In plants, ameiotic parthenogenesis corresponds, from a cytogenetic point of view, to different forms of *apomixis*, here to be understood, according to the botanical tradition, as a form of uniparental reproduction through the seed. These forms are *sporophytic apomixis* (e.g. the *Citrus* species and many orchids), *gametophytic apomixis by apospory* (e.g. *Hypericum perforatum* and *Poa pratensis*) and *gametophytic apomixis by mitotic diplospory* (e.g. *Hieracium* and *Antennaria*) (Section 3.6.2.9).

5.2.3.4 GYNOGENESIS AND PSEUDOGAMY

Gynogenesis, as we have seen in Section 3.6.3, is a form of reproduction similar in many respects to parthenogenesis. As in diploid (or polyploid) parthenogenesis, unreduced eggs are produced that do not need to be fertilized. However, in the case of gynogenesis the egg develops only if somehow 'activated' by a male gamete (Figure 5.26). Gynogenesis, together with hybridogenesis (Section 5.2.3.5), can be seen as a form of *sperm-dependent parthenogenesis.*

In gynogenesis the male gamete simply comes into contact with the egg cell, or penetrates it, but does not contribute any genetic material to the genome of the organism that develops from it. If the gynogenic species is gonochoric, it will be composed of females only and the stimulus to the development of the eggs has to be found in the mating with males of closely related species. In contrast, in hermaphrodite species (examples among flatworms, nematodes, annelids and molluscs), sperm-dependent parthenogenesis does not imply the need for a different sperm-donor species. However, in the hermaphrodite species the more general term of **pseudogamy** (or **pseudogamous parthenogenesis**) should be preferred to gynogenesis, which literally means 'descent from females' (Beukeboom and Vrijenhoek 1998). To complicate the nomenclature, the term *pseudogamy* is used in the botanical literature also to indicate reproduction by parthenogenesis that requires pollination for the fertilization of the central cell of the megagametophyte, from which the triploid endosperm of the seed will develop (Section 3.5.4.2). However, in general, this process is not associated with any form of activation of the egg cell by a sperm nucleus. As in the case of parthenogenesis in the strict sense (*sperm-independent parthenogenesis*), the effects of reproduction by gynogenesis or pseudogamy on the genetic structure of the population depend on the specific mechanism that produces the unreduced egg (Section 5.2.3.3).

As in other forms of maternal inheritance (Section 5.2.3.1), *paternal leakage* can occur in some gynogenic lines. This originates from the occasional incorporation of paternal genome elements into the genome of the gynogenic progeny.

5.2.3.5 HYBRIDOGENESIS

Hybridogenesis (Section 3.6.5) can be defined as a *hemiclonal* form of sperm-dependent parthenogenesis, halfway between amphigonic and uniparental sexual reproduction. In its simplest form (Figure 5.27), hybridogenic females with chromosome complement *B*/*A* (where *B* and *A* refer to the two parental haploid genomes) produce haploid eggs whose genome consists of only the set of maternal chromosomes (*B*), without these having recombined with the paternal chromosome homologues (*A*). The female gametes are therefore partial clones (*hemiclones*) of the maternal genome. Through fertilization by a male of the amphigonic species A, the diploid condition with the typical *B*/*A* chromosome complement is restored in the zygote. Thus, in each individual, the genome of maternal origin is expressed together with that of paternal origin, yet the latter is not transmitted (it is not inheritable) and is replaced

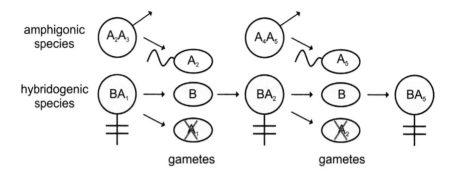

amphigonic species

hybridogenic species

gametes   gametes

Figure 5.27 Schematic representation of the hereditary transmission of nuclear DNA in reproduction by hybridogenesis. In the mode shown, hybridogenic females with *B*/*A* chromosome complement produce haploid eggs whose genome consists of only the set of chromosomes of maternal origin (*B*), which have not recombined with the paternal chromosome homologues (*A*). In each hybridogenic individual the genome of maternal origin is expressed together with that of paternal origin, but the latter is not transmitted to the offspring. Symbol B indicates a chromosome of the hybridogenic species that is transmitted in a clonal manner, while A_1–A_5 represent the homologous chromosomes in the amphigonic species.

in each generation by that of a new male. It could be said that through hybridogenesis a daughter actually manipulates the chromosomes of her father in such a way as to prevent him from becoming the genetic grandfather of her offspring (Avise 2008).

During female gametogenesis, the paternal chromosomes are expunged from the egg cell through a meiosis that proceeds in a particular way and, as a rule, does not allow recombination. Because independent assortment of chromosomes and chromosomal recombination cannot take place, syngamy remains the only source of genetic variation (in addition, of course, to mutation). Alternatively (e.g. in hybridogenic green frogs), paternal chromosomes may be lost before meiosis. The haploid cell then doubles its chromosome complement through an endomitosis followed by a normal meiosis, whose recombination effects are actually frustrated by the perfect identity between the pairing chromosomes. In any case, a hybrid female passes to the next generation the unaltered genome received from her mother, so that, in turn, her offspring will be hemiclones of herself.

As in the case of gynogenesis (Section 5.2.3.4), paternal leakage occurs in some hybridogenic lines, by the infiltration of paternal DNA into the hemiclonal system through sporadic recombination events during oogenesis.

In some species (e.g. the North American salamanders of the genus *Ambystoma*; Figure 5.28) there are very complex genetic systems that cannot be described as strictly gynogenic or hybridogenic. These, for instance, may involve multiple sperm-donor species, whose contribution may or may not be incorporated into the genome transmitted to the offspring, with possible

Figure 5.28 In some species of salamander of the genus *Ambystoma* (here, *A. laterale*) complex genetic systems are present, called kleptogenetic systems. These can involve several sperm-donor species and several recipient ones. The latter can incorporate and transmit to a varying extent the genetic material of the donors.

variations in the ploidy level of the latter. Some authors have suggested the general term **kleptogenesis** to indicate this suite of reproductive modes, where females 'use' very flexibly the sperm 'stolen' from males of related and sympatric amphigonic species (Bogart *et al.* 2007).

5.2.3.6 PATERNAL GENOME LOSS

In some species, despite apparent amphigonic reproduction, the males actually do not transmit to the offspring the chromosomes they have inherited from their father (Figure 5.29). This phenomenon is called **paternal genome loss** (PGL, Section 6.1.3). The chromosomes of paternal origin can be eliminated during spermatogenesis (in this case, only the male gametes lack them, e.g. in some scale insects), or even in the whole early male embryo (in this case the paternal genes can be expressed in the males only in the very early stages of development, or are not expressed at all, e.g. in some mites) (Beukeboom and Vrijenhoek 1998).

5.2.3.7 ANDROGENESIS

Androgenesis is a rare form of reproduction in which the offspring, typically diploid, carry the nuclear DNA of the male parent only. This can occur through different cytogenetic mechanisms, with different effects on transmission genetics (McKone and Halpern 2003; Schwander and Oldroyd 2016). For instance, two male pronuclei can meet and merge in the cytoplasm of an egg in which the maternal genome has degenerated or has been lost (e.g. in some stick insects of the genus *Bacillus*; Mantovani *et al.* 1999; Figure 5.23), producing offspring of both sexes. In this case, the effects on the genetic variation of the nuclear genome will be similar to those of self-fertilization.

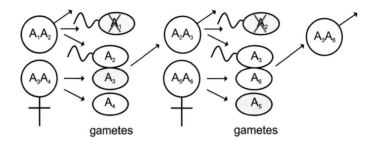

Figure 5.29 Schematic representation of the hereditary transmission of nuclear DNA in reproduction with paternal genome loss. The male produces sperm that contain only the set of chromosomes of maternal origin. Symbols A_1–A_6 indicate homologous chromosomes (for simplicity, female chromosomes are depicted as non-recombinant as well).

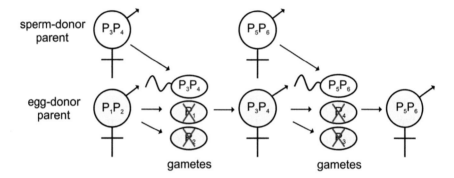

Figure 5.30 Schematic representation of the hereditary transmission of nuclear DNA in reproduction by androgenesis. In the mode shown here (clonal androgenesis), a hermaphrodite produces unreduced sperm cells that at fertilization replace the genome of the egg cells of the partner. The nuclear genome of the offspring is an identical copy of the nuclear genome of the sperm donor parent. Symbols P_1–P_6 indicate homologous chromosomes of paternal origin (for simplicity, female chromosomes are depicted as non-recombinant as well).

In contrast, in the case of so-called *clonal androgenesis* (or *ameiotic androgenesis*; Figure 5.30) the unreduced (2n) genome of a sperm can replace the haploid genome of the egg cell (e.g. in *Cupressus dupreziana*; Pichot *et al.* 2001; Figure 5.31). In this case, the nuclear genome of the offspring will be an identical copy of the genome of the sperm-donor parent, and in separate-sex species with chromosomal sex determination (e.g. some hybrid taxa of the stick insect *Bacillus*; Tinti and Scali 1995), the offspring will be all males.

5.2.4 Sexual Leakage

When rare or irregular episodes of amphigonic reproduction are intercalated with asexual or uniparental sexual reproduction, the effects on the genetic structure of the population are described as those of **sexual leakage** (more exactly, an 'amphigonic leakage'). The genetic reshuffling produced by

Figure 5.31 The Sahara cypress (*Cupressus dupreziana*) can reproduce by androgenesis.

amphigony, even if episodic, can have effects of variable duration on the genetic variation of the population, which depend on the form of uniparental reproduction that amphigony punctuates. Under certain conditions, these effects may show a certain persistence, when the rare episodes of genetic exchange occur with a frequency higher than the average time necessary for the disappearance of their effects on genetic variation owing to uniparental reproduction.

The intermittent recurrence of amphigony in a population that routinely reproduces through a form of uniparental reproduction greatly complicates the mathematical models of genetic transmission and genetic structure of populations. This introduces a further element of complexity to one of the most controversial questions about the evolution of reproductive strategies, that concerning 'evolutionary advantages and disadvantages', for example in terms of fitness associated with the different modes of reproduction. Comparison of strategies based on amphigonic and uniparental reproduction is complicated by the existence of the spurious category of the strategies in which cross-fertilization is only occasional (*mixed mating systems*; Section 3.3.2.2). In this specific case, the core of the question can be summarized by the title of a famous article that discusses the value of occasional forms of genetic exchange: *Is a little bit of sex as good as a lot?* (Green and Noakes 1995). Until further experimental data become available, the costs, benefits and evolutionary stability of occasional genetic exchange remain open questions (D'Souza and Michiels 2010).

5.2.5 Special Cases of Sexual Processes in Eukaryotes

Even when we consider only their cytogenetic aspects, the phenomena of sexual reproduction, and of sex in general, are very diverse. Any attempt to generalize clashes at some point with the existence in nature of situations that do not easily fit into the categories we have devised to classify the diversity of observed processes. Here we report on some cases that do not easily fit into the adopted scheme.

5.2.5.1 CILIATE CONJUGATION

In ciliates, as described in Section 1.5, sex occurs through a peculiar process called **conjugation**. From two genetically dissimilar individuals come two individuals genetically different from those that conjugated, but identical to each other.

These protists have two kinds of nuclei, usually a single copy each: a large hyperpolyploid nucleus, called a *macronucleus*, and a diploid nucleus, the *micronucleus*. The macronucleus, also called a *somatic nucleus*, is involved in protein synthesis during the entire life of an individual, while the micronucleus, also called a *germinal nucleus*, has a function only in sexual exchange. In a typical sexual event (Figure 7.2), two individuals (*conjugants*) unite in the region of their respective cytostomes and their micronuclei undergo meiosis. Three of the haploid nuclei that derive from meiosis in each conjugant degenerate, while the fourth divides once more by mitosis. Of the two nuclei thus formed in each cell, one remains in the individual which produced it (*stationary nucleus*), while the other (*migrant nucleus*) moves into the other cell and merges with the stationary nucleus of the recipient cell. There is therefore a mutual fertilization that restores the diploid condition in the micronuclei of the two cells, which soon separate, and from this moment are called *exconjugants*. The macronucleus present before conjugation degenerates, and a new macronucleus is produced starting from a copy of the new micronucleus. Although this process does not directly affect transmission genetics, it is nonetheless important from the point of view of developmental genetics. So let's have a closer look.

In *Tetrahymena thermophila* (Eisen *et al.* 2006) the diploid micronucleus (MIC) contains 5 pairs of chromosomes, while the hyperpolyploid macronucleus (MAC) contains about 225 microchromosomes, most of them at a ploidy level of about 45 copies. The MAC genome is derived from the genome of the

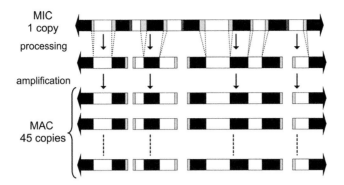

Figure 5.32 Derivation of the genome of the macronucleus (MAC) from a copy of the diploid micronucleus genome (MIC) in the ciliate *Tetrahymena thermophila*. The upper bar represents a portion of one of the five pairs of MIC chromosomes. Sequences that will be found in the MAC are shown in white and black, those that will be eliminated (IESs) in blue, and those of the fragmentation sites (Cbs) in red. Telomeres (green bars) are added at the extremities of the DNA segments (microchromosomes) obtained by fragmentation, after which most of the microchromosomes are replicated about 45 times in an amplification process that produces the highly polyploid genome of the MAC.

MIC, but the two genomes are not identical. When the MAC is produced, a series of programmed DNA rearrangements take place (Figure 5.32). First, from a copy of the MIC, about 6000 specific sequences, called IESs (*internally eliminated sequences*) are eliminated, which reduces the haploid genome of MAC to about 85% of that of MIC. Since repetitive sequences are eliminated in this operation, more than 90% of repeated MIC sequences are not found in the MAC. Subsequently, the chromosomes thus reduced are fragmented at specific nucleotide sites, identified by sequences of 15 bp called Cbs (*chromosome breakage sequence*). During fragmentation, the Cbs and the flanking sequences on both sides, each about 30 base pairs long, are eliminated, while telomeric sequences are added to the new ends. Finally, the microchromosomes thus obtained, ranging in length from a few hundred to a few thousand kilobases, are amplified through successive DNA replications. The microchromosomes containing the sequences encoding ribosomal RNA are exceptionally small (21 kbp) and are replicated until about 9000 copies are obtained, while most of the other microchromosomes are represented in about 45 copies, as mentioned above.

The macronucleus cannot undergo mitosis, and in asexual reproduction it divides by *amitosis* (Section 5.1.3.3).

5.2.5.2 PARASEXUAL CYCLE IN FUNGI

Parasexuality is a peculiar process of some fungi and unicellular eukaryotes that allows the formation of new allelic combinations regardless of whether the organism can reproduce sexually and undergo meiosis. Described for the first time in the mould *Aspergillus nidulans*, parasexuality is based on a cyto-genetic mechanism that involves the fusion of haploid nuclei, mitotic recombination and a subsequent random loss of chromosomes.

Taking a filamentous ascomycete as a model, the fusion of two haploid cells to form a heterokaryon can occur spontaneously in a process called *hyphal anastomosis*. In many fungi this process takes place only between pairs of specific strains, which are said to be *vegetatively compatible*.

Two haploid nuclei in a heterokaryon can merge (karyogamy) to form a diploid nucleus. This is a relatively rare event in vegetative hyphae, and the newly formed diploid nucleus is unstable, because the loss of one of the two chromosomes of each pair is not deleterious, thus escaping purifying selection. During cell proliferation, events of non-disjunction in mitosis progressively lead to the re-establishment of the original haploid condition (*haploidization*) through a series of intermediate steps where the products of mitosis are *aneuploid* nuclei, i.e. with an unbalanced chromosome complement (some chromosomes in a single copy, others in two copies). The aneuploid nuclei are themselves particularly unstable, thus accelerating the process of haploidization.

The chromosomes present in the haploid nuclei that result from this process will represent a random selection of the chromosomes in the two original haploid nuclei. The random loss of chromosomes has therefore the effect of an *interchromosomal recombination*, which in meiosis is instead carried out by the independent assortment of the homologous chromosomes. Moreover, mitotic crossing over can occur between the chromosomes present in pairs, which, as in the analogous meiotic process, can produce *intrachromosomal recombination*. When this occurs, the chromosomes in the final haploid nucleus will be composed of a random combination of chromosome segments of the two original haploid nuclei (Figure 5.33).

In summary, as in the case of ordinary eukaryote sexual processes, para-sexuality makes genome recombination possible, with the production of new genotypes, although the mechanism by which the haploid condition is restored is not meiosis but a series of mitoses.

5.2.5.3 CHIMERISM

In biology, a **chimera** is a multicellular individual made up of different cell lines, each originating from the development of a distinct zygote. At least in principle, *chimerism* is therefore a phenomenon clearly distinct from *mosaicism* (Section 5.1.1.3), in which the genomic diversity between the cells of the same individual has its origin in mutations starting from a single founder cell (although in botany the use of the two terms is anything but rigorous and very often, even in the specialist literature, what are called chimeras are

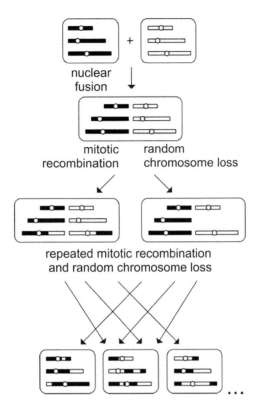

Figure 5.33 Schematic representation of parasexuality in filamentous fungi. Parasexuality is based on a cytogenetic mechanism of genetic reassortment that begins with the fusion of two haploid nuclei and proceeds through repeated cycles of mitotic recombination and random loss of chromosomes. Parasexuality allows recombination between different genomes with production of new genotypes, in the absence of meiosis. White and black in the chromosomes (or parts thereof) indicate their origin from distinct parental nuclei.

actually mosaics). Chimerism has been documented among the representatives of some groups of colonial marine organisms (poriferans, cnidarians, bryozoans and tunicates), but also in the sea cucumber *Cucumaria frondosa* (Gianasi *et al.* 2018), as well as in various species of algae and fungi, and in some mammals (Section 1.4.2).

With regard to transmission genetics, in chimeric organisms situations occur that do not have any equivalent in normal amphigonic reproduction. For example, we have seen (Section 1.4.2) that in marmosets and tamarins a high percentage of pregnancies can result in chimeric littermates, because many cells can be exchanged between siblings during early embryogenesis. In *Callithrix kuhlii* (Figure 5.34) chimerism has also been observed in germline tissues (Ross *et al.* 2007). As a consequence, an individual may not be the genetic parent of its offspring. Its gametes may in fact have derived from its sibling's zygote, so that an offspring of that individual could be genetically

Figure 5.34 In the New World monkeys of the genus *Callithrix*, chimeric individuals may originate from exchange of cells between embryos of the same litter.

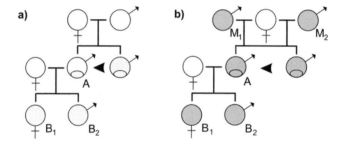

Figure 5.35 Possible effects of chimerism on hereditary transmission in *Callithrix*. (a) Chimerism involves germline cells (oval inside the individual's sex symbol), so that A (a chimeric individual with some germ cells from his brother) is not the genetic father of his children B_1 and B_2. (b) A female mated with two males (M_1 and M_2) and the resulting twins have different fathers, so that M_1 is not the genetic grandfather of his grandchildren B_1 and B_2, even though he is the genetic father of their father A (a chimeric individual). Colours highlight genetic inheritance. Arrowheads indicate cell exchange between brothers.

part of its sibling's progeny (Figure 5.35a). Also, if a female mates with more than one male, the offspring could have different fathers, so that one of these males could not become the genetic grandfather of his grandchildren, even if he were the genetic father of his children (Figure 5.35b). These forms of genetic promiscuity in parent–offspring relationships perhaps lie at the origin of the peculiar and highly cooperative parental-care system in this species.

Chapter 6: Determination of Sex and Mating Type

Acquiring the phenotypic characters specific to a given sex, during development or at some other point during the life cycle of an organism, is usually a complex process. Although the sex of an individual is conventionally defined on the basis of the type of gametes, either eggs or sperm, that it is able to produce (see Section 3.2.1), the phenotype of each sex is generally composed of a multitude of characters. Each of these characters can present a certain degree of independence from other sexual traits in the same organism, be subject to different developmental controls, and show different degrees of sensitivity to the environment. Sexual differentiation is therefore not limited to the development of characteristic reproductive organs and the production of a given kind of gametes, but also extends to the development of the so-called *secondary sexual characters*, morphological, physiological and behavioural, or combinations of these.

Even in eukaryotic species that do not reproduce sexually by means of anisogametes, individuals are generally characterized by belonging to a specific mating type, which determines the compatibility in cross-fertilization with other individuals of the same species (Section 3.5.4). Unlike sexual differentiation, in most instances the differentiation of the mating type simply consists of the synthesis of specific molecules that are exposed on the outer surface of the cell membrane and fix the limits of reproductive compatibility with other members of the same species.

Despite the greater complexity of sexual differentiation, compared to mating type, in most separate-sex species the alternative states of the different sexual characters are closely correlated. In other words, each individual generally presents them all (or almost all) in the version that is typical of what turns out to be its sex, although different sex conditions, which include examples of intersexuality, have been described for many taxa. Thus, even in cases where the development of sexual characteristics is an extremely complex process, sometimes extending over a long segment of developmental time, it is usually possible to identify a key factor that is responsible for the 'developmental decision' of taking one or other of two alternative developmental options, i.e. male or female. For this reason, **sex-determination systems** are traditionally classified based on the nature of the primary causative agent in the specification of an individual's sex. We speak of *genetic sex determination* (*GSD*) when sex is established early in development by genetic factors such as the presence of certain chromosomes, genes or alleles. In contrast, we speak of *environmental sex determination* (*ESD*) when the sex of an individual is established by the values of some environmental parameter such as temperature, which represents signals interpreted by the individual during its development. A third category, which to some extent cuts across the other two, is *maternal sex determination* (*MSD*), in which the sex of the offspring is determined by the developmental environment defined by either the genotype or a physiological condition of the mother. Finally, many organisms have *mixed sex-determination systems*, where genetic and environmental factors are combined to different degrees.

These first sex-determining signals, whether genetic or environmental, can be associated with several different **mechanisms of sex determination**. These are developmental processes that interpret those first signals, and, as we shall see, they can be extremely diverse. This is the case, for instance, with the XY chromosomal system, where the sex of an individual can be determined by a mechanism that depends on the presence of the Y chromosome or, alternatively, by a mechanism that is sensitive to the number of X chromosomes.

Mechanisms of sex determination gradually blend into what we might describe as processes of **sexual differentiation**, which more properly concern the biology of development, rather than the biology of reproduction. This particular aspect of development, although obviously related to the primary determination of sex, nevertheless is to some extent independent of it.

The sex-determination system, the mechanism that implements it and the sexual differentiation that follows may differ greatly even among closely related species, or even within one species. For instance, intraspecific variation in the sex-determination system has been described for the house-fly (*Musca domestica*) and a small rodent, the lemming *Myopus schisticolor* (Marin and Baker 1998). As one can appreciate from the phylogenetic distribution of sex-determination systems in plants and animals (Figure 6.1), sex determination and sexual differentiation are very labile in evolution (for a recent comprehensive review of vertebrate sex determination in a phylogenetic context, see Pennel *et al.* 2018). For plants in particular, Pannell (2017), in consideration of the fundamental modularity of plant development, proposes a relaxation of the distinction often made between genetic and non-genetic sex-determination systems. Different parts of the same plant, at different times, might be channelled towards being male versus female by means of mechanisms where genetic and non-genetic causes cannot actually be disentangled.

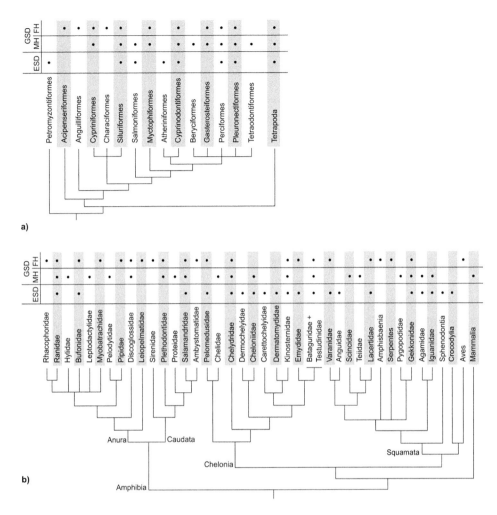

Figure 6.1 a,b Phylogenetic distribution of sex-determination systems (a) in teleosts at the level of orders and (b) in tetrapods at the level of families. Taxa homogeneous for sex-determination system are collapsed at higher rank level (e.g. mammals). Systems are classified as environmental sex determination (*ESD*) and chromosomal sex determination (*GSD*) with male (*MH*) or female heterogamety (*FH*). Only the most common systems for each terminal taxon are shown. (Data from Kraak and Pen 2002)

Therefore, the categories that we use in the following account should not be interpreted too rigidly. On the one hand, evolution produces systems and mechanisms of sex determination that do not necessarily conform to the boundaries identified by these categories. On the other, many of these systems and mechanisms occur in a large number of minor variants that cannot all be adequately treated in these pages.

One last comment. The question of sex determination applies to sequential hermaphrodites, but it does not apply to simultaneous hermaphrodites.

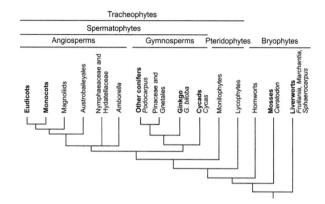

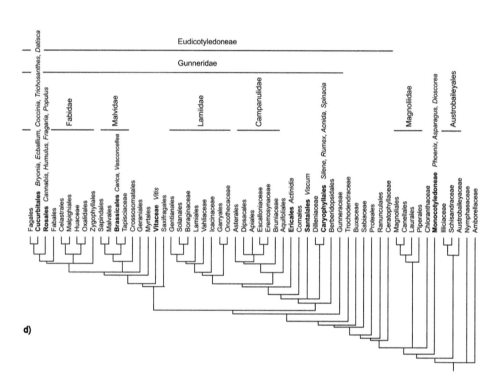

Figure 6.1 c,d Phylogenetic distribution of chromosomal sex-determination systems (c) in embryophytes and (d) in angiosperms. Taxa with names in bold include dioecious species with heterochromosomes. (Data from Ming *et al.* 2011)

However, in species with a *mixed breeding system* (i.e. with unisexual individuals, male and/or female, and hermaphrodite individuals, Section 3.3.3) we find *sex-condition determination systems*, male vs. hermaphrodite (in androdioecious species), or female vs. hermaphrodite (in gynodioecious species), or male vs. female vs. hermaphrodite (in trioecious species). To a large extent, these systems are based on the same principles as sex determination with male vs. female options. To avoid unnecessary complication, in much of the following treatment, unless otherwise specified, sex determination has to be understood in a broad sense, as determination of the sexual condition, which in some cases may also apply to hermaphrodites.

6.1 Genetic Sex-Determination Systems

In **genetic sex determination** (also called **genotypic sex determination**), a gene, a complex of genes or the entire chromosomal complement is responsible for initiating a cascade of developmental events that produce the phenotypic characteristics associated with each sex. Very often, genes involved in the selection of this developmental option are located on a single pair of homologous chromosomes (*sex chromosomes*), which occur in two distinct versions, characterized by a different set of genes or by a different allelic constitution at homologous loci.

The genetic determination of sex has evolved many times independently in animals, generally starting from gonochoric ancestors with environmental sex determination (Section 6.2), and for extant species of animals it represents the most common mode of sex determination (Beukeboom and Perrin 2014). In plants as well, where the dioecious (gonochoric) condition is found in the sporophyte of quite a small percentage of species (10% of land plants, 6% of angiosperms), different genetic systems of sex determination have evolved independently several times (more than 100 in the angiosperms alone). However, in the case of plants, these systems evolved from ancestors with monoecious (hermaphrodite) sporophytes (Ming *et al.* 2011).

6.1.1 Chromosomal Sex-Determination Systems

In **chromosomal sex-determination systems**, males, females and possibly hermaphrodites (in the case of androdioecy, gynodioecy or trioecy: see Section 3.3.3) present a different chromosomal complement, or *karyotype*. In these systems there are one or more **sex chromosomes**, or **heterochromosomes**, which, unlike normal chromosomes (*autosomes*), are present in unequal combination in the two sexes. In species where chromosomes determine the sex condition in the diploid phase, the sex presenting different or unbalanced heterochromosomes (e.g. just one rather than a pair, as in the other sex) is called the **heterogametic sex**, because it produces different types of gametes, while the other sex is said to be **homogametic**. We can have *male heterogamety* or *female heterogamety*, depending on the species, and the gametes produced by the heterogametic sex are those that

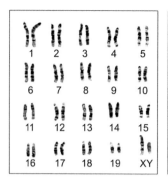
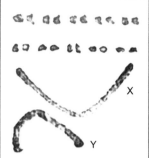

Figure 6.2 Sex chromosomes. Left, male karyogram of a mouse (*Mus musculus*) with 19 pairs of autosomes and X and Y sex chromosomes. Right, male karyogram of the chrysomelid beetle *Alagoasa bicolor*, with a pair of giant sex chromosomes, a distinctive characteristic of some species of this family.

determine the sex condition of the zygote originating from the fusion with the gamete of the opposite sex. In species where chromosomes determine the sex condition in the haploid phase, males and females have distinct sex chromosomes, usually in single copy.

When the different sex chromosomes of an organism are morphologically indistinguishable from each other, they are said to be *homomorphic*, while they are said to be *heteromorphic* in the opposite case. In the most common diploid chromosomal systems, such as XY and ZW (see Section 6.1.1.1), the chromosome that is present in two copies in the homogametic sex (X or Z) is generally of a size comparable to an average autosome, whereas the one found exclusively in the heterogametic sex (Y or W) is generally smaller, but not always (Figure 6.2).

Also, in the case of heteromorphic heterochromosomes, which have a significantly different genetic make-up, the two sex chromosomes (e.g. X and Y) have more or less extensive homologous regions, which allow their pairing at meiosis, so that in a normal meiosis each gamete (or spore) receives exclusively one or other of the two.

During the life cycle, chromosomal sex determination usually takes place during a change in the chromosomal constitution that can result in two alternative configurations, each specific to one or other of the two sexes. In most species with chromosomal sex determination in the diploid phase (diplontic or haplodiplontic), sex is established at syngamy, as a direct consequence of the chromosomal constitution of the participating gametes (see below in this section for some exceptions). In most species with chromosomal sex determination in the haploid phase (haplontic or haplodiplontic), however, sex is established at meiosis, as a consequence of the segregation of sex chromosomes (Figure 6.3 and, in greater detail, Figure 6.4).

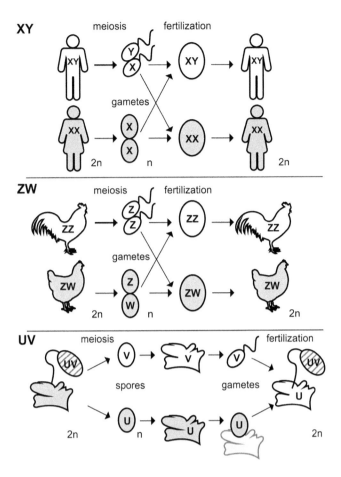

Figure 6.3 Schematic representation of the main chromosomal sex-determination systems: XY, ZW and UV. White, males; grey, females; striped, sexually indeterminate diploid phase of the UV system. In both the XY and ZW systems, sex determination occurs in the diploid phase, in the UV system in the haploid phase. In the XY system the heterogametic sex is the male, in the ZW system the female.

Sex chromosomes are found in a very large number of species of animals and plants, although in the latter they have evolved many times independently and quite recently, compared to animals (Charlesworth 2002, Abbott *et al.* 2017). Hermaphrodite species do not have sex chromosomes, even if vestiges of heterochromosomes can be found in cases where hermaphroditism has recently evolved from the gonochoric condition, as in certain isopods.

Diplontic species with chromosomal sex determination are presumed to be unable to control offspring sex ratio because of Mendelian segregation. However, there is now increasing evidence that species with sex chromosomes can strategically adjust offspring sex ratio, although the underlying mechanism is general not well understood. In a few species of social spiders of the genus *Stegodyphus*, the heterogametic sex (the male) can bias gamete

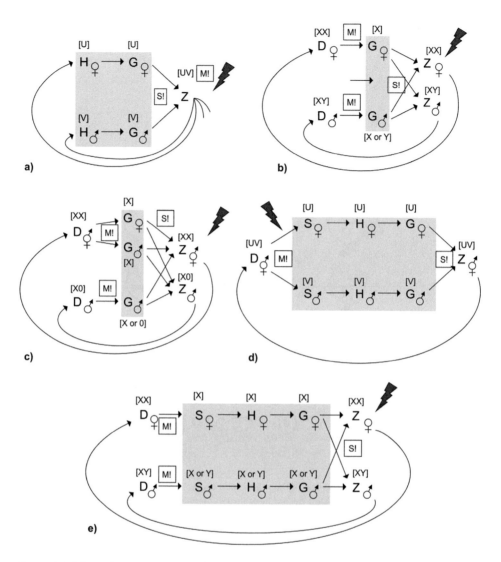

Figure 6.4 Schematic representation of the most common systems of chromosomal sex determination in haplontic, diplontic and haplodiplontic life cycles. In square brackets, sex-chromosome karyotype for the simple systems XY, X0 and UV. H, individual of the haploid phase (e.g. gametophyte); D, individual of the diploid phase (e.g. sporophyte); G, gamete; S, (meio)spore; Z, zygote; M!, meiosis; S!, syngamy. The lightning-bolt symbol indicates the moment at which chromosomal sex determination occurs. Haploid phases (n) on coloured background.

(a) Haplontic species with anisogametes, sex determination at meiosis, zygote sexually indeterminate. Example: the green alga *Volvox carteri*.
(b) Diplontic gonochoric species with sex determination at syngamy, zygote male or female. Example: *Homo sapiens*.
(c) Diplontic androdioecious species with sex determination at syngamy, zygote male or hermaphrodite. Example: the nematode *Caenorhabditis elegans*.
(d) Haplodiplontic species with monoecious (hermaphrodite) sporophyte and unisexual gametophyte, sex determination at meiosis, zygote monoecious. Example: the liverwort *Marchantia polymorpha*.
(e) Haplodiplontic species with unisexual sporophyte and gametophyte, sex determination at syngamy, zygote male or female. Example: the willow *Salix alba*.

production by making significantly more X-carrying (female-determining) sperm (Vanthournout *et al.* 2016).

Female heterogamety is less common than male heterogamety in both animals and plants. Among the animals it is found in lepidopterans, thysanopterans, birds and some representatives of dipterans, crustaceans (*Artemia salina*, some copepods and isopods), fishes, amphibians and snakes. In the fish *Poecilia sphenops* some breeds have male heterogamety, others female heterogamety (Volff and Schartl 2001). In land plants female heterogamety is found in 10% of cases that have been studied, for example in *Ginkgo biloba* and the poplar *Populus trichocarpa*, as well as in *Fragaria elatior* and *Potentilla fruticosa* (both belonging to the Rosaceae).

Sex chromosomes are often heteromorphic, but among the vertebrates, for instance, heterochromosomes are poorly differentiated in fishes and are homomorphic in some amphibians and boid snakes. Heteromorphic sex chromosomes are found in all the main evolutionary lineages of land plants, with the exclusion of hornworts (traditionally classified with the bryophytes), lycopods and ferns, while homomorphic sex chromosomes are known only for gymnosperms and angiosperms (Ming *et al.* 2011).

The main chromosomal sex-determination systems in the diploid phase are the XY male heterogamety systems and the ZW female heterogamety system, including the derived forms X0 and Z0, respectively, where a sex chromosome specific to the heterogametic sex is missing. The UV system is instead the most common chromosomal sex-determination system in the haploid stage. In addition to these 'simple' systems, there is a large number of more complex systems, often rare or even known for only a single species, such as certain systems with multiple sex chromosomes. Let's examine the chromosomal sex-determination systems in more detail.

6.1.1.1 XY AND ZW SYSTEMS

In the **XY** and **ZW systems** (also referred to as **XX/XY** and **ZW/ZZ**) a pair of chromosomes of the diploid complement occurs in two alternative versions, distinguishable for the genes they contain and often also for their size and shape (Figure 6.5). In the case of male heterogamety, females have two sex chromosomes of the same type, XX, while males have two different chromosomes, XY (Figure 6.4b,e). In contrast, in the case of female heterogamety, males have two sex chromosomes of the same type, ZZ, the females two sex chromosomes of different types, ZW. In the heterogametic sex, the genes that are not common to the two heterochromosomes are present in a single copy, and that sex is therefore said to be *hemizygous* at those specific loci.

In spite of the apparent similarity of the genetic systems based on heterochromosomes, the chromosomal constitution of an individual (or of a cell) is

a)   b)

Figure 6.5 Conventional chromosomal sex-determination systems. (a) Mammals (here, *Giraffa reticulata*) typically have an XY system, (b) birds (here, *Bubo africanus*), a ZW system.

translated into a developmental decision (or, for a cell, differentiation), toward one sex or the other, through genetic mechanisms that can be very diverse. These can be grouped into two main categories, those with a *dominant heterochromosome* and those with *genetic balance*.

In sex-determining mechanisms with a **dominant heterochromosome**, the sex chromosome that is found exclusively in the heterogametic sex (Y or W) carries genes specifically involved in sex determination. In mammals, for instance, the Y chromosome carries a gene that is absent in the X chromosome (*testis-determining factor, Sry*), which induces the development of the gonads as testes, rather than ovaries (Section 6.5). Thus, individuals with an aneuploid chromosome complement XXY (Y present) are male, while X0 aneuploid individuals (Y absent) are female.

In contrast, in mechanisms with **genetic balance** the sex chromosomes determine the sex of an individual without the chromosomes Y or W having specific determinants for the heterogametic sex. For instance, in *Drosophila*, which shares with mammals an XY system, sex is determined by the number of X chromosomes in relation to the ploidy level of the autosomes, so that the ratio between the number of X chromosomes and the number of autosome sets (X:A, Table 6.1) is generally a good predictor of the sex of an individual (Erickson and Quintero 2007; Box 6.1).

Individuals with an X:A $\geq$ 1 ratio are female and those with X:A $\leq$ 0.5 are male, while individuals with an X:A ratio between 0.5 and 1 (e.g. XX associated with a triploid complement of autosomes, X:A = 2/3) are sterile and present both male and female features (*intersexual individuals*, or *intersex*; see Section 6.5). Contrary to the case of mammals, therefore, fruit flies with an unbalanced XXY chromosomal complement (X:A = 1) are female, while

Table 6.1 Chromosomal arrangements of sex chromosomes and autosomes illustrating the chromosomal mechanism of sex determination in *Drosophila melanogaster*. Each A indicates a set of autosomes (from Russell 2010)

Heterochromosomes	Autosomes	X:A ratio	Sexual condition
XX	AA	1.00	female
XY	AA	0.50	male
XXX	AA	1.50	female (sterile)
XXY	AA	1.00	female
XXX	AAAA	0.75	intersex (sterile)
XX	AAA	0.67	intersex (sterile)
X	AA	0.50	male (sterile)

Box 6.1 X:A in *Drosophila*

To precisely establish the cause–effect relationships in a group of related phenomena is generally not a simple task, at least not in biology. Appearances are sometimes deceptive, and it is possible to mistake the association of two phenomena determined by a common cause for a cause–effect relationship between the two. A very instructive case is offered by the study of the X:A ratio in *Drosophila*.

Based on the results of classical experiments done by Calvin Bridges over 80 years ago and confirmed by many subsequent experiments, it was believed (and in many manuals it is still stated) that in *Drosophila* the relationship between the number of X chromosomes and the number of homologous sets of autosomes (the X:A ratio) is the *determining factor* in the specification of sex in this insect. According to this model, every cell of the early embryo (*cellular blastoderm* stage) would 'read' its X:A value by measuring the dose of gene products of the X chromosome compared to those encoded by the autosomes, thereby to appropriately regulate the expression of a key sex-determining gene, *Sex-lethal* (*Sxl*).

However, this model does not fit with what is known at the level of the molecular mechanisms that regulate the expression of the genes involved. Activation of *Sxl* depends on the level of expression of four other genes found on the X chromosome: *Sisa*, *Scute*, *Runt* and *Unpaired*, collectively known as *X-linked signal elements* (*XSE*). Under normal conditions, the expression levels

continues

Box 6.1 (cont)

of the *XSE* transcripts from the two X chromosomes in the female are sufficient to activate *Sxl* (which characterizes the female sex), while the *Sxl* level of expression that is reached starting from the single X chromosome in the male does not allow activation of *Sxl* (which establishes the male sex). This mechanism of regulation, based exclusively on the dosage of products of genes on the X chromosome, would seem independent of the expression levels of autosomal genes. But how to reconcile what is known about the activation of *Sxl* with the observation that an aneuploid XA karyotype (with only one X, as in a normal male) is female, while an XXAAA karyotype (with two Xs, as in a normal female) is intersex?

To address this incongruity, Erickson and Quintero (2007) examined in detail early embryo development in haploid individuals (XA) and partial triploids (XXAAA) of *D. melanogaster* and found that ploidy level does not affect sex directly through the X:A ratio. Instead, sex indirectly depends on the increase (in haploids) or decrease (in triploids) of the number of mitotic cycles that precede the formation of cellular blastoderm. Since sex is established almost autonomously in each nucleus before the cellular blastoderm is formed, the delay in cellularization in haploids allows a greater accumulation of *XSE* transcripts, sufficient for the feminization of all the embryo nuclei. On the contrary, the earlier cellularization in triploids does not allow the *XSE* transcripts to reach a concentration sufficient for the feminization of all the nuclei, and a sexual mosaic, or an intersex individual results.

In conclusion, while the X:A ratio is certainly a *predictor* of sex for many karyotypes, it is not actually the signal that determines the sex of the individual. More exactly, the *instructive signal* is provided by the dose of *XSE* transcripts in the early embryo, while the X:A ratio is only a value correlated to the amount of *XSE* transcripts.

This suggests that we should be cautious in assessing cases of sex determination based on the X:A ratio in other organisms. In the absence of direct experimental data, the hypothesis that this relationship constitutes an effective signal in itself must be accepted at least with reservations.

---

individuals with aneuploid X0 karyotype (X:A = 0.5) are male. The genes on the *Drosophila* Y chromosome therefore have no effect on the determination of sex, although they are necessary for sperm differentiation and therefore for male fertility.

In animals, both types of mechanism are well represented, in both XY and ZW systems, and it is not unusual for closely related species to exhibit different mechanisms. Among the plants with XY systems, there are both mechanisms based on the X:A ratio (e.g. *Rumex acetosella*) and mechanisms based on a dominant Y (e.g. *Silene dioica*).

In many taxa, the sex chromosome found exclusively in the heterogametic sex (Y or W) is a degenerate version of the homologous chromosome (X or Z, respectively). The degeneration of one of the two partners in a pair of sex chromosomes lies at the origin of most cases of heteromorphic heterogamety. Generally, the Y and W chromosomes of plants show a lower degree of degeneration than observed in animals. This is due partly to the relatively recent origin of sex chromosomes in plants, and partly to the fact that during the haploid phase of the life cycle (the gametophyte) the genes of the sex chromosomes are expressed in hemizygous condition and are thus significantly exposed to purifying selection, which limits their degeneration (Charlesworth 2002).

Finally, to stress the extreme evolutionary lability of the sex-determination systems, we mention the case of the Japanese amphibian *Glandirana rugosa* (formerly known as *Rana rugosa*), where different populations have different chromosomal sex-determination systems, either ZW or XY, the latter with either heteromorphic or homomorphic sex chromosomes (Janousek and Mrackova 2010).

6.1.1.2 X0 AND Z0 SYSTEMS

In the **X0** and **Z0 systems** (also referred to as **XX/X0** and **Z0/ZZ**), there is only one kind of heterochromosome, present in single copy in the heterogametic sex and in double copy in the homogametic sex (Figure 6.6). In the case of male heterogamety, females have two chromosomes of the same type, XX, while males have only one sex chromosome (X0). Vice versa, in the case of female heterogamety, males have a ZZ karyotype, while females have an unbalanced Z0 karyotype. In either case, the heterogametic sex presents all

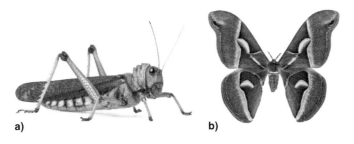

a)   b)

Figure 6.6 Chromosomal sex-determination systems with only one heterochromosome. (a) Many orthopterid insects (here, the blue-winged grasshopper, *Tropidacris collaris*) typically have an X0 system; (b) some lepidopterans (here, the saturniid *Samia cynthia*) have a Z0 system.

the genes of the one heterochromosome in a single copy, and is therefore hemizygous at all the loci on X or Z.

In the nematode *Caenorhabditis elegans*, hermaphrodites are XX and males are X0 (Figure 6.4c). As in *Drosophila*, the sex condition of this worm is determined by the relationship between the number of X chromosomes and the number of pairs of autosomes.

It is believed that the evolutionary origin of X0 and Z0 systems is frequently found in the loss of the Y chromosome in an XY system and, similarly, in the loss of the W chromosome in a ZW system, as a terminal phase in the degeneration of these chromosomes. This would have occurred in many different groups independently (X0: numerous orthopterans and cockroaches; Z0: some lepidopterans). In plants, which have relatively young sex chromosomes, no species with a Z0 system are known (a putative case with an X0 system has not been confirmed by subsequent studies; Janousek and Mrackova 2010).

The inverse evolutionary transition, for instance from X0 to XY with the formation of a new Y, seems to be rarer. A lost Y chromosome can be regenerated when an autosome fuses with the X chromosome and the resulting karyotype is fixed in the population. During meiosis in male gametogenesis, the free homologue of the autosome that has fused with X continues to pair with the homologous sequences now integrated into the new X chromosome (which is referred to as *neo-X*), and at the first meiotic division goes to the opposite pole of the spindle, behaving as a Y chromosome (hence, *neo-Y*). Being subjected to the same selective regime, the neo-Y can face the same process of degeneration that led to the loss of the old Y.

A neo-Y has been described for many orthopterans, stick insects, dipterans and beetles. Similarly, the ZW system that is found in 98% of lepidopteran species has evolved from an ancestral Z0 system (which is found in basal groups of lepidopterans and in trichopterans, the sister group of the Lepidoptera) through the formation of a neo-W in different evolutionary lineages. In some clades of this group, the W chromosome was lost again, thus reverting once more to the Z0 system (Kaiser and Bachtrog 2010).

6.1.1.3 UV SYSTEM

A chromosomal sex-determination system in which the sex is established during the haploid phase of the life cycle is the **UV system** (Figure 6.7). Females (haploid) are characterized by the possession of a sex chromosome U and males (also haploid) by a sex chromosome V. Here sex is not determined at fertilization, but at meiosis (Figure 6.4a,d). This system is typically found in haplontic and haplodiplontic organisms with anisogamety and heterospory, as in some algae and in bryophytes.

Figure 6.7 Chromosomal sex-determination systems in the haploid phase. The haplodiplontic brown alga *Ectocarpus*, with monoecious sporophyte, heterospory and unisexual gametophyte, has a UV system.

Some features of the transmission genetics of this system might not be intuitive, owing to our better familiarity with chromosomal sex-determination systems acting in the diploid phase. For instance, despite the presence of heterochromosomes, in the haploid phase there is no distinction between a homogametic and a heterogametic sex, while the diploid stage, which is hermaphrodite, is invariably heterosporous. In the diploid phase of the life cycle, the heterochromosomes U and V are hemizygous at all loci. Neither chromosome tends to degenerate, because both are subject to purifying selection during the haploid phase. In many bryophytes the U and V chromosomes are heteromorphic, with U generally larger than V (but the opposite is true in liverworts). The UV system allows the accumulation of *sexually antagonist alleles*, i.e. alleles at the same locus that have different relative fitness in the two sexes, or even are beneficial in one sex and detrimental in the other (Bachtrog *et al.* 2011).

6.1.1.4 SYSTEMS WITH MULTIPLE HETEROCHROMOSOMES

Chromosomal sex-determination systems where more than one type of sex chromosome is found in the homogametic sex and/or more than two in the heterogametic sex are called **systems with multiple heterochromosomes** (Figure 6.8). By analogy with the XY and ZW systems, these systems are named using formulae that indicate the set of non-homologous or partially homologous chromosomes found in the heterogametic sex, for instance X_1X_2Y, $X_1X_2X_3Y$ or X_1X_20 (which respectively correspond to the

Figure 6.8 Chromosomal sex-determination systems with multiple heterochromosomes. The echidnas (here, *Tachyglossus aculeatus*) have a system with 5 X and 4 Y.

homogametic karyotypes $X_1X_1X_2X_2$, $X_1X_1X_2X_2X_3X_3$ and $X_1X_1X_2X_2$ in the opposite sex). Most of them originated through chromosomal mutations (translocations) involving sex chromosomes and autosomes. These systems have been studied particularly in animals, where they can be classified on the basis of the mutations from which they originated (White 1973).

Multiple sex chromosomes derived from XY or ZW systems through *central fusions* between the earlier heterochromosomes and autosomes are common in orthopterans, but are also known in coleopterans (Table 6.2), crustaceans, squamates and mammals. In the beetle genus *Cicindela* some species have an X_1X_2Y system, others $X_1X_2X_3Y$. Yet again among the coleopterans, the case of the darkling beetle *Blaps polychresta* stands out, with a system $X_1X_2X_3X_4X_5X_6X_7X_8X_9X_{10}X_{11}X_{12}Y_1Y_2Y_3Y_4Y_5Y_6$ (12 X, 6 Y and 9 pairs of autosomes!), although in this case, as in other examples with many heterochromosomes, the contribution of additional kinds of chromosomal mutation cannot be excluded. The isopod *Jaera marina*, with female heterogamety, has a ZW_1W_2 system. Among the squamates, some lizards of the genus *Anolis* have males with heterochromosomes X_1X_2Y, while in the snake *Bungarus caeruleus*, with female heterogamety, females are Z_1Z_2W and males are $Z_1Z_1Z_2Z_2$. Examples among the mammals are the rat-kangaroo *Potorous tridactylus*, the wallaby *Protemnodon bicolor*, the shrews *Sorex araneus* and *S. gemellus*, the gerbil *Gerbillus gerbillus* and several leaf-nosed bats (Phyllostomidae), all with an XY_1Y_2 system, while in the mouse *Mus minuteides*, the barking deer *Muntiacus muntjak* and several species of mongoose of the genus *Herpestes* the system is X_1X_2Y. The most complex systems in mammals are found among the monotremes: $X_1X_2X_3X_4X_5Y_1Y_2Y_3Y_4Y_5$ in the platypus (*Ornithorhynchus anatinus*)

Table 6.2 Sex-determination systems with multiple sex chromosomes evolved from central fusions in some species of tenebrionid beetles (based on data in White 1973)

Species	Male sex chromosomes	Pairs of autosomes
Blaps lusitanica	X_1X_2Y	8
Blaps lethifera	X_1X_2Y	17
Blaps waltli	$X_1X_2X_3Y$	15
Blaps mortisaga	$X_1X_2X_3Y$	16
Blaps mucronata	$X_1X_2X_3Y$	16
Blaps gigas	$X_1X_2X_3X_4Y$	15
Blaps polychresta	$X_1X_2X_3X_4X_5X_6X_7X_8X_9X_{10}X_{11}X_{12}Y_1Y_2Y_3Y_4Y_5Y_6$	9
Canoblaps nitida	X_1X_2Y	16

and $X_1X_2X_3X_4X_5Y_1Y_2Y_3Y_4$ in the echidnas (genera *Tachyglossus* and *Zaglossus*). It should be noted that the X and Y chromosomes of the monotremes are not homologous to the heterochromosomes of the placentals, suggesting that the sex chromosomes of the latter evolved after the separation of the two evolutionary lineages, about 165 million years ago (Ellegren 2008).

Multiple sex chromosomes resulting from *reciprocal translocation* between the X chromosome and an autosome in an earlier X0 system are very common in mantises, among which the X_1X_2Y system has repeatedly evolved independently in different lineages.

A third category consists of multiple sex-chromosome systems produced by *fragmentation* of the original sex chromosomes, for instance from XY to X_1X_2Y or from XY to XY_1Y_2. These systems are known in insects among heteropterans, homopterans, dermapterans and coleopterans (Table 6.3).

Regardless of the chromosomal mutations that gave rise to them (which are not always known), multiple sex-chromosome systems are also found in other

Table 6.3 Sex-determination systems with multiple sex chromosomes evolved by dissociation (fragmentation) in some species of assassin bugs (Heteroptera Reduviidae) (from White 1973)

Species	Male sex chromosomes	Pairs of autosomes
Acholla multispinosa	$X_1X_2X_3X_4X_5Y$	10
Arilus cristatus	$X_1X_2X_3Y$	11
Harpactor fuscipes	$X_1X_2X_3Y$	12
Sinea confusa	$X_1X_2X_3Y$	12
Sinea rileyi	$X_1X_2X_3X_4X_5Y$	12
Coranus fuscipennis	X_1X_2Y	12
Sycanus collaris	$X_1X_2X_3Y$	12
Fitchia spinulosa	X_1X_2Y	12
Rocconata annulicornis	X_1X_2Y	12
Pselliopus cinctus	$X_1X_2X_3Y$	12

groups, in addition to those already mentioned – for instance among the ostracods (e.g. X_1X_2O, $X_1X_2X_3O$, and from $X_1X_2X_3Y$ to $X_1X_2X_3X_4X_5X_6Y$) and nematodes (different X_nO type systems). Most spiders (85% of the cases studied) have multiple sex chromosomes, typically X_1X_2O (probably the primitive condition), but also $X_1X_2X_3O$. No species of spider has a Y chromosome. More complex systems have been described for the copepod *Diaptomus castor* ($Z_1Z_2Z_3W_1W_2W_3$) and the centipede *Otocryptos sexguttatus* ($X_1X_2X_3X_4Y_1Y_2Y_3Y_4Y_5$).

Systems with multiple chromosomes are also found in plants: in some conifers of the genus *Podocarpus* (X_1X_2Y), in a variety of hop (*Humulus lupulus* var. *cordifolius*, $X_1X_1Y_1Y_2$), in sorrel (*Rumex acetosa*, XY_1Y_2) and in the mistletoe *Viscum fischeri* (female $X_1X_2X_3X_4$, male $Y_1Y_2Y_3Y_4Y_5$). The UV system has also occasionally evolved into a multiple sex-chromosome system, as in the liverwort *Frullania dilatata*, where the female gametophyte has a U_1U_2 karyotype while the male is V (Ming *et al.* 2011).

Finally, the poeciliid fish *Xiphophorus maculatus* has a peculiar chromosomal system with three homomorphic sex chromosomes, X, Y and W: males are XY or YY, females XX, WX or WY. It has been hypothesized that *X. maculatus* is in a phase of transition from an XY to a ZW system (Janousek and Mrackova 2010).

The great variety of chromosomal sex-determination systems escapes rigid classifications. It is therefore not surprising that, regardless of the classification adopted, there is still a group of cases that are placed in a 'miscellaneous other' group that collects all the systems not corresponding to otherwise established classification criteria. The classification adopted here is no exception. In addition to the systems already described, there are others, currently considered rare, sparsely distributed among major taxonomic groups, which are particularly difficult to interpret, possibly in terms of transmission genetics, or for the cytological mechanisms that produce them, or with respect to the evolutionary processes that may have originated them. Let's look at some of them.

In the New Zealand frog *Liopelma hochstetteri* females have a 0W karyotype and males a 00 one. In rodents, many species of voles of the genus *Ellobius* have XX females and XX males (Matveevsky *et al.* 2016), while in *E. lutescens* and in the rat *Tokudaia osumensis* both sexes are X0. In another vole, *Microtus oregoni*, females are X0, males XY. In the female germline of this rodent, during gametogenesis, the XX condition is restored in oogonia owing to a selective lack of disjunction (only the X chromosome) in mitosis, with the production of XX and 00 oogonia, of which only the XX endure to give eggs. The same mis-disjunction occurs in the male germline, with the production of XXY and 0Y cells, of which only the latter differentiate into spermatogonia that will give 0 and Y sperm (Charlesworth and Dempsey 2001). Two species of lemming, *Myopus schisticolor* and *Dicrostonyx torquatus*, have XX and XY females and XY males (Bull and Bulmer 1981).

In most species with chromosomal sex determination in the diploid phase, sex is established at the time of fertilization, but in some cases the sex-specific karyotype of an individual is assembled through the peculiar behaviour of the sex chromosomes during the early stages of embryonic development. For instance, in the fungus gnat *Sciara* (Figure 6.9) all zygotes have the same genotype, with three X chromosomes and two homologous series of autosomes (XXXAA). The loss of one or two paternal X chromosomes determines whether the zygote will develop into a female (XXAA) or a male (XAA) (Sánchez 2008). Similar mechanisms are known for haplodiploid sex determination (Section 6.1.3).

In their evolution from an ancestral autosome, the Y chromosome in the XY systems and the W chromosome in the ZW systems may have undergone considerable degeneration, losing many of the original genes that are still found in the sexual partner chromosome (X or Z, respectively). Therefore, in XY and ZW sex-determination systems many genes on the X or Z chromosome

Figure 6.9 Non-conventional chromosomal sex-determination systems. In the fungus gnat *Sciara*, the male (X0) or female (XX) karyotype is established during the early stages of embryonic development through the loss of a different number of copies of the three X chromosomes present in the zygote. Females maintain two Xs, males only one.

are found in double copy in the homogametic sex, like the genes on the autosomes, whereas they are in single copy in the heterogametic sex. Clearly, in X0 and Z0 systems all the genes on the sex chromosome are affected by genetic imbalance. Since many of these genes, in some cases most of them, require identical levels of expression in the two sexes, because they are not involved in sexual differentiation, sex-determination systems are generally associated with a **dosage-compensation system**, able to balance the level of expression of these genes in the two sexes.

Different systems have evolved in different taxa independently. In the female embryos of eutherian mammals, at the stage of a few hundred cells and in each cell independently, one of the two X chromosomes at random is epigenetically silenced, i.e. made incapable of transcription. Under the microscope, these highly and irreversibly condensed chromosomes assume a typical shape and are known as *Barr bodies*. Each cell line descending from these cells will have only one active X (as in males), so that the females are actually genetic mosaics (Section 5.1.1.3) for the genes on the X chromosome. In marsupials, however, it is the X chromosome of paternal origin that is inactivated in all the cells, owing to a particular epigenetic chromatin marking (Leeb and Wutz 2010). An additional component of mammal dosage-compensation systems is the overexpression of the only active X in the two sexes, thus balancing the level of expression of genes on X with the average level of expression of genes on the autosomes, which are in double copy (Oliver 2007).

In *Drosophila*, dosage compensation is achieved in a way that is opposite to that of mammals. In males there is an augmented transcription (*hypertranscription*) of the genes on the X chromosome, so as to balance the level of expression that in females is achieved by two copies of the same genes. Still another system is found in the nematode *Caenorhabditis*, where dosage compensation is accomplished through a reduction of the transcription level (*hypotranscription*) in both X chromosomes of the hermaphrodite compared to that of the only X chromosome present in the male.

However, dosage compensation is not an imperative among species with a genetic sex-determination system. A very inefficient dosage compensation is characteristic of birds and lepidopterans, where many genes on the Z chromosome are expressed at a significantly higher level in males (ZZ) than in females (ZW). The recent finding that a gene on the Z chromosome, *Dmrt1* (homologous of the sex-differentiation genes *doublesex* in *Drosophila* and *mab-3* in *Caenorhabditis*; see Box 6.2), controls sex determination in birds through a dose effect (Koopman 2009), can perhaps explain why an inclusive dosage-compensation system (at the level of the whole chromosome) has not evolved in birds. However, this does not explain why more selective compensation systems have not evolved for other genes located on the Z chromosome. In this regard, the hypothesis has been recently advanced that, unlike the genes on the X chromosome, many of the genes on chromosome Z are adapted to optimize function and thus fitness of the homogametic sex, in this case the male, so that their reduced expression in females is necessary, rather than problematic, for development (Naurin *et al.* 2010).

Although the XY and ZW systems may appear to be merely reversed with respect to the heterogametic sex, they are actually not equivalent in every respect. Sex chromosomes 'spend' a different fraction of their evolutionary time in the genomes of males and females, owing to their different segregation in the two sexes. For instance, a Z chromosome will be found two out of three times (67%) in a male (ZZ) and one out of three times (33%) in a female (ZW). The Y, Z, X and W chromosomes then spend 100%, 67%, 33% and 0% of their evolutionary time in males, respectively, and complementary percentages of time in females. As sexual selection is usually stronger in males than in females, it follows that chromosome Z (which is found in a male two out of three times) will more easily tend to accumulate sexually antagonistic mutations, i.e. advantageous for one sex and disadvantageous for the other, compared to the X chromosome (which is found in a male one out of three times). Still because of the greater sexual competition among males, which tends to reduce the actual size of the male population, larger effects of genetic drift of the Y chromosome are expected, which therefore tend to degenerate more rapidly than the W chromosome (which is found only in the females).

Box 6.2 *Dmrt* Genes and Sex Development in Animals

Despite the great disparity in the development of sexual traits in animals, recent studies are revealing interesting shared elements at the level of the genetic control of sexual differentiation among all metazoans. We report here the salient features of this system, mainly based on Kopp's (2012) review.

A significant step forward in the exploration of the developmental gene networks involved in sex determination and sexual differentiation in metazoans was the discovery of a family of transcription factors, the **Dmrt genes** (*doublesex/mab-3 related*). Although these genes show considerable disparity in their regulatory domain (through which they interact with the transcription complex of target genes), they share a highly conserved DNA-binding domain called DM. To this family belong the genes *doublesex* (*dsx*) of *Drosophila*, *mab-3* of *Caenorhabditis* and *Dmrt1* and its paralogues of vertebrates. *Dmrt* genes are specifically expressed during the development of the gonads of almost all bilaterians hitherto studied, despite the profound structural and functional differences across the group. The only exception seems to be the nematode *C. elegans*, for which, however, *Dmrt* genes are necessary for the sexual differentiation of extragonadal somatic tissues.

Outside the bilaterians, *Dmrt* gene expression has also been observed in the stony coral (anthozoan) *Acropora millepora*, with a peak that coincides with the season of sexual reproduction. The primary function of *Dmrt* genes in the gonads is to promote the differentiation of male-specific traits and, at the same time, to repress those specific to the female. Thus, to a large extent, in an initially undifferentiated gonad, in arthropods as in vertebrates, *Dmrt* genes promote the development of the testes, while repressing the development of the ovaries.

The involvement of *Dmrt* genes in the development of the gonads in animals as diverse as vertebrates, arthropods and molluscs suggests that they may have played a role as selector genes, between ovaries and testes, in the development of the gonads of the most recent common ancestor of all bilaterians, if not even in an older ancestor. Starting from this ancestral role, changes in the regulation of transcription of *Dmrt* genes may have led to their different role in the gonad developmental network in different taxa. Besides, *Dmrt* genes have been independently co-opted in different evolutionary lineages for the development of other sexually dimorphic organs and structures.

Although the involvement of *Dmrt* genes in the development of metazoan sexual characters is very conserved, independent of the sex-determination system (genetic or environmental, with male or female heterogamety) and the type of sexual differentiation (dominated by autonomous cell behaviour, or

---

Box 6.2 (cont)

depending on the systemic effects of circulating hormones), the genes involved in the signalling systems at the top of the cascade of regulatory events of the same developmental processes are instead very diverse. For instance, the gene *testis-determining factor* (*Sry*), the key factor in male development, present on the Y chromosome of mammals, is not found outside this group. And the gene *Sex-lethal* (*Sxl*), fundamental in sex determination in *Drosophila*, does not play any role in the determination of sex outside the drosophilid dipterans.

Changes at the top of a hierarchy of regulatory interactions between sex-determination genes and sexual-differentiation genes can occur more easily than changes further downstream, at the level of their target genes. Mutations in the latter are more likely to have deleterious effects, due to the multiple phenotypic effects (pleiotropy) that characterize genes that are relatively more downstream. For instance, in the simple case where sex is primarily determined by a single gene at the top of a hierarchy, any gene that takes control of the expression of this gene could easily take the role of a new 'primary' sex-determining gene. Intraspecific variation in the sex-determination system in *Musca domestica* can be explained in the context of these dynamics. Phylogenetic analyses show that genes at the top of the hierarchical systems of sexual-character regulation have been co-opted relatively recently in their role compared to downstream genes (Marin and Baker 1998).

In more complex chromosomal systems, such as those with multiple het-erochromosomes, the problem of dosage compensation (where this exists) becomes obviously more complex. As an example, we only mention the case of the platypus, where the genes of the five X chromosomes show partial and variable dosage compensation, like the genes on the Z chromosome in birds (Deakin *et al.* 2008).

6.1.2 Genic Sex-Determination Systems

In **genic sex-determination systems** (or **multiple-allele sex-determination systems**), there are no sex chromosomes, but males and females have distinct alleles at specific loci, the number of which varies from species to species. This category of systems gradually fades into that of chromosomal sex-determining systems, through the incipient evolutionary phases of sex chromosomes and the most extreme cases of homomorphism between sex chromosomes.

Figure 6.10 Genic (or multiple-allele) sex-determination systems. The green swordtail (*Xiphophorus helleri*) does not possess sex chromosomes and sex is established by the alleles present at different loci distributed on multiple chromosomes. The male (above) carries a long anal fin that is used as a copulatory organ.

The study of these systems has seen a considerable development in recent years, thanks to the application of modern genomics techniques that allow the identification and characterization of the regions of the genome involved in sex determination even in the absence of sex chromosomes. These are not necessarily single or dominated by a single locus. For instance, in a small freshwater fish, the gonochoric poeciliid *Xiphophorus helleri* (Figure 6.10), which does not possess sex chromosomes, sex is genetically determined by a set of factors with masculinizing or feminizing effects, none of which is prevailing, distributed on several chromosomes, and whose collective balance leads to the development of one sex or the other. This is an example of a **polygenic** (or **polyfactorial**) **sex-determination system** (Penman and Piferrer 2008). More frequently, polygenic sex determination, where different factors may also exhibit reciprocal epistatic effects, is associated in the same species with other sex-determination systems (either chromosomal or environmental) in the so-called *mixed sex-determination systems* (Section 6.4). Polygenic determination of sex is considered a transitory state, inherently unstable, in the evolution of genetic sex determination, and, although rare, it is also considered to be the primitive condition in fishes (Penman and Piferrer 2008).

6.1.3 Haplodiploid Sex-Determination System

In hymenopterans, thysanopterans, and some representatives of other animal groups, haploid males develop from unfertilized eggs produced by diploid females through a form of parthenogenesis called *haploid parthenogenesis* or *arrhenotokous parthenogenesis* (Section 3.6.2.2). This reproductive mode, which is not found in plants, is associated with a peculiar system of sex determination, the **haplodiploid sex-determination system**, which relies on a differential ploidy level between the two sexes: females are diploid, males haploid. The number of males in a population therefore depends on the number of unfertilized eggs that are laid. Males produce haploid sperm through a type of spermatogenesis that does not involve a reduction in the number of chromosomes; despite different cytological details, the process actually corresponds to a mitosis (White 1973).

The haplodiploid sex-determination system may seem extremely simple, but this apparent simplicity actually hides a difficulty at the level of the genetic mechanisms that implement it. How is it possible that two copies of the same genome determine the development of the embryo into a female, while a single copy of the same genome determines the development into a male? No gene present in one sex is lacking in the other, and the ratio between the genes involved in sex determination and the other genes is the same in both diploid and haploid conditions.

Unfortunately, almost nothing is known about the genetic basis of haplo-diploid sex determination other than in hymenopterans. To explain the mechanism in this group, in the first half of the last century a model was developed, based on the discovery of diploid males in highly inbred populations of the parasitoid *Habrobracon hebetor* (formerly known as *Bracon hebetor*) (Whiting 1943). In this model, called **complementary sex determination** (CSD), heterozygotes at a multiallelic sex-determination locus develop as females, whereas hemizygotes (and possibly homozygotes resulting from inbreeding) develop as males (Figure 6.11).

Since then, this sex-determination mechanism has been confirmed for more than 60 species of hymenopterans (Asplen *et al.* 2009). In many cases it is

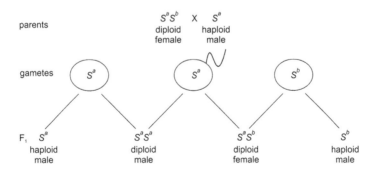

Figure 6.11 Haplodiploid sex-determination mechanism in the wasp *Habrobracon*, according to Whiting's model (1943). At least nine alleles are found at a locus of complementary sex determination (S^a, S^b, S^c, ...). Diploid individuals heterozygous at this locus (e.g. $S^a S^c$ or $S^b S^d$) are females, diploid individuals homozygous for any of these alleles (e.g. $S^a S^a$ or $S^c S^c$) are generally sterile males, and haploid individuals with any allele (e.g. S^a or S^d) are fertile males. The figure shows the case of an $S^a S^b \times S^a$ cross, which is unlikely to occur in a population unless highly inbred.

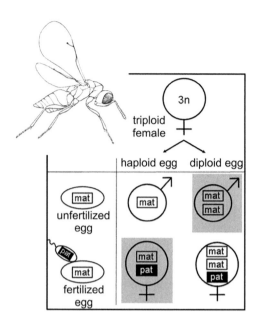

Figure 6.12 Crossing experiment involving a triploid line of the small parasitic wasp *Nasonia* to illustrate the mechanism of haplodiploid sex determination in this insect. The triploid females lay both haploid and diploid eggs. In the absence of fertilization, these will develop into haploid and diploid males, respectively. Fertilized eggs develop instead into diploid and triploid females, respectively. The paternal effect in the determination of sex is evident, since no females develop in the absence of paternal genomic contribution, even under identical ploidy conditions (compare the two cases in grey). pat, paternal chromosome set; mat, maternal chromosome set.

indeed based on a single locus, as in Whiting's original model, while in others it is based on a multi-locus system. In the honey bee (*Apis mellifera*), the gene responsible for complementary sex determination, called *complementary sex determiner* (*csd*), has been cloned and sequenced.

However, the CSD model does not explain sex determination in those hymenopteran species where diploid males are not observed even after prolonged inbreeding, as in cynipids and chalcidoids. Starting from the observation that in the chalcidoid *Nasonia vitripennis* the female sex is determined by the presence of a paternal genome, while the male sex is determined by its absence (Figure 6.12), recent evidence suggests that in this insect one or more loci receive a different genetic imprinting (a different epigenetic modification) in male vs. female germlines. The imprinting must be reversible, because a paternal allele in a generation becomes a maternal allele in the next generation. However, the nature of maternal imprinting is still unknown, and it is also not known to what extent the epigenetic sex-determination mechanism of *Nasonia* is widespread among other species with haploid parthenogenesis (Beukeboom and van de Zande 2010).

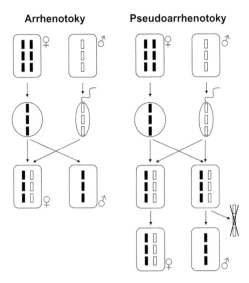

Figure 6.13 Comparison between two haplodiploid sex-determination mechanisms. In arrhenotoky, haploid males develop from unfertilized eggs produced by diploid females through arrhenotokous parthenogenesis. In pseudoarrhenotoky, both males and females develop from fertilized eggs, but the elimination or functional inactivation of the whole set of paternal chromosomes during early embryogenesis results in the development of haploid males, while females will develop if the diploid condition is maintained. Dark and light rectangles indicate maternal and paternal chromosome sets, respectively.

Even for haplodiploid sex-determination systems, cases are known in which the sex-specific karyotype of an individual is obtained through transformations of the chromosome complement during the early stages of embryonic development. In many species of scale insects the elimination or the heterochromatinization of the whole set of paternal chromosomes during early embryonic cleavage results in a functionally haploid blastula, which will develop into a male. On the contrary, if the diploid condition is maintained during development, the embryo will develop into a female (Khosla *et al.* 2006). This genetic mechanism, based on *paternal genome loss* (PGL), also known as **pseudoarrhenotoky** (Figure 6.13), is known for some other taxa, including some mites (Section 3.6.2.2).

A sort of dosage compensation, in this case due not to the imbalance of the sex chromosomes, but to the reduced DNA content of the male nuclei compared to the female ones in a haplodiploid system, is observed in the muscular tissues of the males of most hymenopterans. Here, nuclear DNA undergoes a duplication not followed by the division of the nucleus (*endoreduplication*, or *polytenization*), which re-establishes the DNA content of the diploid condition, probably necessary for the metabolism in these tissues (Aron *et al.* 2005).

6.2 Environmental Sex-Determination Systems

In many species, the sex of an individual is not established by the genetic make-up of its founding cell (e.g. the zygote, or the spore), but is defined through one or more stages of its embryonic or post-embryonic development. The sex that will develop depends on factors external to the organism, such as the temperature to which it is subject during a given developmental phase, or signals coming from conspecific individuals. In all these cases we speak of **environmental sex determination** (or, more rarely, of **metagamic sex determination**, as opposed to the *syngamic*, i.e. genetic, *sex determination*).

In many cases, sex is established irreversibly during a restricted, usually early, phase of development during which the organism is sensitive to a specific environmental signal that can select the developmental option for one sex or another. In other cases, the environmental determination of sex is associated with forms of sequential hermaphroditism, so that the sex of an individual changes, sometimes more than once, in the course of its life.

Until recently, in developmental biology, environmental effects have been greatly underestimated (Gilbert and Epel 2015), acknowledged to have only a *permissive role*: suitable environmental conditions allow normal development, unsuitable conditions forbid it. This is partly due to the fact that, in this regard, the hitherto studied model species in developmental biology are scarcely representative of most biological diversity, precisely because they have been chosen by virtue of their ability to develop in the laboratory in a highly repeatable way (i.e. in a way that is as independent as possible from external factors). However, the environment can have an *instructive role* in development, and in recent times this role in the normal development of various organisms has been increasingly appreciated. Environmental sex determination is a form of **phenotypic plasticity**. This is defined as the correspondence of one genotype to more than one distinct phenotype, the production of which depends on the environmental conditions in which the individual develops (Fusco and Minelli 2010).

6.2.1 Temperature-Dependent Sex Determination

Environmental temperature is among the factors involved in phenomena of phenotypic plasticity, and among the most notable traits that can be influenced by temperature is the sex of an individual, in a system known as **temperature-dependent sex determination** (TSD).

In oviparous amniotes with TSD (tuatara, all crocodiles and many testudinates, lizards and snakes), sex is irreversibly established by the incubation temperature during the middle third of embryogenesis. Three different types

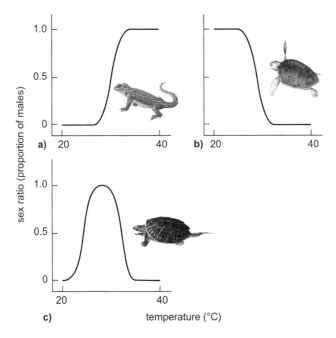

Figure 6.14 Temperature-dependent sex determination: a schematic representation of the relationship between sex ratio and incubation temperature in reptiles. (a) Females develop at low temperatures, males at high temperatures (example: *Sphenodon punctatus*). (b) Males develop at low temperatures, females at high temperatures (example: *Caretta caretta*). (c) Females develop both at low and high temperatures, males at intermediate temperatures (example: *Chelydra serpentina*). In all three cases, non-extreme sex ratios are limited to narrow temperature ranges.

of reaction norm can be identified (Figure 6.14). In some lizards, some crocodiles and in the tuatara (*Sphenodon punctatus*), eggs incubated at low temperatures develop exclusively into females, while eggs incubated at high temperatures all develop into males. The opposite is observed in many turtles, where only males are obtained at low temperatures and only females at high temperatures. In other turtles (e.g. *Chelydra serpentina*) and in other crocodiles, incubation at intermediate temperatures will only give males, while high and low temperatures will only give females. In all cases, there are only narrow temperature ranges within which individuals of both sexes can develop, in different proportions. For instance, in the freshwater turtle *Trachemys scripta* incubation temperatures below 28 °C give only males, temperatures above 31 °C only females and for temperatures between 28 and 31 °C both females and males develop, in direct and inverse proportion to temperature, respectively.

The fact that at intermediate temperatures no hermaphrodite or intersex individuals develop, but only individuals with a well-defined sex, although in variable proportions, suggests that the temperature acts only as a selector switch between the alternative developmental pathways for the two sexes, in particular by inducing one of the two developmental options in gonadal tissues. Recent genetic studies have identified a few candidate temperature-sensitive proteins (e.g. CIRBP and TRPV4) that could be involved in these regulatory events. However the molecular mechanisms of temperature-dependent sex determination have yet to be elucidated (Gilbert and Barresi 2016).

Temperature-dependent sex determination has been described for more than 60 species of fishes, distributed in eight families. In the fishes the norm of reaction of the sex ratio as a function of temperature can assume the same three forms described above for the amniotes, although generally without such a steep transition between the prevalence of one sex and prevalence of the other, and with a more frequent occurrence of the model that sees males developing at higher temperatures than females. In fishes, as in amniotes, TSD operating at early developmental stages is generally irreversible (Penman and Piferrer 2008).

Other cases of temperature-dependent sex determination restricted to small taxa are reported for different metazoan groups, such as the marine lamprey *Petromyzon marinus* and numerous species of biting mosquito (culicid dipterans), among which is *Aedes stimulans* (Cook 2002).

6.2.2 Sex Determination Through Interaction with Conspecifics

The simple spatial proximity of an individual relative to other members of its own species, or the way in which an individual interacts with conspecifics, may constitute a factor capable of determining its sex, reversibly or irreversibly, depending on the taxon. In **sex-determination systems based on interactions with conspecifics** (or **social sex-determination systems**) the signal that evokes the developmental response for one sex or the other can be chemical (i.e. pheromones) or based on other communication channels (e.g. tactile or visual). A chemical signal can directly evoke the developmental response in the recipient individual, whereas other types of signal are generally received through the sensory system of the recipient organism and properly transduced into 'internal' chemical signals through the hormonal system or, in some animals, through the neuroendocrine system.

The sex of *Bonellia viridis*, a marine annelid, is irreversibly established during early development, depending on where the planktonic larva settles to start metamorphosis to the subsequent, and definitive, adult sedentary phase. Larvae that settle on a sea-floor area far from other individuals of the same species develop into females. In contrast, if a larva contacts the 'proboscis'

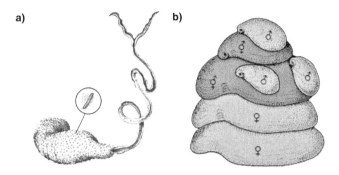

Figure 6.15 Environmental sex determination through interaction with conspecifics. (a) At the larval stage, the sex of the echiurid annelid *Bonellia viridis* is indeterminate. If it metamorphoses on a substrate where no conspecific is present within a short distance, the larva will grow into a female, but if it comes into contact with the body of a mature female it will metamorphose into a male. The adult female can measure in excess of a metre; the male, miniaturized and simplified in anatomy, measures no more than 1–3 mm. It lives as a symbiont in the body of the female, whose eggs it fertilizes. (b) The sex of the larvae of the marine gastropod *Crepidula fornicata* is also indeterminate. These larvae tend to settle on the shell of a metamorphosed individual of the same species, eventually forming a stack of several individuals. Young specimens just after metamorphosis are male, but later they will mature as males or females depending on the composition of the stack and their position within it. The two darker individuals are in a transition phase from male to female.

(prostomium) of an adult female, it begins to develop into a male through the effect of a masculinizing pheromone produced by the female. After a few days the male, which remains small (1–3 mm) compared to the female (up to 1 m, including the prostomium) and in comparison to the latter also has a very simplified body organization, will enter the body of the female, where he will no longer lead an independent life, but will find nourishment and only have to provide for the fertilization of his partner's eggs. From a functional point of view, the male of *Bonellia* is nothing more than a symbiotic producer of sperm (Figure 6.15a).

A different situation is found in the marine gastropod *Crepidula fornicata*, where sex differentiation is not necessarily definitive. In the transition from the planktonic to the sedentary phase, the larvae tend to settle on the shell of an already settled conspecific, forming stacks of individuals of different ages (Section 3.3.2.2). Recently metamorphosed young individuals are male, but this phase is followed by a period of *sexual lability*, accompanied by the degeneration of the male reproductive system. The individual can then become male or female, depending on the composition of the stack. If it fixes on a female it will become a male, whereas if it detaches from the stack it will become a female. In the presence of numerous males in the stack, some of these may become females. In any case, the female condition is not reversible (Figure 6.15b).

Many sequential hermaphrodite fishes, either protandrous or protogynous, can also change sex on the basis of social interactions, which are mediated by the neuroendocrine system. In a small number of species, among which is the goby *Trimma okinawae*, social interactions can induce an individual to change sex several times (*alternating hermaphroditism*). The triggering of the process of sex change is frequently associated with the activity of a stress hormone such as *cortisol*, but further downstream in the process of differentiation, the mediation of the *aromatase* enzyme, which can convert *testosterone* (male sex hormone) into *oestrogen* (female sex hormone; see Section 6.6) can also be found. Changes in the composition of the social group, perceived by the nervous system through the sense organs, can modify the hormonal levels of an individual within a few hours or even minutes. The subsequent phenotypic changes affect a large number of characters, morphological and behavioural alike. Looking at the temporal schedule of the process of sexual transformation, changes in behaviour generally precede those in the gonads.

In the homosporous fern *Ceratopteris richardii* the gametophyte generation is androdioecious, i.e. the same spore can develop into a male gametophyte (with antheridia), or into a monoecious gametophyte (with antheridia and archegonia). Monoecious gametophytes develop from spores in the absence of signals from conspecific gametophytes. These secrete a pheromone (antheridiogen) which induces the development of male gametophytes from the spores exposed to its action. However, the male sex condition of the gametophyte is not definitive, since a decrease in the concentration of the pheromone in the environment can transform a male prothallus into a monoecious prothallus (Juarez and Banks 1998).

6.2.3 Other Environmental Sex-Determination Systems

More rarely, and in only a limited range of taxa, the sex of an organism is established on the basis of other types of environmental signal, very often in combination with other factors, either environmental or genetic. A few examples are discussed here.

6.2.3.1 PHOTOPERIOD

The sex of the amphipod crustacean *Gammarus duebeni* is determined by the photoperiod to which the animal is exposed in an early, sensitive phase of development. The individuals born first in the year develop into males. These, going through a longer growth phase than females, will arrive at the breeding season with a larger average size than the latter. In this species, male reproductive success depends on body size, so this character is under strong sexual selection (McCabe and Dunn 1997).

Figure 6.16 Environmental sex determination dependent on water acidity. The cichlids of the genus *Apistogramma* (here, *A. agassizii*) develop mainly as males in waters with pH lower than 6.2, more frequently as females if the pH is in excess of 7.0.

6.2.3.2 WATER PH

In the South American cichlid fishes of the genus *Apistogramma* (Figure 6.16) and in some poeciliids, water acidity, which varies with atmospheric precipitation, has an effect on sex determination. Individuals developing in acid waters (pH 5.0–6.2) are mainly male, those developing in neutral or slightly basic waters (pH 7.0–7.8) mainly female (Penman and Piferrer 2008).

6.2.3.3 NUTRITION

The adults of the mermithid nematodes are free-living, but their juvenile stages are parasitic on insects. During the juvenile phase, the growth of these worms depends on the available resources represented by the host's tissues. A high density of individuals within a single host and/or a host of small size induces the development of males, the opposite situation the development of females. In *Mermis subnigrescens*, a parasite of locusts, with fewer than 5 individuals per host there is a strongly unbalanced sex ratio with a prevalence of females, with over 15 individuals the sex ratio is decidedly unbalanced in the opposite direction, and for densities of 5–15 individuals per host there are intermediate sex ratios (Bull 1983).

6.2.3.4 BODY SIZE

Although in most dioecious plants sex is determined genetically, cases of environmental determination and instability in the sexual phenotype are also found. For instance, in many plants of the genus *Arisaema* (Araceae), the sex depends on the size of the plant. Small plants have only male flowers, large plants only female flowers, while intermediate-sized plants may have both male and female flowers. This is therefore a case of sex determination in the context of a form of protandrous sequential hermaphroditism with the interposition of a gonochoric phase.

More often, in plants the phenomenon of sex change takes the form of an instability in the suppression of the gynoecium in the flowers of male plants (*subandroecy*, Table 7.1), which thus develop complete flowers and become capable of self-fertilization (Janousek and Mrackova 2010). One form of environmental sex determination, relatively early in development and potentially definitive, has been described in spinach (*Spinacia oleracea*). Plants that germinate from large seeds are more likely to develop as males, whereas those that germinate from small seeds will develop more frequently as females (Freeman *et al.* 1994). However, how and to what extent environmental factors interact with genetic sex determinants has not yet been clarified.

6.2.3.5 PARASITES

Several maternally inherited endosymbionts of arthropods, known as *reproductive parasites*, have developed strategies to convert non-transmitting male hosts into transmitting females through feminization of genetic males and induction of parthenogenesis (Cordaux *et al.* 2011 and references therein).

Alphaproteobacteria of the genus *Wolbachia* are endocellular parasites of numerous arthropods, in which they can be transmitted from mother to offspring through the cytoplasm of the egg cell. In these hosts the bacterium has evolved a form of reproductive parasitism that allows it to increase its diffusion, manipulating the sex ratio of the host in the direction of a greater proportion of females (see also Section 3.6.2.8). Among the different strategies adopted by *Wolbachia* there is the feminization of males. In species with heterochromosomes, these individuals have a female phenotype despite their male karyotype. In the butterfly *Eurema hecabe*, with a ZW system, there are *gynogenic* females, which produce an excess of females because of the feminizing effect of *Wolbachia*. More complex is the situation in the land isopod *Armadillidium vulgare*, where the sex ratio in the progeny of a single female is quite different in different populations. These can include, in different proportions, *monogenic* females which generate offspring of one sex only (thus *gynogenic* or *androgenic*, if the offspring are female or male, respectively) and *amphigenic* females, which generate males and females in equal proportions. Here the effects of *Wolbachia* infection, which qualifies as a *cytoplasmic factor of sex determination*, add up to those of another non-Mendelian feminizing factor (*f*) with a different transmission pattern. These effects are counteracted by a masculinizing gene *M*, antagonist of factor *f*, and by a gene complex *R* that induces resistance to the transmission of *Wolbachia* (Verne *et al.* 2012).

There are other parasites, able to transmit vertically between host generations, that can alter host sex ratio. The bacterium *Cardinium* is able to feminize the males of the mite *Brevipalpus phoenicis*. The unicellular fungus *Nosema granulosis* (Microsporidia) induces the feminization of 'genetic males' in the amphipod crustacean *Gammarus duebeni*.

6.3 Maternal Sex-Determination Systems

In **maternal sex determination** the sex of the offspring depends on the mother. This category in some ways cuts across the previous systems, because here genetic and environmental factors can interact in different ways and to different degrees in establishing the sex of the individual. Maternal sex determination can take very different forms. At one end of the spectrum there are some forms of strict genetic determination, where however it is not the individual's genotype that determines its own sex, but that of its mother. At the other extreme there are cases in which the phenotype or a particular physiological condition of the mother determines the sex of the offspring. In this case, the mother is the source of the specific environmental signal that determines the offspring's sex.

An example of maternal sex determination of the genetic type is seen in the dipterans *Chrysomya albiceps* (Figure 6.17) and *C. rufifacies* (Calliphoridae). In these insects there are two types of females: *androgenic females*, which generate only male offspring, and *gynogenic females*, which produce exclusively female offspring. Gynogenic females are heterozygous (Ff) at a locus F which encodes a maternal factor that accumulates in the oocytes during oogenesis, and determines the female sex condition in the zygotes that originate from them. At the same locus, males and androgenic females are homozygous recessive (ff) and do not produce the maternal factor (Sánchez 2008). The sex of an

Figure 6.17 Genetic type of maternal sex determination. In the calliphorid fly *Chrysomya albiceps*, the sex of the individual depends on its mother's genotype at a specific diallelic locus.

individual is therefore established at syngamy. However this is not the syngamy that produced the zygote from which the individual develops, but instead the syngamy that produced the zygote from which its mother developed.

In contrast, an example of maternal sex determination of the environmental type is provided by the cecidomyid midge *Heteropeza pygmaea* (Figure 2.17). The sex of the offspring (X0 male, XX female) depends on the nutritional condition of the mother. If this is good, females will be born, males otherwise (Cook 2002; see Section 2.8). In response to the nutritional conditions, a factor produced in the mother's brain is secreted into the haemolymph, and from there it reaches the ovaries. In the gonad, this factor determines the course of oogenesis, which is a form of ameiotic parthenogenesis, with or without the loss of an X chromosome.

Maternal sex determination is also possible in species with heterogonic cycles (Section 2.3), with the seasonal passage from parthenogenetic to amphigonic reproduction. In *Daphnia* (Figure 6.18), in response to specific environmental signals such as the shortening of the photoperiod or an increase in population density, parthenogenetic females switch from the exclusive production of females to the generation of males and females, which in the case of these crustaceans are genetically identical to each other and to the mother

Figure 6.18 Environmental type of maternal sex determination. In the heterogonic cycle of the water flea *Daphnia*, in the transition from parthenogenetic to amphigonic reproduction, the sex of the offspring is established by endocrine signals secreted by the mother in response to specific environmental signals.

(Section 2.3). Environmental signals are received by parthenogenetic mothers and transduced into endocrine signals (juvenile hormone). These have an effect on the maturation of the oocytes, which will result in eggs that will develop without being fertilized. Recent studies have shown that the signal is received by the immature oocyte before the beginning of embryogenesis (Ignace *et al.* 2011).

In most aphids with heterogonic cycles, in autumn, in response to environmental signals anticipating the adverse season, a generation of parthenogenetic females is followed by a generation of sexuparous females, which, also by parthenogenesis, produce individuals that will reproduce by amphigony (Section 2.3). Aphids have an X0 sex-determination system, but in parthenogenesis the sexual genotype is established under the control of the maternal hormones through the peculiar behaviour of the chromosomes during the maturation of the oocytes. The eggs that will give males lose an X chromosome, those that will give females keep it. Since the only functional sperm cells produced by the males contain the X chromosome, only females (XX) are born from fertilization, the *founders* of the generations of parthenogenetic females of the following season.

A case verging on the limit of this category is provided by many species of parasitoid hymenopterans. The mother is able to control the fertilization of the eggs by the sperm she has in store since her only mating, thereby determining the sex of the offspring, even if the most proximate cause of sex determination in the latter is the ploidy level of their genome (Section 6.1.3). Fertilized eggs (which will develop into females) are preferentially laid into hosts of large size, whereas the unfertilized eggs (which will develop into males) are generally laid into smaller-size hosts.

Maternal sex determination belongs to a category called **progamic sex determination** (as opposed to *syngamic sex determination*), where sex depends on the condition of the egg before fertilization. The polychaete *Dinophilus gyrociliatus* produces both larger eggs, which will give females, and smaller eggs, from which males will grow. The development of sex characters is regulated by genes whose expression is modulated by the quantity of the trophic resources present in the egg.

6.4 Mixed Sex-Determination Systems and Random Sex Determination

So far we have qualified sex-determination systems as either genetic or environmental, irrespective of whether these operate in the individual itself or in its mother. This simple classification should not however obscure the fact that there are many **mixed sex-determination systems**, in which genetic and

Figure 6.19 Mixed systems of sex determination, here chromosomal/environmental. In the Iberian ribbed newt (*Pleurodeles waltl*), a ZW chromosomal sex-determination system operates in the temperature range 16–24 °C, while at higher temperatures individuals with female karyotype (ZW) and male phenotype develop.

environmental factors combine to different degrees. Thus, strictly genetic and strictly environmental systems can be considered extremes of a continuous spectrum of variation in the modes of sex determination.

In the American salamanders of the genus *Pleurodeles*, with a chromosomal ZW system, from eggs incubated at a temperature between 16 and 24 °C individuals develop with a sex phenotype concordant with the sex karyotype, and the sex ratio is balanced (1:1). From eggs incubated at higher temperatures, in *Pleurodeles poireti* individuals can be generated with male karyotype and female phenotype (*ZZ thermoneofemale*), whereas in *P. waltl*, at the same temperatures (Figure 6.19), individuals develop with female karyotype and male phenotype (*ZW thermoneomales*) (Dournon *et al.* 1990).

In the fish *Menidia menidia* (Atheriniformes) sex is determined by the interaction of several genetic factors (*polygenic sex determination*, Section 6.1.2) with water temperature, but the relative importance of these determinants varies with latitude: at high latitudes genetic factors prevail over environmental factors, whereas water temperature is the dominant factor in sex determination in populations living at low latitudes (Kraak and Pen 2002).

In the skink *Bassiana duperreyi*, genetic, environmental and maternal sex determination are highly integrated. At normal temperatures the sex of this reptile is determined by the sex chromosomes, but at low temperatures the effects of the environment prevail over those of the karyotype and predominantly males develop, although with a typical female sex-chromosome complement, XX. Furthermore, at low temperatures *sex determination via yolk allocation* is observed: smaller eggs, poor in yolk, will develop into males, while larger eggs with more abundant yolk will develop into females (Radder *et al.* 2009).

If the mixed sex-determination systems demonstrate the existence of a *bipolar* continuum between genetic and environmental systems, according to

Figure 6.20 Mixed systems of sex determination, here genic/random. The zebrafish (*Danio rerio*) has a polygenic sex-determination system with added elements of stochasticity.

some authors a third pole should also be recognized, i.e. **random sex determination**; with this addition a *tripolar* continuum of sex-determination systems is obtained (Perrin 2016). According to recent studies, slight random fluctuations in developmental processes (produced by so-called *developmental noise*), an important but too often neglected component of phenotypic variation, would be able to direct the development of sexual characters towards one out of several alternative conditions. The ciliate *Tetrahymena thermophila* provides an example of a totally random system of mating-type determination. But sex-determination systems traditionally considered strictly environmental or multifactorial may also show elements of stochasticity, such as those of the cladoceran *Daphnia magna*, the copepod *Tigriopus californicus*, the seabass *Dicentrarchus labrax* and the zebrafish (*Danio rerio*; Figure 6.20).

6.5 Notes on Sexual Differentiation

The attainment of a given sex condition is an aspect of the development of an organism that will not be dealt with in detail in these pages, and for which we refer the reader to developmental biology textbooks (e.g. Gilbert and Barresi 2016). However, some information will be useful to complete our discussion.

In mammals, for instance, we distinguish between *primary* and *secondary sex determination* (although, for the use we have made of these terms in the chapter, the latter should more properly be described as *sexual differentiation*).

Primary sex determination consists in the differentiation of the gonad into ovary or testis. It is a developmental process that proceeds from an embryonic precursor of the gonad, called a *bipotential* or *undifferentiated gonad*. As we have seen, the Y chromosome is a key factor for the development of the testis, and in its absence the gonad develops into an ovary, although a second

X chromosome is also necessary for the complete formation of the female gonad.

Secondary sex determination concerns all sexually dimorphic phenotypic characters except for the gonads. These characters include *elements of the reproductive system other than the gonads*, such as gonoducts, accessory glands and external genitalia, and *secondary sexual characters*, i.e. features outside the reproductive system such as colour and colour patterns (*sexual dichroism*) or anatomical structures (*sexual dimorphism* in the strict sense) specific to one or the other sex (Section 3.3.1.2). In mammals, these extragonadic characters are usually determined by hormones (typically, steroids) and by paracrine factors secreted by the gonads. In the absence of gonads or their products, the female phenotype develops. Ovaries produce *oestrogen*, a hormone necessary for the development of Müllerian ducts into oviducts, cervix and part of the vagina. Testes produce a paracrine factor, called *anti-Müllerian factor* (*AMF*), which prevents the formation of the uterus and oviducts, and *testosterone*, which masculinizes the fetus, inducing the formation of penis, scrotum and vas deferens, while inhibiting the development of mammary glands (Gilbert and Barresi 2016).

However, even in mammals, there are secondary characters that can be directly controlled by factors produced by the genes of the sex chromosomes, rather than by the circulating hormones. In the wallaby *Macropus eugenii* the marsupium and the mammary glands in the female and the scrotum in the male begin to form before the gonads, at a time when those sex hormones are not yet in circulation. In recent years more and more data are accumulating to show that in other mammals there are sexual characters that are under the control of direct products of heterochromosome genes, rather than being subjected to the hormones produced by the gonads. For instance, significant between-sex differences in gene expression before gonadal differentiation have been documented in the mouse. These new data could lead to a rethinking of the classical model of sexual differentiation in mammals (Arnold 2012).

In other organisms sexual differentiation depends less strictly, or not at all, on circulating hormones. In *Drosophila*, and in insects in general, each cell is sexually determined by its own genotype, in a way substantially independent of external inductive signals from neighbouring cells or gonadal secretions. Despite some exceptions to this rule (e.g. the external genitalia in *Drosophila*), this is still a general principle of sexual development common to many metazoans.

As a consequence of this kind of sexual differentiation, the sex of an individual depends on the independently determined sexual identity of each of its cells. Despite the fact that all body tissues are exposed to the same hormones, these do not have significant effects on the development of sexual characters.

Figure 6.21 Bilateral gynandromorphism in the butterfly *Papilio glaucus*. The left side expresses a male phenotype, the right side a female phenotype.

Sexual differentiation not regulated at a systemic level explains the occurrence of **gynandromorphism**. *Gynandromorphs* are genetically mosaic (Section 5.1.1.3) or chimeric (Section 5.2.5.3) individuals which present abnormal phenotypes characterized by having both anatomical parts with male characters and parts with female characters. They are found in nature, but can also be induced experimentally. Many cases have been described, mostly among insects (Figure 6.21), chelicerates (spiders and mites), decapod crustaceans and birds. Depending on the stage of development in which the anomaly occurs, the division between 'masculine tissues' and 'feminine tissues' may run along the midline of the body, producing a *bilateral gynandromorph*, or appear as a mosaic composition throughout the body. This abnormality can be produced by chromosomal mutations (aneuploidy) affecting the sex chromosomes, or by accidents at syngamy. For instance, during cleavage mitoses in early embryogenesis, disturbances in chromosome segregation in a cell with XY genotype may produce two aneuploid cells with X0 and XYY karyotypes, which in many insects correspond to male and female karyotypes, respectively. Their proliferation will give rise to tissues with karyotype and phenotype of either sex.

In the case of birds, as we have seen (Section 6.1.1.6), the *Dmrt1* gene on the Z chromosome is a fundamental determinant of the male phenotype through a dosage-related mechanism. However, *Dmrt1* is significantly expressed only in the gonads and therefore has little chance of directly determining the sexual identity of other organs and tissues. Individuals that for any reason developed as genetic mosaics (with a mixture of ZZ and ZW cells) can simultaneously exhibit phenotypic characters of both sexes, for instance in the colour of the plumage, but also in neural and behavioural characters, such as those associated with song (Barske and Capel 2010).

Gynandromorphism should not be confused with **intersexuality**, which occurs when genetically uniform individuals (possibly affected by some genetic anomaly: see Section 6.1.1) present parts of their tissues with either a sexual phenotype opposite to their genetic sex or an intermediate sexual phenotype. Intersex individuals are relatively common in some groups of crustaceans, especially among isopods and amphipods, where they are often sterile or functionally female (an exception being the amphipod *Corophium volutator*, where intersexes can instead reproduce as males; McCurdy *et al.* 2008). Individuals with both male and female genital appendages (*gonopods*) are the rule in some species of freshwater crayfishes (Parastacidae) from Oceania and South America. In *Parastacus defossus*, all individuals are intersex, but nearly half of them have a female gonad, half a male gonad, and only 1.5% have a gonad producing both eggs and sperm (*ovotestis*) (Noro *et al.* 2007).

6.6 Determination of Mating Type

In many eukaryote species, for instance in those with isogamety, the sex of the individual is indeterminate. Nonetheless, the genetic exchange associated with sexual reproduction (or with sex in general) cannot take place between all possible pairs of individuals in a population. There are in fact forms of sexual compatibility (and incompatibility, Section 3.5.5) such that an individual, although not assignable to a specific sex, nevertheless belongs to a particular **mating type** which allows it to have sexual exchanges only with individuals belonging to a different mating type. Depending on the tradition of the studies in each group, mating types are indicated by numbers, letters, their combinations, or, if there are only two of them, sometimes simply with the symbols + and –. The number of mating types in a single species can vary from two to a few thousand. Among the ciliates, for instance, there are almost always only two mating types in the species of the genera *Paramecium* and *Blepharisma*, but 5–12 in *Euplotes* and about 100 in *Stylonychia mytilus*. The record, however, seems to belong to the basidiomycete fungus *Schizophyllum commune*, with about 28,000 mating types.

Individuals belonging to distinct mating types present different alleles at a usually small number of specific loci, called **mating-type loci**. Virtually all unicellular eukaryotes, many fungi and algae, but also the cellular slime mould *Dictyostelium discoideum*, have mating-type loci whose products prevent sexual exchange between individuals with the same genotype. In algae with isogametes and alternation of generations, such as the green alga *Cladophora vagabunda*, the mating type is established in the spore at meiosis and is maintained throughout the haploid generation of the gametophyte, until the formation of the gametes.

Fungi which do not present mating types (a condition more frequently found in ascomycetes) are called *homothallic*, while those in which there are two or more distinct mating types are called *heterothallic*. Ascomycetes have only two mating types, generally indicated by MAT1-1 and MAT1-2, but the two alternative sequences at the single mating-type locus are called *idiomorphs*, rather than alleles, because the coded proteins are strongly dissimilar and are thought not to be derived from a common ancestral sequence. In the yeast *Saccharomyces cerevisiae* the two mating types, indicated by a and α, are determined by two idiomorphs at the same locus (*MATa* and *MATα*). In contrast, in basidiomycetes one species can exhibit several thousand distinct mating types. Despite considerable differences in reproductive biology (Sections 7.5.4 and 7.5.5) and the number of mating types between these two groups of fungi, it seems clear that many components of the associated genetic regulation system are shared and highly conserved (Casselton 2002).

In a process analogous to the change of sex in response to an environmental signal, some yeasts, including *Saccharomyces cerevisiae*, can change their mating type. *Mating-type switching* is a programmed DNA rearrangement that occurs in haploid cells and converts a *MATa* idiomorph into a *MATα*, or vice versa. During the transformation, the sequence at the *MAT* locus is removed and replaced by a sequence copied from a different locus, either *HMR* or *HML*. These are silent loci that store a-specific and α-specific sequences, respectively, but which are not transcribed owing to a chromatin marking. Among yeasts, the ability to change mating type evolved at least twice independently, once in the Saccharomycetaceae, which includes *S. cerevisiae* and *Kluyveromyces lactis*, and once in the Schizosaccharomycetaceae, which includes *Schizosaccharomyces pombe*. Mating-type switching is a strategy that allows an isolated haploid cell to have diploid descendants. The cell divides by mitosis asymmetrically (budding) and the larger cell changes its mating type and then fuses with the smaller cell (syngamy), producing a diploid zygote that will continue to divide by mitosis (Gordon *et al.* 2011).

Environmental mating type determination has been described for some green algae such as *Protosiphon* and *Hydrodictyon* (Fritsch 1935).

Recent studies are revealing that what most differentiates mating-type determination from sex determination is nothing more than the association, or not, with anisogamety. As we have seen (Section 6.1.2), not only can the sex of an individual depend on its allelic constitution at specific loci, exactly as in the case of mating types, but the differences between mating types can sometimes appear as early stages in the evolution of heterochromosomes. In many fungi and algae, the flanking regions of the mating-type locus do not recombine and

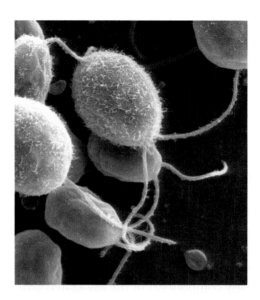

Figure 6.22 Determination of the mating type. In the unicellular green alga *Chlamydomonas reinhardtii* the regions flanking the mating-type locus present specific sequences associated with one or the other of the two mating types (mt+ or mt–). These differences are such that recombination in that region of the genome is very unlikely, and therefore the two alternative mating-type alleles function as if they were on distinct sex chromosomes.

present specific sequences associated with only one of the alternative alleles (or *idiomorphs*). In the unicellular green alga *Chlamydomonas reinhardtii* (Figure 6.22) there are specialized regions of the genome associated with the two mating types, mt– and mt+, which include inversions, translocations, gene duplications and mutations. These structural differences have the effect of suppressing recombination, similar to what happens between the hetero-chromosomes of the XY and ZW systems. In the basidiomycete *Microbotryum violaceum* (Ustilaginales) the chromosomes with mating-type loci appear as heteromorphic heterochromosomes, with the chromosome carrying the *a*1 allele partially degenerated with respect to the chromosome that carries the alternative *a*2 allele. Finally, in *Chlamydomonas*, at syngamy the chloroplasts show uniparental inheritance, these organelles being inherited only from the mt+ parent. Similar instances of uniparental inheritance of mitochondria have been reported for several fungi, including *Neurospora tetrasperma* and *Aspergillus nidulans* (Fraser and Heitman 2004). These cases are obviously suggestive of the asymmetric organelle inheritance frequently observed in anisogamety (Section 5.2.3.1).

Three forms of mating-type inheritance are known in the conjugation of ciliates (Section 5.2.5.1), although the term 'inheritance' is rather inappropriate, as there is no reproduction, but just sex. All three are known for different species of the genus *Paramecium* (Phadke and Zufall 2009).

- *Synclonal inheritance.* Both ex-conjugants express the same mating type, as an effect of having the same micronuclear genome.

- *Cytoplasmic inheritance.* The two ex-conjugants express distinct mating types, corresponding to those expressed before conjugation, although they share an identical micronuclear genome. This type of inheritance is explained by epigenetic influences of the old (degenerating) macronucleus on the new-forming macronucleus.

- *Caryonidal inheritance.* The mating type of each ex-conjugant depends on the independent formation and differentiation of its macronucleus, under the influence of genetic, epigenetic and environmental factors.

Forms of reproductive incompatibility, similar to those of organisms with indeterminate sex conditions, can overlap with those already in place in organisms with anisogamety. In the genome of many angiosperms there is a multiallelic *self-sterility locus*, sometimes with hundreds of different alleles, and cross-fertilization is obligate, because the pollen carrying a certain allele cannot develop a pollen tube through the carpel tissues of a sporophyte which carries the same allele, and therefore cannot fertilize an egg cell from the same plant that produced it (Section 3.5.4).

Chapter 7: Reproduction: a Taxonomic Survey

This final chapter illustrates the most characteristic aspects of reproduction in different taxonomic groups. A few of the taxa discussed below are not monophyletic, but they are nevertheless accepted here because the reader is likely to be familiar with them; these taxa will be flagged when mentioned.

In this chapter, we discuss only the eukaryotes; for the prokaryotes (eubacteria and archaea), see Sections 3.1.1.1 and 5.2.1. The classification of the eukaryotes has been completely revolutionized during the last half century thanks to rapid progress in microscopy and, more recently, in sequencing the long and informative molecules of DNA, RNA and proteins, accompanied by the use of rigorous methods of phylogenetic analysis. In particular, this work has led to a rearrangement of groups traditionally ascribed to protozoa and algae. The latter term remains in use only as an informal collective word referring to the photosynthetic unicellular and multicellular organisms, almost all aquatic, other than land plants or embryophytes. Two large phyla of macroscopic algae (the red algae and the brown algae) are still considered natural groups, and we treat them in distinct sections of this chapter. Another section is devoted to the green algae (which also include single-cell forms such as *Chlamydomonas*), although today they are divided into two phyla, Chlorophyta and Streptophyta, the second of which also includes the land plants, which we treat separately. Of the large and heterogeneous set of the remaining unicellular algae, we discuss here only some of the better-known groups and those with more remarkable phenomena of reproduction and sexuality; alongside these, we mention some groups of autotrophic unicellular eukaryotes once classified among the protozoans.

7.1 Protists (Unicellular Eukaryotes)

All groups of unicellular eukaryotes (for a recent overview, see Archibald *et al.* 2017) can be referred to collectively as protists.

Most protists reproduce by binary division, but multiple division is also widespread, especially among the parasitic taxa. In this case, the nucleus is divided repeatedly, giving rise to up to hundreds of nuclei. Binary division takes place according to mechanisms that in many cases diverge significantly from the mitosis of metazoans or flowering plants. In fact, the nuclear envelope often remains intact during the separation of the chromosomes, in which case the mitotic spindle can be internal to the nuclear envelope, or external; in turn, the spindle can have the usual conformation, or be divided into two differently oriented half-spindles (Section 5.1.5).

There are several types of multiple division, which often occur at different stages of a species' life cycle; these types include budding and schizogony (Section 3.1.1.2). In some groups of protists, sexuality has never been observed. However, residual traces or indirect evidence of sexuality have been found in these species. For example, a more or less complete set of meiotic genes has been found in *Entamoeba* spp. (Stanley *et al.* 2005), *Giardia intestinalis* (Ramesh *et al.* 2005) and *Trichomonas vaginalis* (Malik *et al.* 2008), for which in the past a derivation from evolutionary lines that had never had sexuality (at least in its usual forms) had been hypothesized; in *Toxoplasma gondii*, sexual mechanisms of still uncertain nature are most likely responsible for the remarkable genetic variability observed in this species (Ajzenberg *et al.* 2004).

Sexual reproduction is also unknown in many groups of 'algae', mostly unicellular ones such as Glaucophyta, a large part of Cryptophyta, Bolidophyceae, Dictyochophyceae, Pelagophyceae, Pinguiophyceae, Eustigmatophyceae, Picophageae, Synchromophyceae, Aurearenophyceae, Phaeothamniophyceae, Xanthophyceae (except for the Vaucheriales, which are oogamous; Section 7.1.6), Schizocladiophyceae, Mesostigmatophyceae, Chlorokybophyceae and Klebsormidiophyceae, and it is probably lacking in individual genera or species of other groups. The life cycles of many protists are simple, including only binary divisions; others, however, are very complex, given the alternation of asexual and sexual phases, or of active and quiescent phases. Particularly complex are the cycles of many parasitic protists, in which alternation of phases (which sometimes translates into a conspicuous polymorphism) and reproductive modes are accompanied by the passage from one host to another, or by a change in the nature of the relationship with the host.

The groups discussed in the following sections all belong to the Chromalveolata (except for Hypermastigina, Choanoflagellata and Mycetozoa). In addition, we should briefly mention the Chrysophyceae (Heterokonta), which reproduce either asexually (by binary division or release of zoospores), or sexually (by isogamy or anisogamy s.s., with production of quiescent zygospores) and undergo zygotic meiosis, so that their cycle is haplontic; the Coccolithophyceae (Haptophyta), in which, as a rule, there is alternation between a diploid and a haploid generation, both unicellular; and the Raphidophyceae (Heterokonta), diplonts, which reproduce sexually by binary division and sexually by isogamy.

7.1.1 Foraminifera

Forams reproduce both asexually and sexually, often with more or less regular alternation between a haploid (gamont) and a diploid (schizont) generation, morphologically similar. The gamont produces gametes, usually biflagellate; the schizont is multinucleated and undergoes meiosis, producing new gamonts.

7.1.2 Apicomplexans (Sporozoans)

The apicomplexans reproduce by multiple division, with different modes (merogony, sporogony and gametogony) in combinations that generally display considerable complexity.

The life cycle of *Plasmodium*, the protist responsible for malaria in mammals and birds, is illustrated in Figure 7.1. We can explore it by starting from the *sporozoite*, the stage in which the parasite is injected into the bloodstream of the vertebrate host with the saliva of an infected *Anopheles* mosquito. As soon as it settles in the cells of the liver parenchyma, the sporozoite through several cycles of *schizogony* (here called *merogony*) gives rise to thousands of *merozoites*

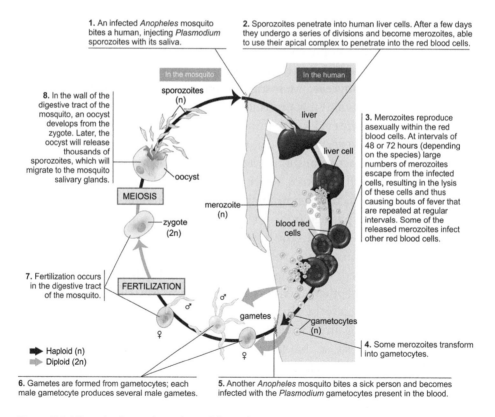

1. An infected *Anopheles* mosquito bites a human, injecting *Plasmodium* sporozoites with its saliva.

2. Sporozoites penetrate into human liver cells. After a few days they undergo a series of divisions and become merozoites, able to use their apical complex to penetrate into the red blood cells.

8. In the wall of the digestive tract of the mosquito, an oocyst develops from the zygote. Later, the oocyst will release thousands of sporozoites, which will migrate to the mosquito salivary glands.

3. Merozoites reproduce asexually within the red blood cells. At intervals of 48 or 72 hours (depending on the species) large numbers of merozoites escape from the infected cells, resulting in the lysis of these cells and thus causing bouts of fever that are repeated at regular intervals. Some of the released merozoites infect other red blood cells.

7. Fertilization occurs in the digestive tract of the mosquito.

4. Some merozoites transform into gametocytes.

➡ Haploid (n)
➡ Diploid (2n)

6. Gametes are formed from gametocytes; each male gametocyte produces several male gametes.

5. Another *Anopheles* mosquito bites a sick person and becomes infected with the *Plasmodium* gametocytes present in the blood.

In the mosquito · In the human

sporozoites (n)
liver
liver cell
oocyst
MEIOSIS
merozoite (n)
blood red cells
zygote (2n)
FERTILIZATION
gametes
gametocytes (n)

Figure 7.1 Life cycle of an apicomplexan (*Plasmodium*).

344

that penetrate the host's erythrocytes and, after passing through the *trophozoite* stage, become small *schizonts*; these in turn produce a new generation of merozoites. As the infestation continues, however, some trophozoites are transformed into *microgametocytes* or *macrogametocytes*, characterized by their different behaviour: when ingested by a mosquito with a blood meal, macrogametocytes do not undergo divisions, while each microgametocyte gives rise, through three mitotic cycles (*gametogony*), to eight flagellate microgametocytes. A zygote (*ookinete*, then *oocyst*) forms by the union of a microgametocyte with a macrogametocyte. The zygote undergoes meiosis and multiple fission (*sporogony*), producing a new generation of sporozoites.

The life cycle of the gregarines is almost as complex. These are large parasites commonly found in the digestive tract of soil invertebrates (earthworms, myriapods, insects) and in those of the marine benthos (polychaetes, ascidians, crustaceans). Characteristic of these sporozoans is the stage of *syzygy*, in which two identical-looking gametes, encased in a *gamontocyst*, synchronously undergo many mitotic cycles from which a large number of nuclei are produced; a bit of cytoplasm (*cytomer*) remains around each of these nuclei, up to the separation of the same number of haploid cells that function as gametes, some of which are flagellated and sometimes smaller than the others, which have no flagellum. Within the gamontocyst many zygotes are thus formed that undergo meiosis and a subsequent mitosis, giving rise to eight sporozoites; of these, four will differentiate into male gamonts (which will produce male gametes), four into female gamonts (which will produce female gametes).

7.1.3 Dinoflagellates (Dinophyceans)

In dinoflagellates, asexual reproduction alternates with sexual reproduction. Vegetative cells, generally covered by a solid case, release naked gametes. In the mitotic division, each of the two daughter cells generally inherits some of the plates that formed the case of the mother cell and produces the remaining ones from scratch; in some species, however, both daughter cells form a completely new case. The zygote formed by the fusion of two gametes undergoes meiosis, of the products of which only a haploid vegetative cell survives, to multiply by mitosis. In *Oxyrrhis* the mitotic spindle forms within the nuclear envelope, which persists during cell division.

7.1.4 Ciliates

In ciliates, sex is not associated with reproduction (Figure 7.2). In these protists, which have cilia arranged in often very complex patterns that differ from species to species, sex is not accomplished through the fusion of two cells (gametes, or gamonts) to form a zygote, but by an exchange of nuclei between

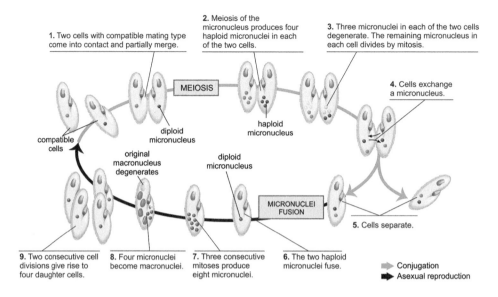

1. Two cells with compatible mating type come into contact and partially merge.

2. Meiosis of the micronucleus produces four haploid micronuclei in each of the two cells.

3. Three micronuclei in each of the two cells degenerate. The remaining micronucleus in each cell divides by mitosis.

MEIOSIS

4. Cells exchange a micronucleus.

diploid micronucleus

haploid micronucleus

compatible cells

original macronucleus degenerates

diploid micronucleus

MICRONUCLEI FUSION

5. Cells separate.

9. Two consecutive cell divisions give rise to four daughter cells.

8. Four micronuclei become macronuclei.

7. Three consecutive mitoses produce eight micronuclei.

6. The two haploid micronuclei fuse.

Conjugation
Asexual reproduction

Figure 7.2 Life cycle of a ciliate (*Paramecium*).

two partners (conjugants). The partners retain their cortical organization (the series of cilia and the complex of fibres of the underlying infraciliature) almost intact even when they are conjoined by the cytoplasmic bridge through which nuclear exchanges take place.

The ciliates also have a nuclear dualism, due to the presence – in its simplest form, for example in *Chilodonella uncinata* – of one micronucleus and one macronucleus. The micronucleus is diploid, whereas the macronucleus contains multiple copies of at least a part of the genetic material present in the micronucleus. It is from these more or less long segments of DNA that the transcription of genetic information occurs. Therefore, the macronucleus can be regarded as a 'metabolic' nucleus that is not involved in sexual processes. Sex instead involves the micronuclei of two partners, called *gamonts* or *conjugants*, between which an exchange of haploid nuclei takes place. As a rule, the conjugants are morphologically indistinguishable and retain independence and cytoplasmic identity until the end of conjugation: at this point they become separated, as *ex-conjugants*, with renewed chromosomal sets, identical between the two. Details of the cytogenetic aspects of ciliate conjugation can be found in Section 5.2.5.1.

The number of divisions the micronucleus of a conjugant undergoes before the nuclear exchange with the partner is sometimes different from three (the two steps of the meiotic division, plus the subsequent equational division of the surviving haploid nucleus). In *Euplotes*, for example, there are four divisions, due to the presence of an equational division before meiosis. Different again are those ciliates that already contain more than one micronucleus when in a vegetative state. Following the onset of the conjugation process, all these micronuclei undergo meiosis; however, only one or two haploid nuclei survive, which will eventually undergo the usual equational division that precedes nuclear exchange. Similarly, the macronucleus may also have a different destiny. For example, in *Paramecium*, at the end of conjugation there may be two or four macronuclei, which will be divided among the same number of daughter cells at the end of one or two mitotic cycles, restoring the presence of a single macronucleus per cell.

Tracheloraphis phoenicopterus, which belongs to the karyorelict ciliates, a group with many primitive traits, has six micronuclei when in a vegetative state; many of the products of their pre-conjugative meiotic division undergo a further equational division, and the subsequent exchange with the partner leads to the formation of numerous diploid nuclei in each conjugant, but only one of these survives.

The number of post-conjugative nuclear divisions that precede the first (reproductive) cell division of each ex-conjugant also varies, from one to four. There are two such divisions in *Euplotes*, between two and four in *Paramecium*. In *Tracheloraphis phoenicopterus* there are four: of the 16 diploid nuclei thus produced, four degenerate, six give rise to macronuclei, the other six to micronuclei.

The situation is more complex in ciliates with gamonts of different sizes, a *microgamont* and a *macrogamont*. These are sessile forms, such as the peritrichs and the suctors, in which the microgamont is the only mobile phase in the life cycle. In these ciliates, conjugation involves the complete fusion of the two gamonts, an extremely rare event in the majority of ciliates, in which the gamonts are identical. The result of the fusion of micro- and macrogamont is a conventional zygote. The haploid micronuclei contributed by the two gamonts have different origins: the nucleus contributed by the microgamont is one of the products of a meiosis, while the nucleus contributed by the macrogamont is one of the two haploid nuclei formed by the usual post-meiotic equational division. The diploid zygotic nucleus gives rise through three mitotic cycles to eight diploid nuclei, one of which will represent the micronucleus, while the other seven will give rise to macronuclei.

Finally, in some ciliates there is *autogamy*, with the fusion of the two haploid nuclei derived from the same diploid nucleus (Section 3.2.2).

7.1.5 Diatoms

Diatoms are one of the few groups of unicellular algae in which the diploid condition distinctly prevails over the haploid one (Figure 7.3). An unbroken series of mitotic divisions can continue for months or years, but this leads to

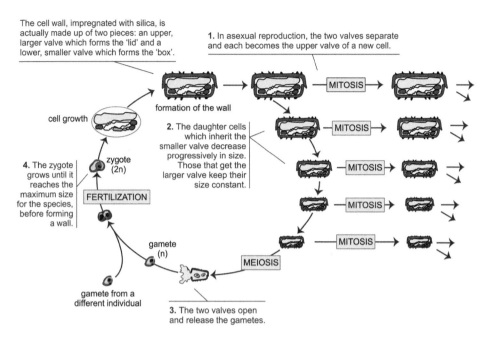

Figure 7.3 Life cycle of a diatom (*Tabellaria*).

a progressive reduction in cell size. This is a consequence of the mechanics of cell division in these unicellular organisms. Each diatom has two siliceous valves of slightly different sizes and, at division, each of the two daughter cells inherits one of the two valves, which ends up in each of them as the major valve, while the minor valve is built from scratch. The return of the population to its initial cell size requires sexual reproduction. The zygote matures into a diatom cell with two valves of the maximum size for the species, which will start a new sequence of asexual reproduction events.

Sexual reproduction is more often by oogamy, sometimes however by anisogamy s.s. or even isogamy. Oogamous diatoms produce small male gametes (flagellate, except in *Rhabdonema*) and large egg cells. Male gametes are the product of a series of mitoses unaccompanied by the formation of new valves, followed by meiosis and the formation of the flagellum, which is however lacking in pinnate diatoms; in one of these, *Pseudostaurosira trainorii*, sex pheromones have been described and the male gamete emits viscous strands that can capture an egg cell (Sato *et al.* 2011). One or two products survive from meiosis. The zygote has the value of an auxospore: that is, it grows in size, often protected by siliceous scales; when maximum size is reached, it transforms into a vegetative cell, with the usual two valves. However, some diatoms (*Achnantes, Bellerochea, Biddulphia, Ditylum,*

Grammatophora and *Rhabdonema*) can produce auxospores without resorting to sexual reproduction. These auxospores are the diatoms' resistance stage.

7.1.6 Xanthophyceans

The Xanthophyceae are a group of heterokont algae widespread especially in fresh water, but with some marine species and others living in terrestrial environments. They can have very different organization, either unicellular, with uninucleate or multinucleated cells, or multicellular and filamentous.

We mention here only *Vaucheria*, a filamentous alga in the form of a long tube with a single large central vacuole surrounded by a thin layer of peripheral cytoplasm populated by thousands of nuclei. During asexual reproduction, the cytoplasm at one end of the tube is cut off through a transverse wall and pairs of flagella are formed near each nucleus; then the old wall breaks down and the multiflagellate cells disperse into the water. As soon as their flagella are lost, these cells begin to grow at both ends, forming a new tube. In sexual reproduction, biflagellate spermatozoa are formed by a process similar to the production of the flagellated zoospores involved in asexual reproduction. The egg cell is formed instead as a small multinucleated protrusion of the tube, which subsequently becomes isolated through a transverse wall beyond which remains a large uninucleate cell. After fertilization, the zygote forms a thick wall and separates from the tube from which it originated. After a period of dormancy, the zygote undergoes meiosis and germinates, forming a new tube full of haploid nuclei.

7.1.7 Hypermastigines

The Hypermastigina are symbionts of cockroaches and termites. Sexuality is documented in a few species only; of these, some are isogametic, others anisogametic s.s. In the latter, the macrogamete differentiates a fertilization cone to which the microgamete attaches. Karyogamy does not always occur immediately, but sometimes not until 5–6 days after plasmogamy. In *Urinympha*, meiosis is accomplished through a single division, a peculiarity shared by flagellated protists of other groups, such as *Oxymonas* and *Saccinobaculus* among the Metamonadida and *Leptospironympha* among the Parabasalia.

7.1.8 Choanoflagellates

Of the different groups of protists, many of which are imperfectly known from the point of view of reproductive and sexual processes, particular attention should be paid to the Choanoflagellata, which recent studies of molecular

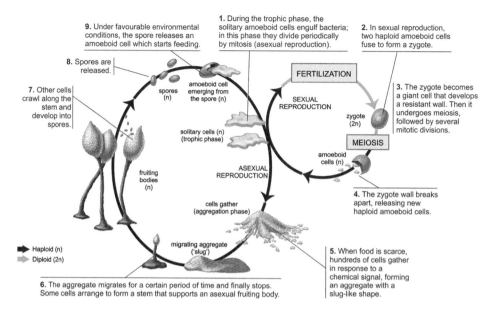

9. Under favourable environmental conditions, the spore releases an amoeboid cell which starts feeding.

1. During the trophic phase, the solitary amoeboid cells engulf bacteria; in this phase they divide periodically by mitosis (asexual reproduction).

2. In sexual reproduction, two haploid amoeboid cells fuse to form a zygote.

8. Spores are released.

FERTILIZATION

spores (n)

amoeboid cell emerging from the spore (n)

SEXUAL REPRODUCTION

3. The zygote becomes a giant cell that develops a resistant wall. Then it undergoes meiosis, followed by several mitotic divisions.

7. Other cells crawl along the stem and develop into spores.

zygote (2n)

solitary cells (n) (trophic phase)

MEIOSIS

amoeboid cells (n)

fruiting bodies (n)

ASEXUAL REPRODUCTION

cells gather (aggregation phase)

4. The zygote wall breaks apart, releasing new haploid amoeboid cells.

➡ Haploid (n)
➡ Diploid (2n)

migrating aggregate ('slug')

5. When food is scarce, hundreds of cells gather in response to a chemical signal, forming an aggregate with a slug-like shape.

6. The aggregate migrates for a certain period of time and finally stops. Some cells arrange to form a stem that supports an asexual fruiting body.

Figure 7.4 Life cycle of a cellular slime mould (*Dictyostelium*).

phylogenetics indicate as the sister taxon of metazoans, within the broader group of the Opisthokonta. Choanoflagellates reproduce asexually, by mitosis, and we have only limited clues as to their sexuality. In a single species, *Salpingoeca rosetta*, the alternation between haploid and diploid conditions has recently been demonstrated and anisogamy s.s. has been observed, i.e. the fusion of two types of gametes of different sizes, both flagellated (Levin and King 2013).

7.1.9 Mycetozoans

Many forms of slime mould belong to the Mycetozoa. These are amoeboid single-celled organisms presenting life cycles that alternate solitary and aggregation phases. They reproduce both sexually and asexually. Among the mycetozoans there are the dictyosteliids or cellular slime moulds (Figure 7.4) and the myxogastrids or plasmodial slime moulds (Figure 7.5). Their life cycles are described in Section 2.6.

7.2 Brown Algae (Phaeophyceans)

Typical of the brown algae (see Fritsch 1945; Lee 2008; de Reviers *et al.* 2015), all of which are multicellular, is a multigenerational life cycle with alternation between a haploid gametophyte and a diploid sporophyte. Among the brown algae there are isogamous, anisogamous s.s. and oogamous taxa; fertilization occurs in water, except in oogamous forms.

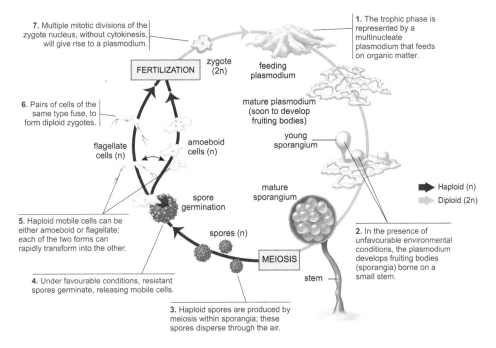

7. Multiple mitotic divisions of the zygote nucleus, without cytokinesis, will give rise to a plasmodium.

1. The trophic phase is represented by a multinucleate plasmodium that feeds on organic matter.

FERTILIZATION   zygote (2n)

feeding plasmodium

6. Pairs of cells of the same type fuse, to form diploid zygotes.

mature plasmodium (soon to develop fruiting bodies)

flagellate cells (n)

amoeboid cells (n)

young sporangium

spore germination

mature sporangium

Haploid (n)
Diploid (2n)

5. Haploid mobile cells can be either amoeboid or flagellate; each of the two forms can rapidly transform into the other.

spores (n)

MEIOSIS

2. In the presence of unfavourable environmental conditions, the plasmodium develops fruiting bodies (sporangia) borne on a small stem.

4. Under favourable conditions, resistant spores germinate, releasing mobile cells.

stem

3. Haploid spores are produced by meiosis within sporangia; these spores disperse through the air.

Figure 7.5 Life cycle of a plasmodial slime mould (*Physarum*).

In the different groups of brown algae, the relative importance of the two generations varies greatly. In the Fucales, Ascoseirales and in *Syringoderma abyssicola* (Syringodermatales) the gametophyte is very small and remains included in the sporophyte. In extreme cases the gametophytic generation is suppressed (diplontic cycle, for example in *Fucus*). The gametophytes of the Laminariales are also tiny, but in this case they are free, with separate sexes; the sporophytes of this group are perennial and sometimes of enormous size (Figure 7.6). In the Scytosiphonaceae, on the contrary, the sporophyte is extremely reduced. Gametophyte and sporophyte are similar in the Asterocladiales, Nemodermatales and Ectocarpales.

Sexuality has apparently been lost in Discosporangiales and other brown algae. In the diploid sporophytes of *Ectocarpus* some cells of the lateral branches grow in size and eventually become unicellular sporangia. The nucleus divides by meiosis, followed by a series of mitoses that lead to the formation of 32 or 64 haploid zoospores. After a dispersal phase, the zoospores attach to the substrate and develop into haploid gametophytes that produce multicellular gametes. There are two types of gametes: the female ones attach themselves to the substratum and secrete a pheromone that attracts the male gametes, which are mobile. The zygote develops into a new sporophyte.

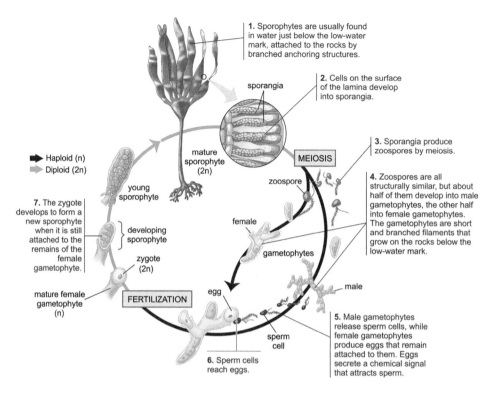

1. Sporophytes are usually found in water just below the low-water mark, attached to the rocks by branched anchoring structures.

sporangia

2. Cells on the surface of the lamina develop into sporangia.

Haploid (n)
Diploid (2n)

mature sporophyte (2n)

MEIOSIS

3. Sporangia produce zoospores by meiosis.

zoospore

4. Zoospores are all structurally similar, but about half of them develop into male gametophytes, the other half into female gametophytes. The gametophytes are short and branched filaments that grow on the rocks below the low-water mark.

young sporophyte

7. The zygote develops to form a new sporophyte when it is still attached to the remains of the female gametophyte.

developing sporophyte

female

gametophytes

zygote (2n)

mature female gametophyte (n)

FERTILIZATION

egg

male

sperm cell

6. Sperm cells reach eggs.

5. Male gametophytes release sperm cells, while female gametophytes produce eggs that remain attached to them. Eggs secrete a chemical signal that attracts sperm.

Figure 7.6 Life cycle of a diplontic brown alga (*Laminaria*).

7.3 Red Algae (Rhodophyceans)

Almost all red algae (see Fritsch 1945; Lee 2008; Kamiya *et al.* 2017) are multicellular, and some of their life cycles are among the most complex. A 'typical' life cycle cannot be identified – but here we describe *Polysiphonia* (Figure 7.7), whose cycle is similar to that of other taxa. In this red alga, the meiosis originates from four haploid spores (*tetraspores*), all viable, each of which produces a male or female gametophyte, multicellular and haploid. At maturity, special structures (*spermatangia*) are formed on some specialized branches of the male gametophyte. Spermatangia contain male gametes, in this case called *spermatia*. Similarly, on some specialized branches of the female gametophyte *carpogonia* are formed, unicellular gametangia that are in fact equivalent to the egg cell. In red algae there are no flagellated cells, and the sperm, carried by water, reach the carpogonia passively. The latter have a filamentous cytoplasmic expansion, the *trichogyne*, which receives the male gametes, allowing fertilization. A composite structure is then formed, the *carposporophyte*, to which both the haploid cells deriving from

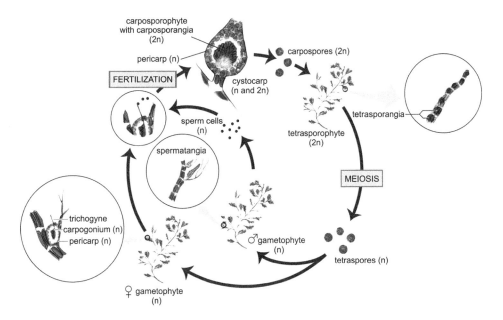

Figure 7.7 Life cycle of a red alga (*Polysiphonia*).

the female gametophyte and the diploid cells originating from the zygote by mitosis contribute. The carposporophyte represents a generation that is additional to those of the gametophyte and sporophyte typical of other algae and land plants. The carposporophyte in turn produces diploid spores (*carpospores*), which are released into the water, developing into *tetrasporophytes*, also diploid. Some cells of the tetrasporophyte undergo meiosis and give rise to the *tetraspores*.

The events following fertilization differ from one order to another. In many species, the fertilized carpogonium produces another long filament that carries the diploid nucleus outside and deposits it in an auxiliary cell, where mitosis begins. In orders with a more derived organization there are two types of auxiliary cells with which the fertilized carpogonium fuses: the nourishing auxiliary cells supply the nutritious substances supporting the development of the carposporophyte, while the generative auxiliary cells originate the filaments of the *gonimoblast*, at the top of which are differentiated the carposporangia that produce the carpospores.

Asexual reproduction is also known, through spores capable of developing into a thallus similar to the one from which they originated. Monosporangia, each of which produces only one spore, are distinguished from parasporangia, which produce more than one spore each.

7.4 Green Plants (Viridiplantae)

Green plants include the paraphyletic group of the green algae and the monophyletic group of the land plants or embryophytes. We will consider them
here under five headings: green algae, bryophytes, pteridophytes, gymnosperms and angiosperms; all of these, except the last, correspond to popular
polyphyletic or paraphyletic groups.

7.4.1 Green Algae

The green algae (see Fritsch 1935; Lee 2008; Neestupa 2015; Leliaert *et al.*
2015a, 2015b) are an informal group which includes two main clades, the
Chlorophyta in the strict sense, and those members of the Streptophyta that
do not belong to the lineage of the land plants. Green algae are a group of
eukaryotes with an extraordinary diversity of reproductive mechanisms and
life cycles, which only partially matches the obvious disparity in their structural plans. The latter range from unicellular (e.g. *Chlamydomonas*; Figure 7.8)
to colonial (e.g. *Volvox*), plasmodial (e.g. *Caulerpa*) and multicellular with
filamentous aspect (simple or branched; e.g. *Spirogyra, Oedogonium*) or laminar
(e.g. *Ulva*), etc. Important differences in reproductive modes are often observed
even among forms that are phylogenetically very close. We give some
examples below. The unusually complex life cycles of some green algae are
described in Box 7.1.

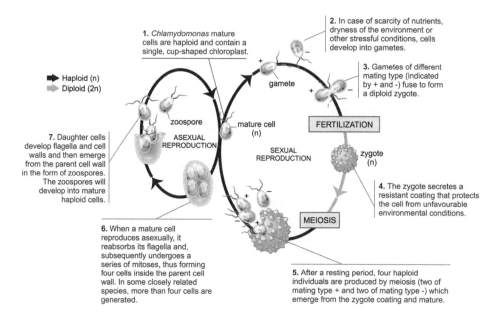

Figure 7.8 Life cycle of a haplontic unicellular green alga (*Chlamydomonas reinhardtii*).

Box 7.1 The Complex Life Cycles of Some Green Algae

The filamentous green algae of the genera *Oedogonium* and *Bulbochaete* (Oedogoniales) can reproduce asexually by fragmentation or by multiflagellate zoospores, which after a period of free life become attached to the substratum by rhizoids and begin to divide, starting the formation of a new filament; or sexually, through strongly dimorphic gametes. The male gametes are multiflagellate and similar to the zoospores, but smaller; the female gamete (*oogonium*) is large and spherical. In these genera, some species are monoecious, others dioecious, and two different sexual cycles are recognized. In the **macrandrous** mechanism, present for example in *Oedogonium cardiacum*, the antheridium is a series of small discoidal cells at the tip of a parental cell; in each anteridium ripen two or four spermatozoa, which are attracted to a pheromone released by the oogonia, which they reach and fertilize. In the **nannandrous** species, such as *Bulbochaete hiloensis*, there is the release of an *androspore*, of intermediate diameter between a spermatozoon and a zoospore, which attaches to an algal filament near the mother cell of the oogonium which has attracted it – and there it germinates, as if it were a zoospore, but generates a dwarf male filament in the apical cells, from which the sperm cells differentiate. Freshly released, these can fertilize the adjacent oogonium. Androspores and dwarf male filaments can also de-differentiate, taking vegetative structure and functions. Meiosis is zygotic.

The life cycle of *Hydrodictyon* is probably the most complex and the most plastic of all green algae. In the vegetative state, this alga looks like a large-mesh network, consisting of a thousand elongated cells that meet in groups of three to form the nodes of the network. In a mature colony, each cell can grow up to 1 cm in length. Within a single cell, a repeated series of divisions gives rise to cells (also called *zooids*) that can behave in three different ways. A first possibility is the initiation of an asexual cycle: the zooids produced by a cell lose the flagella, join together in the network and undergo a series of mitoses, while the wall of the parental cell turns into gel. In the end, the new network is released into the water and its cells undergo a huge increase in size, but will not divide any more. Alternatively, the zooids are released in the form of flagellate cells for which two different fates are open. They can behave like isogametes, giving rise to a zygospore in which meiosis occurs, followed by some mitotic divisions that lead to the formation of 4–8 zoospores; alternatively, the zooid that has not participated in a conjugative event behaves as an azygospore, from the germination of which derives a single zoospore. The further fate of the zoospores, however, is identical, whether they are derived from a zygospore or an azygospore. All zoospores, in fact, are

continues

transformed into uninucleate polyhedra, within which many nuclear divisions occur, leading to the production of a plasmodial (or coenobitic) polyhedron; from the cellularization of this (that is, from the separation of distinct cytoplasmic units containing one nucleus each) a new network is formed, with modes similar to those in the asexual cycle.

A number of different types of asexual reproduction are known in the green algae, including the simple fragmentation of a filament into two or more parts. A more specialized mechanism is the production of flagellate spores (*zoospores*) – most often by non-differentiated vegetative cells, in a few cases instead by specialized structures (*sporangia*) – in a number that is usually a power of two. Some chlorophyceans, such as *Trebouxia*, produce non-flagellate spores (*aplanospores*). Aplanospores of the same shape as the mother cell, produced within it in a number equal to a power of two, are called *autospores*; they are found in many of the unicellular green algae traditionally ascribed to the genus *Chlorella* (which today is known to be polyphyletic). Green flagellate algae organized in the form of colonies of flagellated cells with a fixed number of members (these structured colonies are also known as *coenobia*) generate daughter colonies that reach the final cell number before separating from the mother colony; after leaving the latter, the cells that form a daughter colony will increase in size but will not undergo further mitoses.

A distinction between germ and soma appears within the coenobia. The cells responsible for the asexual reproduction of the colony (*gonidia*) are approximately one-half the cells in a coenobium in *Pleodorina*, but account for just a few units in the large coenobia of *Volvox*. In the latter genus, the release of daughter coenobia following the laceration of the maternal one is soon followed by the death of the latter.

Sexual reproduction in the green algae can also take place according to different modes, above all as regards the nature of the gametes. In some species, these are all morphologically equal (isogametes); in others, one gamete is larger than the other, but both are mobile (anisogametes s.s.); in others (those traditionally ascribed to the genera *Pleodorina*, *Eudorina* and *Volvox*), smaller, flagellated 'male' gametes contrast with much larger 'female' ones without flagella (oogametes). In some unicellular forms, such as *Dunaliella* and *Polytomella*, ordinary vegetative cells may act as gametes, but in all other cases the gametes are specialized cells. In *Phyllocardium* a series of divisions leads to the formation of gametes smaller than ordinary cells; *Dangeardinella* produces gametes of different sizes, but the fusion may involve either gametes of different types or gametes of the same type. In the other isogamous taxa, the tiny gametes are formed by binary divisions within the wall of the parent cell, which often produces 16, but also 32 or 64 gametes. In *Chlamydomonas coccifera* the macrogamete is formed from an ordinary vegetative cell, without the intervention of a nuclear division; the microgametes are formed instead in groups of 16, as a result of four divisions within a cell. An individual of the unicellular *Chlorogonium oogamum* produces either a single naked egg or a large number of spermatozoans. In the oogamous forms, however, gametes usually form in specialized organs (*gametangia*).

The gametes of the chlorophyceans are almost always flagellated, but the Zygnematales produce gametes without flagella. In some cases, gametogenesis is induced by environmental factors, in other cases the presence of two different sexual strains is necessary and the vegetative cells of one of the two strains secrete a substance that induces differentiation into gametes of some cells of the other strain. In some oogamous species the egg produces a sex pheromone capable of attracting male gametes.

Eudorina and *Pleodorina* are generally dioecious. In *Pleodorina*, male coenobia are smaller than the female. The genus *Volvox* includes both monoecious and dioecious species. Monoecious forms can be either protandrous or protogynous. Parthenogenesis, in the form of *parthenospore* production from unfertilized egg cells, has occasionally been observed in *Volvox aureus* and *Eudorina elegans*. The meiotic reduction occurs during the germination of the zygospore. Of the products of meiosis, four may survive (e.g. *Chlamydomonas*, *Gonium*), or only one (*Pleodorina*, *Eudorina*, *Volvox*). In many groups, meiosis occurs immediately after the formation of the zygote. Among the green algae, sexuality is unknown in a number of different groups, such as Pseudoscourfieldiales, Pedinophyceae, Chlorodendrophyceae, Chaetopeltidales and many groups of Sphaeropleales.

Let's now examine the reproductive modes of some groups of green algae in greater detail.

Many green algae ascribed to the **Chlamydomonadales** are unicellular – for example those traditionally attributed to the genus *Chlamydomonas*, which has proved to be strongly polyphyletic; others are colonial/coenobitic, such as those traditionally classified in the genera *Gonium*, *Pleodorina*, *Eudorina* and *Volvox*. While the new classification of the whole group is still problematic, we continue to use the traditional names of genera and species.

Chlamydomonas species are exemplary unicellular haplonts: after fertilization, the zygote immediately undergoes meiosis and usually produces four flagellated haploid spores (*zoomeiospores*), which develop into unicellular haploid individuals; sometimes, however, the spores are formed by the meiotic products through an additional mitotic division (*zoomitospores*). By mitotic division, a vegetative cell can reproduce asexually, but it can also produce two isogametes.

Asexual reproduction and sexual reproduction also coexist in colonial chlamydomonads, including *Volvox*, in which the colonies (*coenobia*) comprise up to a few thousand cells arranged to form a hollow sphere. In *V. carteri*, a young adult asexual colony consists of a monolayer of 2000–4000 small biflagellate somatic cells and about 16 large *gonidia* immersed in the extracellular matrix beneath the somatic cells. At maturity, a gonidium has a volume equal to one-thousandth of a somatic cell and differs from it also in the absence of functioning flagella. While somatic cells seem to be absolutely unable to divide, each mature gonidium undergoes a series of 11 or 12 divisions, giving rise to a new colony formed in turn by somatic cells and gonidia. Sexual reproduction in these green algae involves the production of resistant spores capable of surviving adverse conditions. Eggs and spermatozoa are formed from somatic cells that become germ cells in response to specific environmental signals. Since *V. carteri* is haploid, gametogenesis does not involve meiosis, but it requires a series of asymmetric mitoses that lead to the production of large cells, each of which differentiates into an egg or a packet of sperm cells. In *V. carteri* sexes are separate and genetically distinct, but in other *Volvox* species both types of gametes can be produced by a single clone (Kirk 2001).

The **Cladophorales** produce both zoospores and flagellated gametes; isogamety is the rule. At least in some species of *Cladophora* and *Chaetomorpha* there is alternation between a sexual and an asexual generation, but these are isomorphic. Some *Cladophora* have separate sexes, and heterochromosomes are recognizable. *Cladophora glomerata* is diplontic.

The **Ulvales** are normally dioecious. In *Ulva* and *Enteromorpha* there is alternation between an asexual diploid generation and a sexual haploid generation, morphologically similar (Figure 7.9). Asexual reproduction can occur by simple fragmentation of the thallus or by zoospores, 4–8 from each mother cell. Sexual reproduction is by isogamy, through biflagellate gametes produced in the same way as zoospores. The latter and the gametes are similar.

In the **conjugates** (**Conjugatophyceae** or **Zygnematophyceae**) there are no flagellated stages and sexual reproduction occurs by amoeboid gametes; the zygote is covered with a resistant wall, thus taking on the function of a spore. The conjugates include two morphological types: some (almost all of which belong to the desmid clade) are unicellular and discoidal; the others (almost all of them zygnematals, such as *Spirogyra*) are filamentous and can reproduce by fragmentation.

Characteristic of all conjugates is the mechanism through which sexual reproduction takes place, which is called *conjugation*. In the filamentous

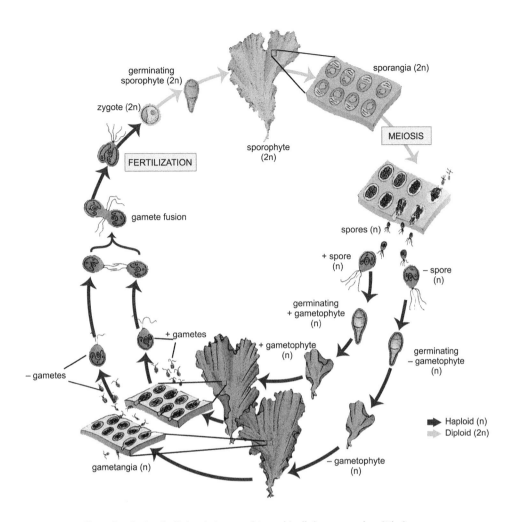

Figure 7.9 Life cycle of a haplodiplontic isomorphic multicellular green alga (*Ulva*).

zygnematals conjugation begins with the formation of a slimy sleeve which holds two contiguous filaments together. In each pair of cells involved (one per strand), one of the cells forms a protruding papilla in the direction of the other filament; as a rule, all the cells that start producing these papillae belong to the same filament, which in some species is the male one, while in others it can also be the female one. Subsequently, the other filament also forms similar protruding papillae. The apices of the opposing papillae come into contact with each other, the cell walls dissolve, and each set of paired cells forms a conjugative tube. The cytoplasm of the male cell detaches from the cell wall and progressively moves into the partner cell. In some species, the zygospore is formed in the conjugation tube, in others in the 'female' cell after it has received the partner's cytoplasm. However, there is no real distinction between donor (male) and acceptor (female) filaments, because the same filament can be simultaneously engaged in conjugative phenomena with two other filaments, and its cells can act at the same time as 'male' towards the one and as 'female' towards the other.

Conjugation can be *lateral* or *scalariform*. In the first case, a tube is formed between two contiguous cells of the same filament, and the content of one cell passes into the other through this tube. In the second case, the conjugation tube is formed between cells of two distinct filaments. As a rule, conjugation involves filaments belonging to the same clone (*homothallic conjugation*), but sometimes only between filaments of different clones (*heterothallic conjugation*). In this case, karyogamy is delayed with respect to plasmogamy, but is immediately followed by meiosis, from which only a haploid nucleus survives; the cell that contains it will give rise to a new algal filament.

The **Charales** are haplonts, either dioecious or monoecious. They reproduce asexually, by means of unicellular or multicellular *bulbils* that are formed on the branches of the rhizoids, and sexually, by oogamy. Gametes are produced in specialized structures (male *antheridia* and female *oosporangia*). At fertilization, the egg cell remains protected by coating cells characteristically coiled in a spiral and surrounded by a calcified envelope. Meiosis occurs immediately before the germination of the fertilized oogonium; the latter will grow into a protonema, from which the primary axes of the alga will form.

7.4.2 Bryophytes

In traditional botanical classifications a division (or phylum) Bryophyta was recognized, divided into three classes: **Hepaticae** or liverworts, **Musci** or mosses and **Anthocerotae** or hornworts (see Frey and Stech 2009). The phylogenetic relationships accepted today, however, indicate the paraphyletic nature of the bryophytes, because the hornworts would be the sister group of the vascular plants ('pteridophytes' and spermatophytes), the mosses the sister group of hornworts plus vascular plants, and the liverworts the sister group of all those clades. For convenience, however, we discuss mosses and liverworts together (Figure 7.10).

Like all embryophytes, the bryophytes are haplodiplonts: in their life cycle there is alternation between haploid and diploid phases of comparable complexity and duration. The green moss plantlets are called *gametophores* and constitute the haploid generation (gametophyte). They sprout up from a common base consisting of a sort of green horizontal stolon, the *protonema*, and comprise a thin stem, sometimes branched, with two or more often three rows of leaflets. Gametophores bear the reproductive organs (male gametangia

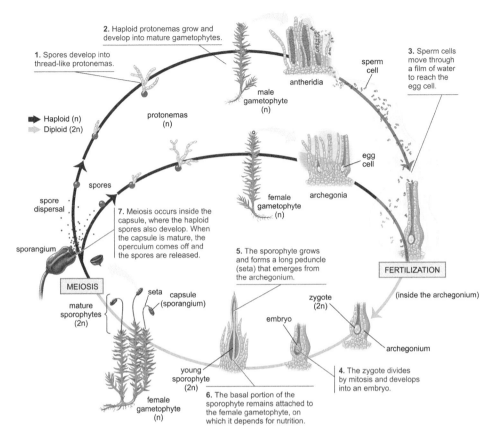

1. Spores develop into thread-like protonemas.

2. Haploid protonemas grow and develop into mature gametophytes.

3. Sperm cells move through a film of water to reach the egg cell.

Haploid (n)
Diploid (2n)

protonemas (n)

sperm cell

antheridia

male gametophyte (n)

egg cell

spores

spore dispersal

7. Meiosis occurs inside the capsule, where the haploid spores also develop. When the capsule is mature, the operculum comes off and the spores are released.

female gametophyte (n)

archegonia

sporangium

5. The sporophyte grows and forms a long peduncle (seta) that emerges from the archegonium.

FERTILIZATION

(inside the archegonium)

MEIOSIS

mature sporophytes (2n)

seta   capsule (sporangium)

zygote (2n)

embryo

archegonium

young sporophyte (2n)

4. The zygote divides by mitosis and develops into an embryo.

female gametophyte (n)

6. The basal portion of the sporophyte remains attached to the female gametophyte, on which it depends for nutrition.

Figure 7.10 The haplodiplontic heteromorphic life cycle of a moss (*Polytrichum*).

or *antheridia* and female gametangia or *archegonia*), which respectively produce mobile (flagellated) male gametes (*antherozoids*) and immobile female gametes (*oospheres*). Since the plant that produces them is also haploid, these gametes are not the result of a meiosis, but of a mitosis, accompanied by a specific and divergent differentiation of the cells that become antherozoids or oospheres, respectively. From the zygote produced by the fertilization of an oosphere by an antherozoid, a diploid multicellular organism is formed, which has the form of a long filament (*seta*) ending with an *urn* containing cells capable of going into meiosis. The products of meiosis are the spores. Filament and urn are therefore the diploid multicellular phase (the sporophyte) of the life cycle of bryophytes. The oosphere, when it is fertilized, remains inserted in the cellular structure of the gametophyte (i.e. it does not detach, as an animal egg would); as a consequence, the sporophyte that derives from it remains inserted on the gametophyte, giving rise to a chimera with cells containing the different genomes of gametophyte and sporophyte.

Some mosses (e.g. *Funaria*, *Pottia*) are monecious, with antheridia and archegonia on the same gametophore, while in other mosses (e.g. *Barbula*, *Polytrichum*, *Rhacomitrium*) there are distinct male and female gametophores (dioecy). When gametophores of opposite sex nonetheless derive from the same protonema, this condition is called *pseudodioecy*. Virtually all mosses are homosporous: all spores are of the same size and seem to have the same value. Some species of *Macromitrium* and *Schlotheimia*, however, produce two types of spores in each capsule. Larger spores develop into large gametophytes with archegonia, smaller spores into dwarf male gametophytes that live epiphytically on the female ones.

Sexual reproduction is common for most monoecious mosses but rare for many dioecious species. Sporophytes are rare or unknown for 69% of the dioecious species of the British moss flora (Haig 2016).

In liverworts, the sporophyte is even less conspicuous than in mosses and is also completely dependent on the gametophyte. As in mosses, liverwort gametophores can be monoecious, producing both antheridia and archegonia, or dioecious. In the former case, the gametangia of the two sexes can be arranged on distinct branches or on the same branch, or even mixed together in a common reproductive structure. All liverworts are homosporous.

In bryophytes self-incompatibility is rare and sexual reproduction can occur by self-fertilization, but parthenogenesis is unknown.

Clonal propagation of bryophytes is essentially restricted to the haploid phase. Many bryophytes have perennial gametophytes and produce asexual propagules. Of a number of species (for example, 18% of English mosses) only the gametophytic phase is known. Vegetative reproduction of the sporophyte is accidental, but known for many species (Haig 2016).

Tiny soil arthropods (springtails, oribatid mites) can facilitate the dispersal of male gametes. The moss *Ceratodon purpureus* emits volatile substances, different in male and female gametophytes, which are attractive to such arthropods, suggesting a mechanism similar to the zoogamous pollination of flowering plants (Rosenstiel *et al.* 2012).

7.4.3 Pteridophytes

The traditional division (or phylum) Pteridophyta is no longer recognized as a natural group. According to the current understanding of phylogeny, the most basal split among the tracheophytes or vascular plants (those provided with vessels for sap transport) is that between the Lycophytina (to which belong the lycopods, the lesser clubmosses *Selaginella* and the quillworts *Isoetes*) and the Euphyllophyta. The latter in turn include the Moniliformes (ferns and horsetails) and the seed plants or Spermatophyta. As with 'bryophytes', however, we

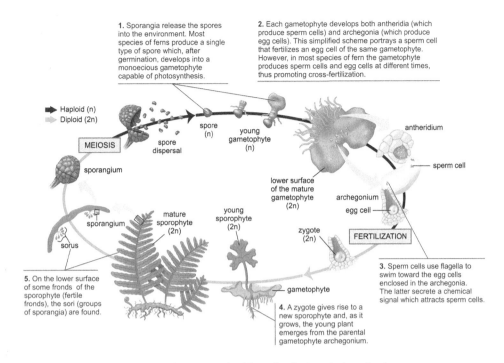

1. Sporangia release the spores into the environment. Most species of ferns produce a single type of spore which, after germination, develops into a monoecious gametophyte capable of photosynthesis.

2. Each gametophyte develops both antheridia (which produce sperm cells) and archegonia (which produce egg cells). This simplified scheme portrays a sperm cell that fertilizes an egg cell of the same gametophyte. However, in most species of fern the gametophyte produces sperm cells and egg cells at different times, thus promoting cross-fertilization.

Haploid (n)
Diploid (2n)

spore (n)
spore dispersal
MEIOSIS
sporangium
young gametophyte (n)

antheridium
sperm cell

lower surface of the mature gametophyte (2n)
archegonium
egg cell

sporangium
mature sporophyte (2n)
young sporophyte (2n)
zygote (2n)
FERTILIZATION

sorus

gametophyte

5. On the lower surface of some fronds of the sporophyte (fertile fronds), the sori (groups of sporangia) are found.

3. Sperm cells use flagella to swim toward the egg cells enclosed in the archegonia. The latter secrete a chemical signal which attracts sperm cells.

4. A zygote gives rise to a new sporophyte and, as it grows, the young plant emerges from the parental gametophyte archegonium.

Figure 7.11 The haplodiplontic heteromorphic life cycle of a fern (*Polypodium*).

continue, for convenience, to group the ancient lycopodials, ferns and horsetails under the collective term of pteridophytes.

Pteridophytes (see Fischer 2009) have a haplodiplontic cycle, but the haploid gametophyte is, as a rule, much less conspicuous than the sporophyte (Figure 7.11). The sporophytes of lycopods are herbaceous plants with creeping rhizomes from which short vertical branches sprout out; these branches bear the sporangia, often in the shape of a small pinecone. All lycopods are homosporous. The gametophytes developing from their spores are monoecious (i.e. they produce both antheridia and archegonia); in some species these gametophytes are green and autotrophic, while in others they are heterotrophic and develop underground. *Selaginella* species, instead, are heterosporous and the female gametophyte develops inside the former spore wall. In this respect, these plants are closer to seed plants than are lycopods, but in *Selaginella* the megaspore is released rather than retained on the sporophyte. The microgametophyte consists of a single vegetative cell, plus the antheridium that produces numerous flagellated sperm cells (*antherozoids*), which are released following the breaking of the microspore wall and reach an archegonium moving in the thin film of water that frequently covers these plants.

Horsetails (*Equisetum* spp.) are homosporous. Some species produce two types of sporophytic stems, one sterile, the other fertile, i.e. with sporangia;

in other species there is only one type of sporophytic stem, with apical sporangia and the lower part covered in whorls of green 'leaves' similar to those of the sterile stems of the other horsetails. Spores germinate on moist soil, developing into tiny green gametophytes which are monoecious or change sex with age (successively monoecious). Male gametes are mobile and have numerous flagella, while female gametes remain attached to the gametophyte. The zygote, which is also sessile, is fed for a while by the gametophyte.

A distinction between fertile and sterile fronds (sometimes very different morphologically, as in *Blechnum spicant* and *Osmunda regalis*) is also found in some ferns, but in most species the same frond performs both photosynthetic and reproductive functions. The sporangia, grouped in sori, are located on the edge of the frond or, more often, on its lower surface. Ferns are homosporous, except for two small groups of water ferns (Marsileales and Salviniales). From the spores are formed small ribbon- or heart-shaped green gametophytes (*prothalli* or *prothallia*), each of which, as a rule, produces both antheridia and archegonia. As in horsetails, antherozoids are also mobile in ferns, while female gametes remain attached to the gametophyte even after fertilization. Vegetative reproduction of the sporophyte is very common, but in some ferns, especially among the Hymenophyllaceae, the gametophyte can also reproduce by propagules. In some phyletic lines among the Pteridaceae and the Hymenophyllaceae, no sporophyte is produced, and populations survive only thanks to the asexual reproduction of gametophytes (Kuo *et al.* 2017).

7.4.4 Gymnosperms

The spermatophytes (or seed plants, or phanerogams) include the gymnosperms, together with the angiosperms or flowering plants (see next section). They have a haplodiplontic life cycle that is similar to that of the pteridophytes, but in the spermatophytes the sporophyte is much more developed than the gametophyte. The sporophyte is the diploid plant with roots, stems, leaves and reproductive organs. Specialized leaves (the male *microsporophylls* and the female *megasporophylls*) bear the *sporangia* (male *microsporangia* and female *megasporangia*), in which spores (*microspores* and *megaspores*, respectively) will originate by meiosis. Only one megaspore survives from the meiosis of a megaspore mother cell, whereas in the case of microspores all four products of the meiosis of a mother cell are viable. *Microgametophytes* and *megagametophytes*, respectively, originate from the two kinds of spores. Spermatophytes are therefore heterosporous with unisexual gametophytes.

Spermatophytes, as the name implies, produce seeds. In gymnosperms (Richards 1997) the seed sits on a modified leaf (one of the scales of a pine cone, or an equivalent structure). The seed is almost always naked (unlike in

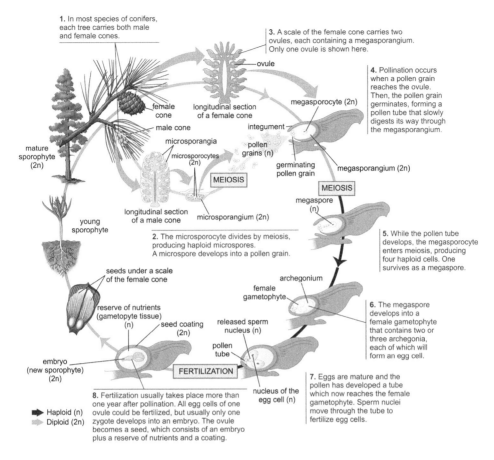

1. In most species of conifers, each tree carries both male and female cones.

3. A scale of the female cone carries two ovules, each containing a megasporangium. Only one ovule is shown here.

ovule

4. Pollination occurs when a pollen grain reaches the ovule. Then, the pollen grain germinates, forming a pollen tube that slowly digests its way through the megasporangium.

female cone

longitudinal section of a female cone

megasporocyte (2n)

male cone

integument

microsporangia

pollen grains (n)

mature sporophyte (2n)

microsporocytes (2n)

germinating pollen grain

megasporangium (2n)

MEIOSIS

MEIOSIS

megaspore (n)

longitudinal section of a male cone

microsporangium (2n)

young sporophyte

2. The microsporocyte divides by meiosis, producing haploid microspores. A microspore develops into a pollen grain.

5. While the pollen tube develops, the megasporocyte enters meiosis, producing four haploid cells. One survives as a megaspore.

seeds under a scale of the female cone

archegonium

female gametophyte

6. The megaspore develops into a female gametophyte that contains two or three archegonia, each of which will form an egg cell.

reserve of nutrients (gametopyte tissue) (n)

seed coating (2n)

released sperm nucleus (n)

pollen tube

embryo (new sporophyte) (2n)

FERTILIZATION

nucleus of the egg cell (n)

7. Eggs are mature and the pollen has developed a tube which now reaches the female gametophyte. Sperm nuclei move through the tube to fertilize egg cells.

➡ Haploid (n)
➡ Diploid (2n)

8. Fertilization usually takes place more than one year after pollination. All egg cells of one ovule could be fertilized, but usually only one zygote develops into an embryo. The ovule becomes a seed, which consists of an embryo plus a reserve of nutrients and a coating.

Figure 7.12 The haplodiplontic heteromorphic life cycle of a conifer (*Pinus*).

angiosperms; see next section) – but in a few cases, such as *Ginkgo*, *Taxus* and *Juniperus*, the cone is completely closed around the seed, forming a fleshy envelope that resembles a fruit. The endosperm in the seeds of gymnosperms derives from the tissues of the gametophyte that produced the female gamete and is called *primary endosperm*, to distinguish it from the endosperm in the seeds of the flowering plants, which is of different origin.

In the **conifers** (Figure 7.12), the main group of gymnosperms, the reproductive structures are the unisexual cones (strobili), on which are found the mother cells of the spores. Each microspore forms a microgametophyte of four cells, only one of which is a gamete.

Like the pollen grains of the angiosperms, gymnosperm microgametophytes are dispersed by wind. The female gametophyte resulting from the germination of a megaspore is much larger, and for a long time it retains a coenocytic or plasmodial organization, i.e. it contains many nuclei (up to 7200) within a common cytoplasmic mass. Only at a later stage of

365

development, which may take up to one year, does the female gametophyte eventually become cellularized and differentiate two or three archegons, in each of which a large egg cell matures.

Pollination occurs before the female gamete is mature, and it may take more than a year between pollination and fertilization. Within a single gametophyte two or three egg cells can be fertilized, but only one zygote develops into an embryo.

A few gymnosperms (the conifers *Picea vulgaris*, *Pinus laricio*, *Abies balsamea* and *A. pindrow* and the cycad *Encephalartos villosus*) are parthenogenetic. The diploid condition is obtained by the fusion of the egg nucleus with that of the ventral cell of the archegonium, in a form of intragametophytic self-fertilization.

Some cases of polyembryony are known, for example in *Pinus*, but generally all the multiple embryos, except one, die during seed maturation. Among the **Gnetales**, archegonia are still recognizable in *Ephedra*, but not in *Welwitschia* and *Gnetum*, where, as in the flowering plants, the whole female gametophyte is reduced to a small number of nuclei (eight, as a rule), one of which is the female gametic nucleus.

7.4.5 Angiosperms

Angiosperms (see Richards 1997) are also heterosporous haplodiplonts, with unisexual gametophytes (Figure 7.13).

In its typical form, the female gametophyte is represented by the embryonic sac, formed by the megaspore through three mitotic cycles. It therefore comprises a total of eight haploid nuclei, distributed into seven cells: the egg cell, three antipodal cells, two synergids and a central cell with two nuclei. In addition to this model (called '*Polygonum* type' by specialists in plant cytogenetics), variants are numerous, depending on the number of megaspores that participate in the formation of the embryo sac (in some cases, two or even four products of the meiosis of a mother cell survive, rather than only one, as usual), the number of mitotic divisions undergone by their nuclei, and the specific nature of the differentiated cells. For example, the '*Oenothera*-type' female gametophyte originates from a single megaspore, but the latter undergoes only two divisions, which give rise to the egg cell, the two synergids and the central cell, while the antipodal cells are missing. In grasses, on the other hand, the antipodals are often numerous, up to 300 in a bamboo species (*Sasa paniculata*).

Bisporic gametophytes (that is, deriving from two megaspores) occur in the '*Allium* type', where each megaspore undergoes only two mitoses, so the gametophyte contains eight cells as in the '*Polygonum* type'. *Tetrasporic*

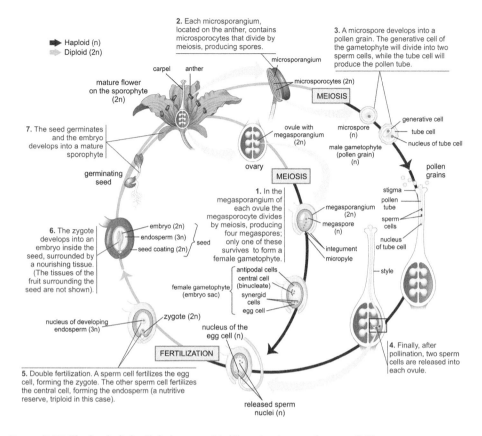

Figure 7.13 The haplodiplontic heteromorphic life cycle of an angiosperm (*Iris*).

gametophytes (formed by four megaspores) characterize the '*Adoxa* type', where each megaspore divides only once, thus giving again a total of eight nuclei, and the '*Fritillaria* type', where three megaspores merge, forming a triploid nucleus; the latter divides twice, as also does the haploid nucleus of the fourth megaspore: eight nuclei are thus obtained, four of which are haploid, the other four tetraploid. In some plants, female gametophytes of different compositions may coexist, even among the flowers of the same plant, up to five different types in *Delosperma* (Mauseth 1988).

The pollen grain is the male gametophyte of the angiosperms. At first it is formed by two cells only: a vegetative cell, which will produce the pollen tube, and a generative cell, which later divides to form two sperm cells. These, passing through the pollen tube, will be involved in a process known as *double fertilization* (Section 3.5.4.2): one of them will fertilize the egg cell, forming the zygote, the other will fuse instead with the nucleus of the central cell, forming a triploid cell from which, by mitosis, originates a tissue called *secondary endosperm* that will provide nourishment to the embryo. Unlike the gymnosperms (Section 7.4.4), the seeds of the angiosperms are formed inside an ovary, which will turn into a fruit with the ripening of the seeds. Some species, no more than 200, are viviparous: the seed, that is, germinates before being released from the mother plant. This phenomenon is widespread among the mangroves (Rhizophoraceae and Avicenniaceae). Some grasses (41 species in 13 genera, among them the so-called var. *vivipara* of *Poa alpina*) are pseudoviviparous: instead of producing flowers, the apical part of their inflorescences generates bulbils or even small plants with well-differentiated leaves, which are abscissed from the mother plant as propagules ready to take root.

Most species of angiosperms are monoecious. Dioecious species, although much less numerous, are present in many families, distributed throughout the whole group. Few families include dioecious species only; among these are the Salicaceae. Dioecy is more common in tropical forests and among the flowering plants of remote ocean islands. In all dioecious species the sex of the individual plant is fundamentally determined by sex chromosomes, but the phenotypic expression of sex can be modified by environmental influences. Male heterogamety (XX/XY system) is the rule, but systems with multiple sex chromosomes are also known, for example in various *Rumex* species (Section 6.1.1).

The distribution of the male and female functions in individuals and populations can be very diverse, as summarized in Table 7.1, and can vary considerably even among closely related species. As an example, nine different types of distribution are recorded in the ca. 60 species of the genus *Urtica* (Grosse-Veldmann and Weigend 2018).

For the various forms of apomixis (parthenogenesis) in flowering plants see Section 3.6.2.9. For vegetative reproduction, including polyembryony, see Sections 3.1.2.3 and 3.1.2.4.

Table 7.1 Temporal and spatial distribution of male and female functions in flowering plants (after Richards 1997 and Renner 2014, modified)

TEMPORAL DISTRIBUTION OF MALE AND FEMALE FUNCTIONS

Dichogamy	Temporally separated male and female function in each bisexual flower or in each inflorescence	Common, more often with protandry, as in carnations (*Dianthus*) and geraniums (*Geranium*), more rarely with protogyny, as in *Parietaria officinalis*

Table 7.1 (cont)

Sequential hermaphroditism/monoecy	In the individual plant, a male phase first, then a female phase, with a possible return to the male condition	ca. 250 species, including *Gurania*, *Psiguria*, *Arisaema*, *Elaeis*, catasetine orchids
Duodichogamy	The individual plant produces seasonally, in succession, male flowers, then female, then male again, asynchronously in respect to other plants in the population	A few species belonging to 5 genera in 4 families: *Acer*, *Bridelia*, *Castanea*, *Cladium*, *Dipteronia*
Heterodichogamy	Populations with two kinds of genetically different individuals, some male first, then female, the others vice versa (individual flowers can be unisexual or bisexual, protandrous or protogynous)	ca. 50 species of 20 genera in 12 families, e.g. a few *Acer*, *Spinacia*
DISTRIBUTION OF MALE AND FEMALE FUNCTIONS BETWEEN FLOWERS OF THE SAME PLANT		
Monoclinous condition	All the flowers of the plant are bisexual	Most flowering plants
Diclinous condition	All the flowers of the plant are unisexual	All dioecious angiosperms and those with unisexual flowers among the monoecious angiosperms

continues

Table 7.1 (cont)		
Andromonoecy	The individual plant produces bisexual flowers and male flowers	Rare, e.g. *Leptospermum scoparium*
Gynomonoecy	The individual plant produces bisexual flowers and female flowers	Common, in Asteraceae especially; e.g. *Aster, Solidago, Ligularia*
DISTRIBUTION OF MALE AND FEMALE FUNCTIONS WITHIN THE POPULATION		
Monoecy with bisexual flowers	All the flowers of all individuals are bisexual	72% of angiosperm species
Monoecy with unisexual flowers	All individual flowers are unisexual, but the individual plant produces both male and female flowers	Somewhat more frequent than dioecy; e.g. maize (*Zea mays*)
Androdioecy	Populations with both monoecious and male individuals	50 species, e.g. *Sagittaria* spp., *Datisca glomerata*, manna ash (*Fraxinus ornus*), *Phillyrea angustifolia*, *Schizopepon bryoniifolius*, a few species in *Potentilla* and *Geum*, rambutan (*Nephelium lappaceum*). Accidental in annual mercury (*Mercurialis annua*) and red campion (*Silene dioica*), in the latter species as a

Table 7.1 (cont)		
		consequence of attack by a pathogen
Gynodioecy	Populations with both monoecious and female individuals	More than 250 angiosperm genera contain at least one gynodioecious species; 59 genera contain both dioecious and gynodioecious species
Subandroecy	Populations with both female and andromonoecious individuals	Very rare: creeping thistle (*Cirsium arvense*) and a few American *Fuchsia*
Subgynoecy	Populations with both male and gynomonoecious individuals	Perhaps 2% of angiosperm species, e.g. cucumber (*Cucumis sativus*)
Dioecy	Populations with both male and female individuals	ca. 15,600 species belonging to 987 genera in 175 families, e.g. hemp (*Cannabis sativa*), common hop (*Humulus lupulus*), sheep's sorrel (*Rumex acetosella*), red campion (*Silene dioica*), spinach (*Spinacia oleracea*), willows (*Salix* spp.), poplars (*Populus* spp.), common grapevine (*Vitis*

continues

Table 7.1 (cont)		
		vinifera), papaya (Carica papaya), squirting cucumber (Ecballium elaterium)
Trioecy	Populations with monoecious, male and female individuals	Atriplex canescens, Carica papaya, Fraxinus excelsior, Pachycereus pringlei
Polygamy	Populations with five kinds of individuals (not all of which are necessarily present in the same population): with male flowers only, with bisexual and male flowers, with bisexual flowers, with bisexual and female flowers, and with female flowers only	Very rare; e.g. Thymelaea hirsuta

7.5 Fungi

Apart from the widespread possibility of multiplying by simple fragmentation of the mycelium, most fungi practise both asexual reproduction by *mitospores* and sexual reproduction by *meiospores* (Nieuwenhuis and James 2016). Several species, however, reproduce only sexually, others only asexually. Furthermore, in some species parasexuality occurs.

The simplest method of asexual reproduction in the fungi is by fragmentation. Some yeasts, which are unicellular fungi, reproduce by simple cell division. In filamentous fungi the mycelium can fragment into segments, each of which grows to form a new individual. In most yeasts and some filamentous fungi, asexual reproduction is asymmetric and is called budding. A bud may produce other buds even before being released, so a chain of cells is formed. These cells, however, eventually become free and function as spores. In addition to serving as reproductive cells, spores also represent a means of dispersal as well as a phase of resistance. Only the spores of the chytridiomycetes have flagella. The asexual spores of chytridiomycetes and zygomycetes are generally produced in sporangia at the tip of hyphae, whereas the asexual spores of ascomycetes and basidiomycetes, called *conidia*, are formed by differentiation of the terminal cells of a specialized hypha (*conidiophore*).

Parasexuality is a peculiar reproductive mode that allows recombination of genomes and production of new genotypes without resorting to meiosis but instead, in essence, relying on a series of mitoses (Section 5.2.5.2). In nature, a parasexual cycle is known both in some ascomycetes, for example in the mould *Aspergillus nidulans*, and in some basidiomycetes, for example in the wheat rust *Puccinia graminis*. Under experimental conditions, the possibility of undergoing a parasexual cycle has been demonstrated in several ascomycetes, among them the moulds *Fusarium moniliforme* and *Penicillium roqueforti*, the plant parasite *Verticillium dahliae* and the pathogen yeast *Candida albicans*. Some fungi are haplonts (for example, many zygomycetes), others diplonts (for example, some ascomycetes), while most, especially among the ascomycetes and basidiomycetes, have a life cycle that can be described as haplodiplonts only in a broad sense, because of the interposition of a dikaryotic (n+n) phase between the haploid (n) and the diploid (2n) phase (Section 2.1). In these fungi, two haploid hyphae (*monokaryotic hyphae* with a single nucleus per cell) can merge (plasmogamy without karyogamy), producing a first binucleate hypha (beginning of the dikaryotic phase) which proliferates, giving rise to a dikaryotic mycelium (formed by *dikaryotic hyphae* with two nuclei per cell). The dikaryotic mycelium proliferates and forms fruiting bodies that carry the reproductive structures specific for each group. In these cases, karyogamy occurs and later, by meiosis, haploid spores are produced that will give rise to the monokaryotic mycelium of the new generation. In the ascomycetes the dikaryotic phase is typically short-lived, while in many basidiomycetes it represents the predominant phase of the life cycle. Fungi are either *homothallic* (when fusion of two hyphae of the same individual is possible) or *heterothallic* (when only hyphae of individuals with a different conjugative type can fuse together).

In many fungi only asexual reproduction is known. In the absence (or at least the apparent absence) of sexual reproductive structures, these fungi were traditionally classified in an artificial taxonomic group called deuteromycetes.

7.5.1 Chytridiomycetes

Chytridiomycetes (see Voigt 2012) are the only fungi that reproduce sexually through flagellated zoospores. Sexual reproduction, known for a few species only, occurs according to different modes, but always with the formation of a zygote that represents the phase of resistance to adverse environmental conditions. Mobile isogametes are formed in *Synchytrium*. Other chytrids are oogamous, while still others form a conjugation tube between two thalli, through which the gametic nuclei can pass and eventually unite.

7.5.2 Zygomycetes

In the zygomycetes (Figure 7.14), the group of fungi to which many common moulds belong (see Voigt 2012), the mycelium consists of multinucleated (coenocytic) branching hyphae. These fungi do not form complex fruiting bodies. Germinating haploid spores produce a hypha that soon becomes multinucleate and branches strongly, producing rhizoids that anchor the fungus to the substrate and help absorb nutrients, like plant roots. Specialized hyphae (*sporangiophores*) grow upwards to form sporangia containing asexual spores. Sexual reproduction begins with the formation of anastomotic bridges between the hyphae of two individuals of compatible mating types. This results in plasmogamy and karyogamy followed by repeated mitoses, with the formation of a large *zygosporangium* containing many diploid nuclei. After a period of quiescence, sometimes lasting a few months, the nuclei contained in the zygosporangium undergo meiosis and a sporangiophore is formed from which the new haploid spores are released.

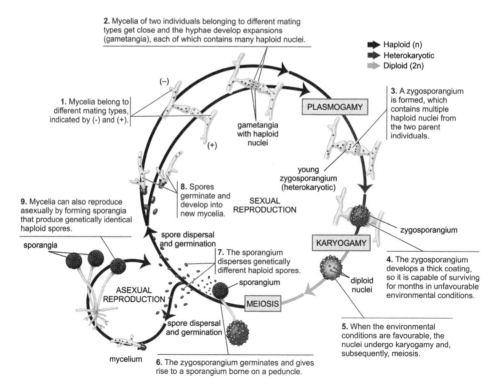

Figure 7.14 Life cycle of a zygomycete (the black bread mould, *Rhizopus stolonifer*).

7.5.3 Glomeromycetes

Glomeromycetes (see Redecker 2012; see also Section 3.2.3) apparently reproduce asexually only. However, indications such as the presence of genes encoding proteins necessary for the meiosis process, in species of the genus *Glomus* (Halary *et al.* 2011), suggest that in these fungi there may be some form of sexuality.

7.5.4 Ascomycetes

Ascomycetes (see Jaklitsch *et al.* 2016) reproduce mainly by asexual means, through mitospores called *conidia* (Figure 7.15). In many species no sexual cycle is known, but there can be parasexuality.

In *homothallic* ascomycetes even hyphae belonging to the same clone can participate in a sexual event, whereas in the *heterothallic* ones the union is limited to pairs of genetically different hyphae, belonging to compatible mating types. In the ascomycetes the mating type is controlled by a single

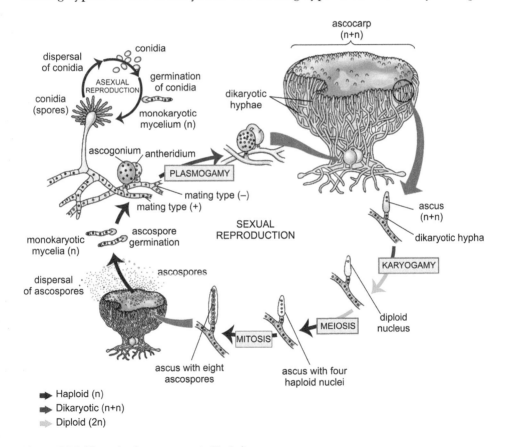

Figure 7.15 Life cycle of an ascomycete (*Peziza*).

locus with two alleles (multiple alleles only in *Glomerella cingulata*). In the true heterothallic species, prevalent in the group, each spore contains only a haploid nucleus with one of the two alleles, but there are also *pseudo-homothallic* species that produce spores with two nuclei, each with one of the two alleles, which immediately give rise to a dikaryotic mycelium. The hyphae are septate (i.e. divided into cellular compartments), and form two types of sexual structures (*gametangia*), the *ascogons*, from which emerges a very thin hypha, the *trichogyne*, which fuses with a gametangium of the other type (*antheridium*) of a mycelium of different mating type. The nuclei of the antheridium migrate into the ascogon. Plasmogamy is not immediately followed by karyogamy, and therefore a dikaryotic hypha is formed with pairs of nuclei, deriving from those of the two hyphae in anastomosis. With each mitosis, the two nuclei divide in a synchronous manner. This dikaryotic hypha (*ascogenous* or *fertile hypha*) is associated with a vegetative mycelium formed by sterile monokaryotic hyphae, giving rise to a fruiting body, the *ascocarp*, endowed with a fertile surface (*hymenium*). Here, at the tip of the ascogenic hyphae, differentiate outgrowths in the shape of a backwards-folded hook. By mitotic division, the two genetically different nuclei hosted in the apical region of the hypha divide, and with the subsequent formation of two walls three compartments are formed in which the four nuclei are distributed: one in the basal portion of the hypha, one in the hook and two (genetically different) in the U-shaped section where karyogamy eventually occurs. Thus a diploid zygote is produced, which develops into an ascus, an elongated capsule the nucleus of which undergoes meiosis; the four resulting haploid nuclei are the nuclei of as many *ascospores*. Dispersed by the movements of the air or 'fired' at a distance when the ascus opens, the ascospores germinate, forming new haploid hyphae. The meiosis can be followed by one or more mitoses before the release of the ascospores; consequently, an ascus may contain eight or more spores, up to a maximum of about 7000 (*Trichobolus*).

7.5.5 Basidiomycetes

The mycelium of most basidiomycetes (Figure 7.16) consists of long septate hyphae (i.e. hyphae that are divided into cellular compartments). The monokaryotic haploid hyphae are usually characterized by a mating type based on allelic differences in one (*unifactorial mating type*) or two loci (*bifactorial mating type*), but in some species the number of mating types can be up to 1000. Two monokaryotic hyphae of compatible mating types unite, allowing plasmogamy: the nuclei of the one migrate into the other mycelium and form pairs with the resident nuclei. However, karyogamy is delayed, so that the compatible nuclei remain in pairs; the resulting mycelium is called *dikaryotic*.

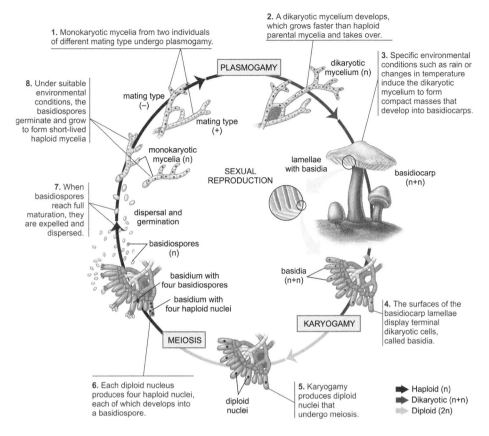

1. Monokaryotic mycelia from two individuals of different mating type undergo plasmogamy.

2. A dikaryotic mycelium develops, which grows faster than haploid parental mycelia and takes over.

PLASMOGAMY

dikaryotic mycelium (n)

3. Specific environmental conditions such as rain or changes in temperature induce the dikaryotic mycelium to form compact masses that develop into basidiocarps.

8. Under suitable environmental conditions, the basidiospores germinate and grow to form short-lived haploid mycelia

mating type (–)

mating type (+)

monokaryotic mycelia (n)

SEXUAL REPRODUCTION

lamellae with basidia

basidiocarp (n+n)

7. When basidiospores reach full maturation, they are expelled and dispersed.

dispersal and germination

basidiospores (n)

basidia (n+n)

basidium with four basidiospores

basidium with four haploid nuclei

4. The surfaces of the basidiocarp lamellae display terminal dikaryotic cells, called basidia.

KARYOGAMY

MEIOSIS

6. Each diploid nucleus produces four haploid nuclei, each of which develops into a basidiospore.

diploid nuclei

5. Karyogamy produces diploid nuclei that undergo meiosis.

➡ Haploid (n)
➡ Dikaryotic (n+n)
➡ Diploid (2n)

Figure 7.16 Life cycle of a basidiomycete (*Agaricus*).

The dikaryotic condition can last for many years, until the fruiting bodies are formed, on which the *basidia* differentiate: these are specialized cells (usually shaped like a club) in which karyogamy takes place. This is immediately followed by meiosis, the products of which are four haploid *basidiospores*, which are usually carried at the tip of the basidium from which they detach and disperse. From a basidiospore a new haploid mycelium is generated. During meiosis, alleles that control mating type are segregated, so that the four spores produced by a basidium belong to two different types in the case of unifactorial mating types and to four different types in the case of bifactorial mating types.

Many species, however, behave other than in this generalized way. First of all, there are *homothallic* basidiomycetes, in which no mating type is recognizable, so that there can be fusion between two hyphae of the same mycelium. In other species, in the course of meiosis two nuclei of different but compatible mating types migrate into each basidiospore, creating the dikaryotic condition as soon as the new mycelium is generated. Many of these species, but not all, form only two spores on each basidium. In other basidiomycetes, when the spores are still on the basidium, meiosis is followed by one or more cycles of mitotic division, so that from each basidium more than four basidiospores are formed – though not necessarily in multiples of four, because of asymmetries in the migration of the nuclei or degeneration of some of them. For example, *Craterellus* often produces six spores on each basidium, and in the common *Agaricus bisporus* the number of spores produced on each basidium ranges from one to four.

Particularly complex are the life cycles of many **rusts** (**Pucciniales**), tiny basidiomycetes parasitic of plants. The transition from one generation to the next is mediated here by different types of spores. The basidia of the rusts are produced by the overwintering, resistant *teleutospores*. The primary mycelium develops from a haploid basidiospore, a product of meiosis, and extends inside a leaf of the host plant, emerging on its upper surface by means of small flask-shaped corpuscles, the *picnidia* (or *spermogons*), which contain the tiny *spermatia*, intermingled with *receptive hyphae*. When they come into contact with a receiving hypha of the opposite mating type, in a spermogonium different from that in which they were formed, the spermatia release their nucleus; the latter, through the hypha that crosses the inner tissues of the leaf, reaches an *aecidium* (a yellow pustule protruding from the lower surface of the leaf), which becomes binucleate. From this the *aeciospores* originate, also binucleate, like the mycelium that forms from them, which develops in the tissues of a new host plant on which it causes the formation of *uredosori*, linear pustules from which the *uredospores* detach, also dikaryotic. At the end of the favourable season, in the uredosori pairs of cells are produced that undergo karyogamy, producing the diploid *teleutospores*. In many species, however, the cycle is less complex; it is simplest in the microcylic species, where it does not include host change or the formation of pycnidia and aecidia.

The **smuts** (**Ustilaginales**) are another group of basidiomycetes parasitic of plants, characterized by blackish pulverulent spores, the *teliospores*; these are initially dikaryotic, but in each teliospore karyogamy occurs, followed by meiosis, right at the time of germination. The life cycles of the ustilaginals are complex and diverse and often involve alternation between a unicellular vegetative phase, comparable to a yeast, and a mycelial phase. The basidio-spores frequently fuse in pairs before being released.

7.5.6 Lichens

Lichens are symbiotic associations between a photoautotrophic organism (usu-ally a green alga, otherwise a cyanobacterium) and a fungus (usually an asco-mycete). The fungal component is nearly always the more specialized of the two partners, which justifies our brief discussion of lichens here. Tripp and Lendemer (2018) list 27 different modes of reproduction in lichens, which include sexual and asexual processes and the possibility of parasexual reproduction.

The fungi involved in lichen symbiosis (see Jaklitsch *et al.* 2016) may have asexual reproduction with mechanisms that do not disrupt the association between the fungal and algal symbionts, but also sexual reproduction, which involves the dissociation of the fungus from the alga and the subsequent reconstitution of the lichenic symbiosis in the next generation. Vegetative reproduction can occur by simple fragmentation or through the production of *soredia*, small masses of hyphae and algal cells. In the lichen *Vezdaea aestivalis*, whose fungal component is an ascomycete, the ascospore does not produce a mycelium, but a short hypha that releases a conidium. The latter will produce the new mycelium. Conidia can also be produced by sexual spores from fungi not engaged in lichen symbiosis (some ascomycete species of the genera *Ascocoryne*, *Nectria* and *Runstroemia* and some basidiomycete species of the genera *Calocera*, *Dacrymyces* and *Exobasidium*).

7.5.7 Microsporidians

Finally, we mention the microsporidians, tiny single-celled organisms, which represent the largest group of eukaryotes without mitochondria and which until 1999 were classified among the protozoans. However, it is universally accepted today that they belong to the fungi. All microsporidians are intracellular parasites, especially of insects but also of crustaceans and fishes and, occasionally, also of humans. Some of them seem to have a purely asexual cycle, while in others there is alternation between sexual and asexual reproduction, generally associated with the passage to a different host. We describe here briefly the cycle of a microsporidium known as responsible for occasional intestinal diseases in humans (*Enterocytozoon bieneusi*). Here a spore (the only non-parasitic phase of the cycle), coming into contact with a host, emits a polar tubule through which the content of the spore (*sporoplasm*) is injected into a host cell. The sporoplasm undergoes binary or multiple division (*schizogony*). Finally, in a special *sporangiophorous vacuole* of the host cell, or freely in the cytoplasm of this, the parasite undergoes sporogenesis; by breaking the host's cell, the spores are released and pass on to infect other cells.

7.6 Metazoans

Metazoans have a diplontic life cycle. Sexual reproduction is known for all the main groups and is often accompanied by different forms of asexual reproduction, of variable importance from group to group.

7.6.1 Sponges

The sponges (Porifera) usually reproduce by sexual means; asexual reproduction is regular only in the freshwater species, through *gemmules* (which are produced, however, also by a small number of marine sponges), covered by a strong spongin wall and by spicules with a characteristic shape. Gemmules survive the death of the sponge, resisting desiccation even for a long time. As a rule, a new sponge will originate from each gemmule. However, young sponges deriving from more than one gemmule (and also more than one larva, in some species) can merge together, giving rise to a single individual.

Sponges have regenerative capacities superior to those of almost all other metazoans; this ability also allows occasional reproduction by fragmentation. Sponges do not have gonads. Freshwater species are mostly gonochoric, while the marine ones are often hermaphrodite, either protandrous or protogynous. The sperm cells are mostly provided with flagella, but those of the calcareous sponges are aflagellate, so they are immobile and are led by carrier cells to contact the female gametes. In sponges with flagellated spermatozoa, fertilization can be external or internal.

7.6.2 Cnidarians

Many representatives of the Cnidaria (see Fautin 1992) have a metagenetic life cycle with alternating generations (polyp and medusa) or a derivative form of this cycle (Figure 7.17). The sperm cell has primitive features including the lack of acrosome.

Most cnidarians are gonochoric; the hermaphrodite species are either sequential or, more often, simultaneous. The latter condition is more frequent in the Anthozoa (many sea anemones and perhaps all the tube-dwelling anemones or Ceriantharia), less so in Scyphozoa (*Nausithoe eumedusoides*, *Chrysaora isoscella*) and Hydrozoa (*Eleutheria dichotoma*).

Asexual reproduction is widespread and may lead to the production of (i) polyps from polyps, (ii) medusae from medusae, or (iii) medusae from polyps. In many anthozoans, the generation of new polyps from a parent polyp, or from a stolon into which a parent polyp extends, is not followed by the detachment of the new polyps, this causing colonies to develop. Colonies are also formed in similar ways in many hydrozoans, but in this case, in addition to a variable number of polyps, the colony frequently also includes units that can be interpreted as modified medusae. In anthozoans, asexual reproduction occurs in very different ways. As an alternative to the binary division of the whole parent polyp, new polyps are sometimes formed by laceration, i.e. starting from a fragment detached from the foot of a polyp (known in *Metridium senile* and *Aiptasia diaphana*). The production of medusae

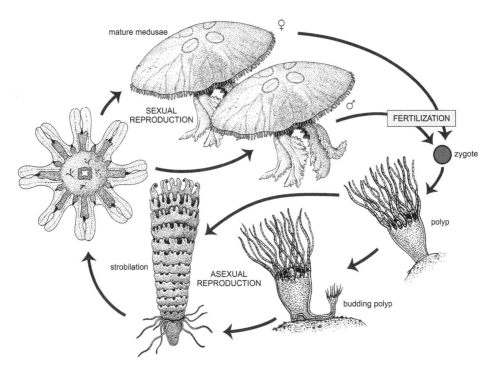

Figure 7.17 The metagenetic life cycle of the medusozoan *Aurelia aurita*.

from other medusae is limited to the Hydrozoa, where it can occur in many ways: new individuals can originate on the manubrium, at the base of the tentacles or near the gonads; in *Cladonema radiatum* the whole body of the mother medusa splits longitudinally. Some species of Anthozoa, e.g. *Haliplanella luciae*, reproduce exclusively asexually (Adler and Jarms 2009).

Sexual reproduction generally involves amphigony, with internal or external fertilization, depending on the species. Parthenogenesis is known in a few species, e.g. in the anthozoans *Actinia equina* and *Cereus pedunculatus*. Most species are oviparous; some however practise parental care, incubating eggs in pockets of the gastrovascular system (some anthozoans and cubozoans) or in other parts of the body. Among the Cubozoa, *Carybdea marsupialis* and *Alatina alata* are viviparous. The anthozoan *Alcyonium digitatum* can be either gonochoric or hermaphrodite, while the populations of *Cereus pedunculatus* can be viviparous and parthenogenetic, or oviparous and gonochoric, or viviparous and hermaphrodite. In freshwater hydras, as well as in *Podocoryne*, *Aurelia* and *Cyanea*, the sexual phase is induced by crowding (high CO_2 concentration). In the cubozoans *Copula sivickisi* and *Tripedalia cystophora* there is a striking sexual dimorphism and males engage in true courtship behaviour. By helping with the tentacles, the male induces the female to ingest a spermatophore, which via the gastrovascular cavity reaches a sort of spermatheca, where it remains for some hours (Marques *et al.* 2015). A few cnidarian life cycles of unusual complexity are described in Box 7.2.

Some cnidarians practise parental care: eggs and larvae of some scyphozoans and the anthozoan *Actinia equina* are released in the water and enter into adult conspecifics, which is in effect a sort of commensalism or intraspecific parasitism. The planula larvae of some scyphozoans are cared for externally by adoptive parents of another species. Scyphozoans living in the open ocean produce large eggs that develop directly into small medusae called ephyrae

---

Box 7.2 Special Cases of Cnidarian Life Cycles

In the marine hydrozoan *Margelopsis haeckeli* the polyp, which is solitary and pelagic, generates by budding the medusa. This produces both subitaneous eggs, which immediately develop into polyps, and durable eggs, whose development into polyps is postponed to the following spring. According to Werner (1955, 1963), durable eggs are non-reduced eggs that develop by ameiotic parthenogenesis, so that, from a genetic point of view, the entire cycle could be described simply as clonal reproduction.

At the base of the polyp of some scyphozoans, such as *Chrysaora isoscella* and *Aurelia aurita*, a kind of capsule can be produced from which emerges a sort of larva (*planuloid*) capable of metamorphosing into a polyp similar to the one that generated it. The formation of a new polyp is similar, starting from a multicellular frustule which detaches from the wall of the polyp body in some hydrozoans, among which is *Craspedacusta sowerbyi*, a freshwater species in which a medusa stage also occurs.

Planuloids also form asexually inside the gastrovascular cavity of some *Actinia*, even in male individuals.

In the hydrozoan *Pegantha smaragdina*, larvae are formed in the gastrovascular cavity of the parent medusa and are subsequently released, also as medusae. In another hydrozoan, *Cunina proboscidata*, large female medusae produce eggs that can develop with or without fertilization. The fertilized eggs give rise to gamma larvae, which are transformed into dwarf male medusae. Non-fertilized eggs develop instead into alpha larvae, which by budding give rise to beta larvae. Both alpha larvae and beta larvae develop into dwarf female medusae. Medusae derived from alpha larvae produce eggs that give rise by parthenogenesis to large female medusae; those derived from beta larvae produce eggs that are fertilized by dwarf males and develop into large male medusae.

(*Pelagia*), or incubate modified polyps (*scyphistomas*), and the stage eventually released into open water is the jellyfish produced by them (*Stygiomedusa*).

7.6.3 Ctenophores

Ctenophores are mostly hermaphrodite and capable of self-fertilization. *Dissogony* has been reported for a few species: the same individual experiences two distinct reproductive periods, one as a juvenile and one later in adulthood. A population of *Mertensia ovum* in the central Baltic Sea basin consists exclusively of larvae and therefore reproduces only by paedogenesis (Jaspers *et al.* 2012). Gametes are usually released in the water, but cases of parental care are known (*Tjalfiella* retains the eggs in a brood chamber), as well as *larvipary*, i.e. the release of the offspring when they are already in a larval stage. Vegetative reproduction is comparable to the laceration of some anthozoans and is limited to two benthic genera with a creeping habit (*Ctenoplana* and *Coeloplana*).

7.6.4 Placozoans, Orthonectids, Dicyemids

Placozoans are benthic marine animals shaped as tiny discs, with a very simple organization. Their reproduction is poorly known. However, they exhibit both asexual reproduction through the detachment of hollow spheres of 40–60 μm, suitable for dispersal, and sexual reproduction: sperm cells and eggs have been observed inside the same individual, but there is no information on meiosis or karyogamy.

Orthonectids are parasites of a number of marine invertebrates. The most conspicuous phase of their life cycle is a plasmodium derived from host cells, which reproduces by fragmentation. Within the plasmodium accumulate some nuclei (*agamonts*) of the parasite, which divide repeatedly, producing cellularized individuals. Depending on the species, a plasmodium can produce males, females or individuals of both sexes. Sexual stages abandon plasmodia and host. In the water, the male adheres to the female, which is much larger and receives the very small sperm cells through a genital pore. In some cases the whole male enters the female to fertilize her. The zygote develops into a ciliated larva, initially endowed with a multicellular coating and free internal cells. Following the laceration of the body wall, the larva abandons the female and attacks a new host. Once the latter is reached, the larva loses its external coating and every internal cell induces in the host the production of a plasmodium.

Adult **dicyemids** are renal parasites of cephalopods. In the characteristic vermiform stage called the *nematogen*, an epidermis is recognized which covers a large axile cell. The nematogen can reproduce asexually, by means of *axoblasts* originating from the axile cell. The young animal abandons the parent following tearing of the body wall. Excessive population density in the excretory organs of the host triggers in the nematogens the production of *rhombogens* with hermaphrodite gonads, in which fertilization takes place.

7.6.5 Acoelomorphs

The **acoels** and the **nemertodermatids**, until a few years ago classified among the flatworms (Platyhelminthes s.l.), are currently included in this group, whose monophyly is still controversial. These are a few hundred species of tiny worm-shaped animals with a very simple organization. Some of them practise asexual reproduction by architomy, paratomy or budding. *Convolutriloba retrogemma*, for example, reproduces by buds formed at the back end of the body. The sexual reproduction of these small hermaphrodite metazoans involves internal fertilization; in some nemertodermatids, this is practised by hypodermic insemination.

7.6.6 Gastrotrichs

Gastrotrichs are hermaphrodite. The protandrous macrodasyids reproduce exclusively by amphigony, while in the chaetonotids there is alternation between parthenogenesis and amphigony (heterogonic cycle), or parthenogenetic reproduction exclusively. In almost all gastrotrichs, mature eggs are released following tearing of the body wall. In some macrodasyids, male reproductive organs do not open at the copulatory organ, but at some distance, so that the copulatory organ must collect the sperm already released to the outside before transferring it to the partner (this is an indirect transmission of the sperm cells, as in dragonflies and spiders). Sperm cells are injected into the partner's body or are attached to its external body surface, grouped in spermatophores.

7.6.7 Gnathostomulids and Micrognathozoans

All **gnathostomulids**, tiny invertebrates of the marine interstitial fauna, are hermaphrodite. Their sperm cells are often aflagellate and fertilization is internal. There are no gonoducts, and the fertilized eggs are released after laceration of the dorsal epidermis.

Of **micrognathozoans**, very small metazoans whose only known species (*Limnognathia maerski*) was first described in 2000 (Kristensen and Funch 2000), only female individuals are known. These possess a paired ovary, in which only one egg cell matures at a time. Two types of eggs are produced, some being subitaneous, the others undergoing diapause.

7.6.8 Synderms

The three groups of tiny aquatic invertebrates traditionally classified as **rotifers** (seisonids, monogononts and bdelloids) have recently been brought together in a phylum Syndermata, together with the **acanthocephalans**, parasitic animals of much larger size. All synderms are gonochoric and practise internal fertilization, but parthenogenesis is very common. Seisonidea are amphigonic and practise sperm transfer by spermatophores. All bdelloids are parthenogenetic.

Many monogonont species have a heterogonic cycle alternating between amphigony and parthenogenesis, as described in Section 2.3 (see also Figure 7.18). During the favourable season, the population consists exclusively of amictic females which reproduce by thelytokous (diploid) parthenogenesis. At the end of the favourable season, mictic females developed under the

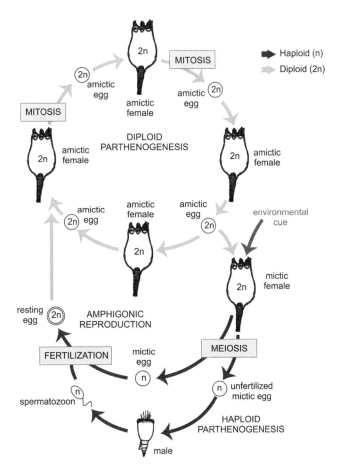

Figure 7.18 The heterogonic life cycle of a monogonont rotifer (*Keratella*).

stimulus of specific environmental signals produce haploid eggs through a regular meiosis. Reduced but unfertilized eggs develop into haploid males, whose sperm cells can fertilize other haploid eggs produced by mictic females. The fertilized eggs will spend the adverse season in quiescence, to give rise to the first generation of amictic females in the new year.

As a rule, a single female of monogonont rotifers produces only eggs of one type (either mictic or amictic), but in *Asplanchna*, *Sinantherina* and *Conochiloides*, females have been described that produced eggs of both types. Some *Asplanchna* are viviparous. The males of the monogononts are much smaller than the females and have no gut, so they do not take food; active sperm cells are found in male rotifers when they are still in the egg.

7.6.9 Flatworms (Rhabditophorans and Catenulids)

Recent developments in zoological systematics have led to a radical revision of the classification of so-called flatworms. Five main distantly related lines were recognized within the traditional phylum Platyhelminthes. Two of these, the acoels and nemertodermatids mentioned in Section 7.6.5, have only remote affinities to the other flatworms. The same applies to the xenoturbellids, sometimes classified with the acoels and nemertodermatids but often assigned in recent years to the deuterostomes; of these 'worms', we only know that *Xenoturbella bocki* is hermaphrodite, while *X. profunda* is gonochoric (Rouse *et al.* 2016). We discuss here the two remaining lines of what were traditionally known as platyhelminths, one of which (**Rhabditophora**) includes the vast majority of flatworms, while the other (**Catenulida**) includes only a small number of species.

Asexual reproduction and/or considerable regenerative capabilities are known for many flatworms. Both architomy and paratomy are known in free-living forms. The latter mechanism can lead to the temporary formation of a chain of individuals (*zooids*), up to a few hundred units, as in the catenulid *Africatenula riuruae*. In some tapeworms (Cestoda) such as *Echinococcus granulosus*, the larva (*onchosphere*) that hatches from the egg transforms into a cystic stage (*hydatid*) that undergoes vegetative reproduction, giving rise to a number, perhaps a high number, of *protoscolices*, each of which represents the primordium of a future adult tapeworm.

Almost all the flatworms are protandrous hermaphrodites. Among the few forms with separate sexes are the flukes (Digenea) of the genus *Schistosoma* and the tapeworms of the genus *Dioecocestus*. Parthenogenesis is widespread, either as the only reproductive mode of individual populations, as in some freshwater planarians (free-living forms belonging to the group Tricladida), or as a reproductive mode that alternates with amphigony in a complex, multi-generational life cycle, as in the digeneans.

Fertilization is internal and very often reciprocal; sometimes it takes place in the ovary, or in the oviduct (planarians). Sperm cells (usually without acrosome but with two flagella, an unusual condition in metazoans; aflagellate in Catenulida and Macrostomorpha) are almost always transferred by means of a penis. In some free-living forms, sperm transfer is by hypodermic insemination (Section 3.5.2.1). Production of spermatophores is rare. Several species self-fertilize, among them the freshwater planarian *Cura foremanii* and those big tapeworms of which the host is home to only one individual, a typical 'solitary worm' such as *Taenia solium*. The genito-intestinal canal of some flatworms may allow the digestion of sperm and/or self-fertilization. In some Polycladida there is *dissogony*, i.e. there are two distinct reproductive periods in the course of a worm's life, one at a juvenile stage, the other as adult (a similar strategy is used by some ctenophorans; see Section 7.6.3).

The life cycle of many parasitic flatworms is very complex and still of uncertain interpretation. In the Digenea (see Galaktionov and Dobrovolskij 2003), two phases are recognized, one with sexual reproduction (*marita*), one with asexual reproduction (often however interpreted as parthenogenetic, hence the name *parthenita*). The asexual phase, in turn, can include more generations of sporocysts and/or rediae. The life cycle of the Monogenea is, instead, monogenerational. Worthy of mention is the genus *Diplozoon*, whose representatives unite in pairs in permanent union, with fusion of tissues and joining of the vagina of one with the deferent duct of the other. All mono-geneans are oviparous, with the sole exception of *Gyrodactylus*, which presents polyembryony accompanied by viviparity (Section 3.1.2.4).

The life cycle of some tapeworms is distinctly bigenerational, due to the intercalation in the sexual cycle of a stage (such as the hydatid of *Echinococcus granulosus* or the multilocular cyst of *E. multilocularis*) from which a number, sometimes very high, of individuals capable of reaching sexual maturity is produced asexually.

7.6.10 Cycliophorans

First described in 1995 (Funch and Kristensen 1995), the phylum Cycliophora includes only two species, both of which live on the buccal appendages of lobsters (*Nephrops* and *Homarus* species). Sexual dimorphism is very strong: females measure almost half a millimetre, while males barely reach 40 μm and are made up of no more than 60 cells. The life cycle is very complex and multigenerational. We can start from asexual reproduction, which involves the formation of a *pandora larva* inside an incubation chamber of what is conventionally considered the adult (*trophic stage*) of the animal. After leaving its parent through the latter's anus, the pandora larva settles in the vicinity of this and turns into a new trophic stage. Shortly before the host completes a moult, the cycliophoran undergoes sexual reproduction: it stops feeding and generates, internally, a primary male (*prometheus larva*) which abandons the parent and attaches itself to an adult still in the trophic phase, inducing in the latter the production of a female primordium. This produces a single oocyte, while the primary male buds off one to three secondary males. After fertilization, the female leaves the parent and adheres to the host's cuticle in the form of a cyst. From the latter a further type of larva (*chordoid larva*) differentiates, which will go in search of a new host, on which it will develop into a trophic stage.

7.6.11 Entoprocts

Entoprocts are small marine invertebrates (only one freshwater species is known), all benthic, that take the form of individual or colonial polyps. They practise both asexual reproduction, through buds formed at the base of the stem or on the oral side of the cup, and sexual reproduction. Loxosomatids are generally protandric hermaphrodites and pedicellinids are mostly simultaneous hermaphrodites, while barentsiids form colonies in which single polypiform individuals are unisexual, but individuals of both sexes coexist in the same colony. Parental care is known: eggs are fertilized when they are still in the ovary and are held in two lateral pockets of the terminal intestine until the larva is released.

7.6.12 Nemerteans

Most of the nemerteans or ribbon worms are capable of regeneration, and some of them (*Lineus* spp.) can reproduce by architomy. Almost all nemerteans are gonochoric; one of the rare hermaphrodite species is *Pantinonemertes agricola*, one of the very few terrestrial ribbon worms. Fertilization is almost always external, but sometimes it does not take place in open water, but within a mucus sleeve surrounding the two partners (*pseudocopulation*). However, there are also viviparous species with internal fertilization, such as *Prosorhochmus claparedii* and *Pantinonemertes agricola*.

7.6.13 Bryozoans, Phoronids, Brachiopods

These are three groups of benthic, sessile animals, all marine except for a few freshwater bryozoans. Except for one genus (*Monobryozoon*), **bryozoans** are colonial animals. Each colony unit (a *zooid*) comprises a *cystid* and a *polypid*, of which the first represents a case within which the latter can be retracted. They are mostly hermaphrodite (exceptions are the Stenolaemata and some of the Cheilostomata): more precisely, almost always the single individuals of the colony are hermaphrodite (sometimes simultaneous, more often protandrous). Fertilization is generally internal. Almost all the bryozoans provide parental care: they host the eggs in the coelom (some Ctenostomata), or in particular structures of the zooid or in specialized zooids. Many species of the Stenolaemata reproduce regularly by polyembryony. The first blastomeres of the embryo separate from each other, but then associate again to form a primary embryo from which secondary embryos are formed by budding; these develop mainly into larvae, but in some species they instead produce numerous tertiary embryos (Zimmer 1997).

The **phoronids** are either gonochoric or protandrous hermaphrodites. Fertilization can be external or internal: in gonochoric species, females use the lophophore to capture the spermatophores released by males; the sperm cells reach the eggs through the intestine or the tentacles, crossing one of the mesenteries that divide the general body cavity into four compartments. In *Phoronis ovalis*, the embryos are incubated in the tube inhabited by the adult.

Most species in the **brachiopods** have separate sexes, but some species are protandrous hermaphrodites. Fertilization takes place in water. Some species practise parental care.

7.6.14 Molluscs

In the large phylum Mollusca only sexual reproduction is known, except for the singular case of the African freshwater bivalve *Mutela bourguignati*, which produces larvae that live parasitically on fishes (a behaviour shared with other freshwater bivalves). The behaviour of the larva is unique: instead of undergoing the expected metamorphosis, it produces a bud – and it is the latter that metamorphoses into the adult (Fryer 1961). This is a case of asexual reproduction bordering on development (see Box 2.3).

Separate sexes are found in all species of cephalopods, caudofoveates and scaphopods (apart from rare hermaphrodites occasionally occurring in a few species) and in the majority of the monoplacophorans (with the exception of *Micropilina arntzi*) and polyplacophorans. There are many hermaphrodites among the solenogasters, gastropods and bivalves. Most bivalves, however, have separate sexes; gonochorism and hermaphroditism can occur even among closely related species. Among the hermaphrodite bivalves, some have distinct ovary and testicle, while others possess a single hermaphrodite gonad. Subsequent hermaphroditism is not uncommon, more often protandrous, more rarely protogynous, or alternate, as in *Ostrea edulis*, which is male in the first sexually mature phase. Mixed breeding systems are also known, as in

Crassostrea virginica, where within the same population there are individuals with a permanent sexual condition, either male or female, and individuals which are sequential hermaphrodites (trioecy). The basal lineages of the gastropods are gonochoric, with some exceptions such as *Crepidula fornicata* (Section 3.3.2.2), and some cases of thelytokous parthenogenesis, such as *Melanoides tuberculata*.

Besides the functional spermatozoa, many gastropods also produce atypical spermatozoa with abnormal chromatinic content, which transport the functional spermatozoa up to the encounter with the egg.

Fertilization is external in caudofoveates, in most polyplacophorans, in the most primitive gastropod lineages and in most bivalves and scaphopods; often, however, fertilization occurs in the mantle cavity and can pave the way for the evolution of parental care. Fertilization is internal in the solenogasters, in most of the gastropods and in all cephalopods, and usually involves the use of copulatory organs, very different in nature and position, or the production of spermatophores, or both.

In the bivalves there are no copulatory organs; fertilization is almost always external, but in some cases it takes place in the mantle cavity.

In gastropods with internal fertilization, the copulatory organs may derive from a cephalic tentacle, the edge of the mantle or the foot; the production of spermatophores is widespread, both in species without a copulatory organ and in species with one. In some groups of gastropods, in addition to the fertile eggs a number, sometimes high, of trophic eggs are produced: in some cases their content is transferred to the fertile eggs during oogenesis, in other cases the embryos that will develop from the latter will swallow the former by oophagy. Heterobranch gastropods (an assemblage roughly corresponding to the Pulmonata plus the Opisthobranchia of traditional classifications) are almost all simultaneous hermaphrodites.

In hermaphrodites, the exchange of gametes can be reciprocal, as in *Helix*, or non-reciprocal, as in *Bulinus globosus*. Some snails (*Helix, Cepaea*) have a calcareous stiletto, the so-called love-dart: in the phase preceding the exchange of gametes, each partner fires its love-dart into the tissues of the partner's foot, where it delivers a pheromone that leads to elimination of any sperm that the receiving individual may have preserved from a previous mating. Sperm cells are often stored in sperm bags, where they are kept alive for up to two years.

Among the Polyplacophora, some species are ovoviviparous, while others provide parental care by holding the embryos in the mantle cavity, as also does *Micropilina arntzi* among the monoplacophorans. Numerous bivalve species retain the eggs in the mantle cavity or in marsupial pockets of the gills, from which they later release larvae ready to disperse in the water. Instances of parental care are also known in some gastropods, in which the eggs (and also the larvae, in some aquatic taxa) are housed in the mantle cavity or under the foot; the marine species *Janthina janthina* produces a float of air-filled bubbles to which the eggs are attached.

Numerous cephalopod species show remarkable sexual dimorphism, evident especially in the smaller size of the male and in the transformation of one or two arms into copulatory organs, which take the name of *hectocotyli*. In some species, the hectocotylus is abscissed from the male's body and autonomously reaches the partner's mantle cavity, following the path traced by the pheromones produced by the female. It has been traditionally held that cephalopods, with the sole exception of the nautilus (which releases gametes once a year, for up to 20 years), are semelparous, that is, they reproduce only once during their lives, although often over a quite long season, up to seven months in the cuttlefish (*Sepia officinalis*) (Rocha *et al.* 2001). An exception, however, is provided by the vampire squid (*Vampyroteuthis infernalis*), at least by the females of the species, which appear to have up to about 20 spawning periods, separated by times of reproductive rest (Hoving *et al.* 2015).

7.6.15 Annelids (Including Pogonophorans, Sipunculids and Echiurids)

Among the traditional metazoan phyla, Annelida is one of the most difficult to circumscribe. Recent developments in molecular systematics have led to placing (or bringing back) several groups within this phylum, some of which lack any evidence of division of the body into segments. These groups were traditionally attributed to different phyla; we will mention them at the end of this section. Furthermore, the traditional distinction between polychaetes (almost exclusively marine, mostly with separate sexes and generally possessing short locomotor appendages, the parapodia) and clitellates (mostly freshwater or terrestrial, hermaphrodite), has also been abandoned. Clitellates, all lacking parapodia, have at maturity a *clitellum*, a specialized region close to the genital openings, rich in glands specializing in the production of mucus used to form both a sleeve around the two partners in the pseudocopulation, and a cocoon for the eggs. The clitellates (to which the earthworms and leeches belong) are currently considered one of the branches of the extensive adaptive radiation of polychaetes, so the latter term, as traditionally used, would refer to a paraphyletic group. Nevertheless, we continue to use 'polychaetes' here in an informal sense (cf. Rouse and Pleijel 2001), applied to the marine groups to which the use of the term was restricted in the past.

In the annelids, asexual reproduction is widespread and can occur by *architomy* (spontaneous fragmentation into individual segments, as in *Dodecaceria caulleryi*, a polychaete, or into groups of several segments, as in

Enchytraeus fragmentosus, an oligochaete, followed in any case by the regeneration of the missing segments) or by *paratomy*. In the latter case, the parent individual turns into a *stolon* consisting of a series of zooids that progressively differentiate and separate from the chain. Some species of the Syllidae produce a series of buds at the posterior end of the body. Architomy is more widespread than paratomy, the latter being restricted to sipunculans, serpulids, aeolosomatids and naidids. No form of regeneration is present in the hirudineans. Regeneration of the posterior segments is widespread; less so the regeneration of the anterior segments, which is unknown in nereids, capitellids, echiurids and many other groups (Zattara and Bely 2016).

Most polychaetes are gonochoric, while the clitellates are simultaneous hermaphrodites. However, hermaphroditism is also present in representatives of numerous clades of polychaetes, with a range of morphological and functional solutions. The Hesionidae of the genus *Microphthalmus* are simultaneous hermaphrodites; sperm cells mature in the anterior half of their body, which is equipped with copulatory organs, while eggs mature in the posterior segments. The tiny dorvilleids of the genus *Ophryotrocha* are protandrous hermaphrodites, but in particular environmental conditions they can change back from female to a new male phase.

Among the gonochoric annelids, some have environmental sex determination (*Bonellia*; Section 6.2), whereas others rely on a progamic mechanism of sex determination (*Dinophilus*; Section 6.3). In this phylum there are no true gonads, and the gametes are often expelled either through the excretory organs (metanephridia), more or less modified, or by simple tearing of the body wall. Rather rare is the presence of copulatory organs, which in polychaetes of the genus *Pisione* (where – as is the rule in polychaetes – there is diffuse gametogenesis on the wall of the coelomic cavities) are present in many pairs (up to 30, one pair per segment); a penis is also present in some leeches, for example in *Haemopis*.

Some cases of extreme sexual dimorphism are known. A classic example is seen in *Bonellia viridis* and related species belonging to a group long treated as phylum Echiura. In *Bonellia*, the dwarf male (1–3 mm) lives inside the body of the female, which can reach a length of about 10 cm, plus an extensible prostomium ('proboscis') that can extend up to 1 m. The *Osedax* polychaetes are also strongly dimorphic. These are forms of deep marine waters, first described in 2004 (Rouse *et al.* 2004), which live on the seabed, in the skeletal remains of cetaceans; the very small males live in large numbers inside the gelatinous tubes produced by the females, up to hundreds in one tube.

Parthenogenesis is widespread among the oligochaetes (about 15 species of the Lumbricidae, such as *Aporrectodea trapezoides* and *Eiseniella tetraedra*, and some Enchytraeidae). In some groups of polychaetes, the body comprises a sterile (atokous) anterior part and a fertile (epitokous) posterior part. At maturity, the epitokous part is detached, and releases the mature gametes.

Fertilization is external in most marine annelids, but it is internal in the few species endowed with penises, many of which (e.g. Pisionidae, Hesionidae and Dinophilidae) are of very small size and live in the interstitial environment, i.e. in the spaces between sand grains on the seabed. Many polychaetes lay their eggs in mucus cocoons and practise parental care.

Earthworms exchange sperm cells by *pseudocopulation*: two individuals, side-by-side but in opposite head-to-tail orientation, place the male genital pore on the opening of the spermatheca of the partner; thus the transfer of sperm is quite easy, helped by the mucous layer in which both worms are covered. Leeches do not have spermathecae; insemination is hypodermic, or achieved by insertion of a penis into the female genital pore of the partner; in any case, fertilization is internal. Through the secretion of glandular cells of the clitellum, a body region enlarged at maturity in oligochaetes and hirudineans, a cocoon is produced that envelops the eggs, immersed in a trophic fluid also produced by cells of the clitellum; in oligochaetes, the gametes meet in the cocoon.

Within the annelids are now classified three groups of metazoans with interesting reproductive biology, traditionally attributed to distinct phyla, the Pogonophora (currently reduced to the status of a family, as Siboglinidae), Sipuncula and Echiura.

Most **siboglinids** have separate sexes, but *Siboglinum poseidoni* is an exception. Fertilization probably takes place inside the tubes in which the females live, or perhaps in their oviducts. It is believed that sperm cells, joined in bundles or even in true spermatophores, can be transferred into the female tubes by means of the tentacles.

Sipunculans are gonochoric but without sexual dimorphism, with the sole exception of *Nephasoma minutum*, which is a protandrous hermaphrodite. One species (*Themiste lageniformis*) reproduces by thelytokous parthenogenesis. Vegetative reproduction is known in two species (*Sipunculus robustus* and *Aspidosiphon elegans*), where it occurs by budding and by transverse division of the body, respectively.

All species of **echiurans** have separate sexes, and only sexual reproduction is known. As a rule, fertilization takes place in open water. The two sexes are morphologically indistinguishable, with the exception of the bonelliids, where the differences between female and male are extreme, with tiny dwarf males. In *Bonellia viridis* the males live on the female's body or inside it. Fertilization takes place in female nephridia. *B. viridis* is the classic example of sex determination by interaction with conspecifics (see Section 6.2.2), but it seems that a genetic component is also in play. In any case, the larvae develop in most cases into females, unless they stay for a few days, during a critical phase of development, on the body of a female, in which case they preferentially develop into males.

7.6.16 Tardigrades and Onychophorans

Almost all members of the **tardigrades** are gonochoric, but those of the genus *Isohypsibius* are simultaneous hermaphrodites. Parthenogenesis, generally ameiotic and often associated with polyploidy, is widespread among the terrestrial and freshwater species. In many species of Echiniscidae, males are unknown. Fertilization is generally internal. In some aquatic forms, however, insemination is external, as the sperm cells are released by the male in the space between the old and the new cuticle of a moulting female: hence the sperm reach the cloaca of the female and then the eggs, which are fertilized inside the mother's body.

The **onychophorans** have separate sexes and are almost all amphigonic, but a thelytokous population of *Epiperipatus imthurni* is known. Fertilization, described for a few species only, occurs by means of spermatophores that the male attaches in the proximity of the genital opening of the female (for example in *Peripatus*) or at any point on her body (*Peripatopsis*). All Peripatidae are viviparous and matrotrophic, as are also some species of Peripatopsidae. In the latter family, other species are also viviparous but not matrotrophic, and still others are oviparous. All oviparous species (for example *Ooperipatus*, *Ooperipatellus*) and some non-matrotrophic viviparous species (*Austroperipatus eridelos*) possess an ovipositor.

7.6.17 Nematomorphs, Priapulids, Loriciferans and Kinorhynchs

All members of the nematomorphs, priapulids, loriciferans and kinorhynchs are gonochoric. Fertilization is external in the priapulids (with the possible exception of the small interstitial species), internal in the other three phyla. Sperm transfer occurs by true copulation in the marine genus *Nectonema*, in gordiaceans instead by *pseudocopulation*, similar to earthworms. In the Kinorhyncha, the length of a sperm cell can be equal to one-fifth of the length of the entire animal. A species of nematomorph, *Paragordius obamai*, is parthenogenetic (Hanelt *et al.* 2012).

7.6.18 Nematodes

In the nematodes, the sexes are generally separate. Apart from the reproductive organs, the two sexes are often practically identical, but in some species the sexual dimorphism is remarkable, up to the extreme case of *Sphaerularia bombi*, a parasite of bumblebees. In the females of this species a peculiar solution is found to the otherwise insurmountable conflict between the limited elasticity of the cuticle that covers the body and the production of a voluminous mass of eggs, as expected in a parasite. At maturity, in fact, there is a prolapse of the enormous uterus, which can reach a volume 300 times that of the worm; the latter ends up as a tiny appendix of its enormous extroflected reproductive system.

Some saprobious and zooparasitic rhabditids are hermaphrodite and practise self-fertilization. In several species of the genera *Caenorhabditis*, *Pristionchus* and *Oscheius* male and hermaphrodite individuals coexist. In *Meloidogyne* there are, besides amphigony, forms of thelytokous parthenogenesis, both meiotic and ameiotic. Some species of freshwater and soil nematodes reproduce exclusively by parthenogenesis, meiotic in *Rhabditis*, ameiotic in the Heteroderidae, by pseudogamy in some Rhabditidae. Fertilization is internal. The aflagellate sperm cells are inserted by the male into the genital opening of the female, and from here they move with amoeboid movements up to the egg cells. Some species are viviparous, among both the free-living nematodes (for example, *Anoplostoma viviparum*) and the parasitic ones, such as *Trichinella spiralis* and the Guinea worm (*Dracunculus medinensis*).

7.6.19 Arthropods

In the largest of the animal phyla, the Arthropoda, asexual reproduction is limited to polyembryony, the few instances of which are mostly found among the hymenopterans (Section 3.1.2.4). Most arthropods are gonochoric and exhibit a very conspicuous sexual dimorphism. The size of the male is often very small compared to the female (the case of the spiders of the genus *Nephila* is emblematic: for example, a body length of 35 mm in the female, no more than 5 mm in the male of *N. plumipes*). The external colouration is often also very different: for this reason, males and females of a number of species were initially described as different taxa. In many insects, only the male is able to fly, while the female has very small and non-functional wings, or is completely wingless: this occurs for example in many fireflies (lampyrid beetles), in many hymenopterans, in the psychid moths and in many homopterans. The sexual dimorphism of the scale insects is remarkable: their females are wingless and larviform and in many cases, after a first active stage, they become permanently fixed on the host plant, taking on the appearance of a shield, with strongly regressed legs and antennae; the males, by contrast, are mobile and generally winged.

Sexual dimorphism is common in crustaceans (Vogt 2016). The larger sex is sometimes the female, sometimes the male. An extreme form of sexual dimorphism is the reduction of males to dwarf males, which has independently evolved in sessile cirripeds and parasitic isopods, copepods and cirripeds. Other extreme forms of sexual dimorphism are observed among the parasitic crustaceans, in particular in some families of isopods and copepods, where the mature female, besides reaching gigantic size, takes on an aspect totally departing from the usual architecture of arthropods.

Hermaphroditism is widespread in the crustaceans: simultaneous hermaphrodites include remipedes and cephalocarids; also, almost all cirripedes are hermaphrodite (but some Thoracica are androdioecious; see Section 3.3.3), as are a few species of other groups (the petrarcid Ascothoracida, some protandrous shrimps and some isopods, either protandrous or protogynous). In some branchiopods, males are accompanied by individuals with hermaphrodite gonads, which behave like females in the presence of males, but as sufficient hermaphrodites in their absence. Among the tanaidaceans, some species are gonochoric, others simultaneous hermaphrodite, others protogynous hermaphrodite with conspicuous male polymorphism. Hermaphroditism is extremely rare in arthropods other than crustaceans.

Thelytokous parthenogenesis is widespread. Among the arachnids, it is found in a small number of scorpions (e.g. *Tityus metuendus*), amblypygids, spiders, palpigrades, pseudoscorpions and harvestmen, but it is common in mites; among the myriapods, it occurs in very few centipedes and millipedes. Among the crustaceans, parthenogenesis is apparently obligate in the darwinulid ostracods, in some cladocerans and in some terrestrial isopods; it alternates with amphigony, in a heterogonic cycle, in most of the cladocerans (Figure 7.19; see also Section 2.3). In many species of cypridid ostracods there is a mixture of amphigonic males and females and parthenogenetic females, whose presence in the population is however not seasonal, as it is in cyclical parthenogenesis (Bode *et al.* 2010).

Among the insects (reviews in Vershinina and Kuznetsova 2016; Gokhman and Kuznetsova 2018), parthenogenesis is known in some mantises, many phasmids (obligate or optional), psocopterans (obligate or optional), some thrips, and homopterans (some scale insects, some aleurodids). Parthenogenesis has at times a geographical character, especially among the curculionid beetles (Section 3.6.2.3). In most aphids it alternates with amphigony, in a heterogonic cycle (Figure 7.20; see Section 2.3).

Fertilization is almost always internal, even in aquatic forms; the only exceptions are represented by Xiphosura and Pycnogonida. The transfer of spermatozoa can take place in very different ways and does not always imply a direct contact between the two partners. Sperm transfer is direct, with the help of gonopods or penises, in many arachnids (spiders, harvestmen, ricinulids, solifuges and some mites), indirect in others, sometimes via spermatophores.

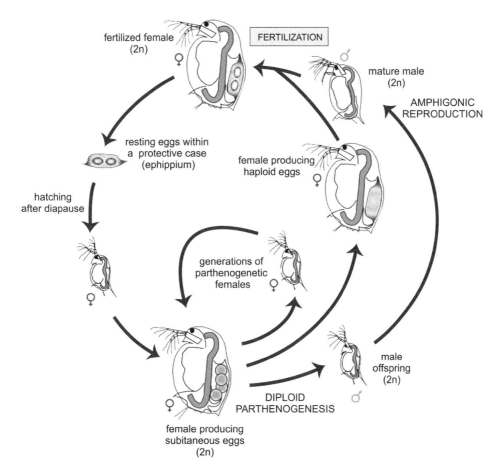

Figure 7.19 The heterogonic life cycle of a cladoceran crustacean (the water flea *Daphnia*).

In scorpions these are complex and are transferred at the end of long courtship rituals. Among the chelicerates, spermatophores are also used by uropygids, amblypygids and pseudoscorpions; among the myriapods, by all centipedes, symphylans, pauropods and the penicillate millipedes; by a large number of crustaceans, whose sperm cells are generally aflagellate; by all primitively wingless hexapods and some groups of winged insects. Spermatophores are not always received by the female's external genitalia. In symphylans, for example, spermatophores are collected with the mouthparts and stored in parabuccal pockets; the eggs will receive the sperm outside the mother's body, immediately after being laid.

The vast majority of arthropods are oviparous, but there is no shortage of lecithotrophic viviparous forms (most of the scorpions, some cockroaches, psocopterans and thrips, the earwigs *Marava arachidis* and *Chaetospania*

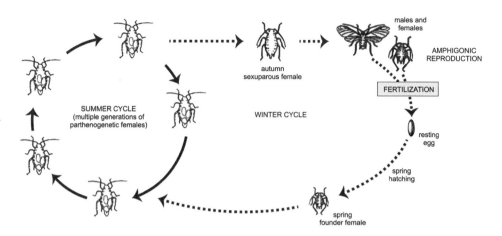

Figure 7.20 The heterogonic life cycle of an aphid (*Aphis*).

borneensis (Kočarek 2009) and polyctenid heteropterans). Some species of scorpions and cockroaches are viviparous and matrotrophic, as are two very specialized genera of dermapterans, *Hemimerus* and *Arixenia*, both epizoic on mammals.

Parental care is very widespread. In pycnogonids, eggs are carried by the male until hatching. In spiders, eggs are laid in silk cocoons that in some families are fixed to the substrate; in others, they are carried around by the female, fixed ventrally to the opisthosoma, or held between the buccal appendages (chelicerae) for the whole duration of embryonic development. After hatching, the young wolf spiders (lycosids) remain on their mother's body until the first moult, those of other spiders (some *Coelotes*, *Theridion* and *Stegodyphus*) also remain on the web of the mother, from which they receive food. In the pseudoscorpions, the fertilized eggs are carried by the female in a bag of secreted material attached to its genital opening, from which pours out a nutritious fluid produced in the ovary, which is sucked by the embryos, sometimes very quickly, thanks to the particular conformation of the embryonic pharynx. In the amblypygids the eggs are kept by the female for 3–4 months in a ventral pouch. The females of some solifuges remain close to their eggs until hatching.

In myriapods, parental care is offered by the mother in three groups of centipedes (craterostigmomorphs, scolopendromorphs and geophilomorphs), and by either the mother or the father in some millipedes.

In crustaceans the eggs are often carried by the female, for example by the branchiopods in an ovisac at the level of the eleventh pair of trunk appendages, by most copepods in one or two bags (some species release the eggs individually into the water), in many ostracods in a dorsal space inside the bivalved carapace. Eggs are carried under the carapace in thermosbaenaceans, between the thoracic appendages (pereiopods) in leptostracans, many stomatopods and some species of euphausiaceans (krill), attached instead to the abdominal appendages (pleopods) in all decapods except the penaeids, which release them in water. The females of the peracarids (mysids, amphipods, tanaidaceans, isopods) possess a marsupium, formed by laminar expansions (oostegites) of the thoracic appendages.

In many insects (for example, in most earwigs, in many heteropterans, embiids and hydrophilid beetles) the female protects the eggs with her body. Prolonged parental care is known in many species of Cassidinae among the leaf beetles or Chrysomelidae, and in some groups of sawflies (the hymenopteran families Tenthredinidae and Pamphiliidae), while the lace bugs of the genus *Gargaphia* and various genera of treehoppers remain with their offspring until they reach adulthood. Protective egg cases are produced by mantises and cockroaches and by the most primitive termites (*Mastotermes*). Very refined and complex forms of parental care are present in the hymenopterans, up to the forms typical of eusociality (bees, wasps, ants), which has also evolved independently in the termites.

The temporal distribution of reproductive effort is very varied within the arthropods (Minelli and Fusco 2013). At one end of the spectrum are the mayflies, fragile insects whose adult life hardly extends beyond a day or a few days, just enough for a single mating, immediately followed in females by egg laying. In many other arthropods the fertile period is much longer and occupies a more or less extended fraction of the final (adult) stage of development, though without further moults intervening after the beginning of the reproductive activity. This condition is common to nearly all the pterygote insects, but also to proturans, most arachnids, some of the millipedes, the pauropods, the copepods and some ostracods among the crustaceans. In many other arthropods, however, reproductive activity is prolonged throughout a more or less long series of stages, morphologically similar to each other but separated by moults. This behaviour is shared by the pycnogonids, some millipedes, the centipedes, the symphylans, many crustaceans including malacostracans, and the hexapods other than proturans and pterygotes. Among the latter, a peculiar condition is found among the mayflies, in which a moult separates the fleeting adult stage from the previous equally fleeting stage of subimago.

Restricted to a few groups is the condition known as *periodomorphosis*, in which the animal is reproductively active in two or more stages, separated by one or more intercalary stages in which gonads and sexual appendages are more or less extensively regressed: this phenomenon is known in the males of some julid millipedes and in some springtails. The behaviour of the females of the isopod crustaceans is similar, with each reproductive period extending over two stages: in the first, the female possesses a ventral space (*marsupium*) for the incubation of the eggs, delimited by laminae (*oostegites*) carried by some limb pairs, and this is the condition in which it can be inseminated; in the second, in which incubation is prolonged, the oostegites are reduced and the female cannot be inseminated. The *parturial moult* that accompanies the release of the offspring from the mother's pouch ushers in a new reproductive cycle.

7.6.20 Chaetognaths

The chaetognaths or arrow-worms are simultaneous hermaphrodites, with an occasional tendency to protandry. In the benthic genus *Spadella*, exchange of spermatophores has been described as reciprocal in *S. cephaloptera* (Ghirardelli 1953), as non-reciprocal in *S. schizoptera* (Goto and Yoshida 1985). Fertilization is internal, but there are no copulatory organs, so sperm cells are simply pressed, in packages, onto the surface of the partner's body, where they move up until they reach the genital pore and eventually the internal reproductive organs. Self-fertilization has been observed in *Heterokrohnia*, where the testes communicate directly with the ovaries (Casanova 1985).

7.6.21 Echinoderms

Most of the echinoderms are gonochoric, but some species of the ophiuroids, asteroids and holothuroids, generally smaller than 1 cm, are hermaphrodite; these are often viviparous or provide parental care. Few species, including the common starfish *Asterias rubens*, are parthenogenetic. Fertilization is almost always external, except in the viviparous sea cucumber *Leptosynapta clarki* and a handful of species of asteroids and ophiuroids.

Parental care is known for a small number of species distributed among all major groups. In some stemless crinoids, embryos develop in incubation pockets (marsupia) in specialized pinnules of the arms. Incubation chambers are also present in some sea urchins (Spatangidae), while other sea urchins (Cidaridae) protect their offspring between the spines. Some sea stars (*Hymenaster, Peltaster*) retain the embryos in the respiratory chamber of the aboral face, others in gastric pockets. In the ophiuroids, development sometimes takes place in incubation spaces known as *genital pouches*. Similarly, in sea cucumbers incubation can occur in different locations: between the tentacles (Dendrochirotida), under the creeping sole (*Psolus*), in various folds of the body surface, and even in the coelom (Synaptidae).

Some asteroid larvae reproduce asexually, either by paratomy (with the detachment of daughter larvae from the tip of the arms), or by autotomy (spontaneous detachment) of the anterior portion of the preoral lobe, or even following the separation from the tip of the arms of individuals similar to early embryos. Some species reproduce asexually also as adults, by splitting of the disc or by regeneration from large body fragments, detached by autotomy, which include at least a part of the central disc.

7.6.22 Hemichordates

The worm-like **enteropneusts** are gonochoric, with external fertilization. The tubicolous **pterobranchs** are poorly known from the point of view of reproduction and sexuality. In the colonial forms, male and female zooids can coexist in the same colony. In *Rhabdopleura normani*, fertilization takes place inside the tubes in which the colony lives and in which embryos are retained until the larva is formed. Vegetative reproduction by budding is widespread, in both the cephalodiscids and the rhabdopleurids.

7.6.23 Cephalochordates

The cephalochordates are gonochoric. The gametes are released after laceration of the walls of the peribranchial space above the gonads and reach water through the atrial pore. Fertilization is external.

7.6.24 Tunicates

The tunicates (or urochordates) are hermaphrodite, with the sole exception of a species of appendicularian (*Oikopleura dioica*). Both sexual and asexual reproduction are practised (Gasparini *et al.* 2015). The sea squirts (Ascidiacea) are simultaneous hermaphrodites. Fertilization often takes place in the peribranchial space, where the whole embryonic development takes place. Matrotrophic viviparity is limited to rare cases (*Botrylloides*, *Hypsistozoa*): here, the fertilized eggs are retained in the terminal segment of the oviduct, where they receive nourishment from the parent.

Asexual reproduction can produce the so-called *social ascidians*, in which the zooids derived from a single larva are joined by a stolon, and the *synascidians*, in which the integration between the zooids extends as far as sharing the tunic and circulatory system and sometimes also the formation of a cloacal space common to all their atrial siphons.

Among the Thaliacea, the salps and doliolums have a metagenetic life cycle with alternation of sexual and asexual generations. Among the salps, *Thalia democratica* is a sufficient hermaphrodite with internal fertilization (Boldrin *et al.* 2009). The solitary oozoid is produced sexually; it generates asexually a ventral stolon, which in turn produces a chain of 25–50 pairs of blastozooids.

The stolon divides by strobilation into 1–3 such chains of blastozooids. The latter, protogynous hermaphrodites, detach one by one from the stolon. The ovary produces only one egg, which is fertilized by a sperm cell produced by the same blastozooid.

Among the colonial tunicates, a number of asexual reproduction mechanisms are known (Brown and Swalla 2012):

- *Aggregated stolonial budding*, a very modified budding mechanism evolved in pelagic tunicates (salps) in which a lateral chain (stolon) develops, at the end of which the buds of new zooids are progressively differentiated.

- *Peribranchial budding*, as found in the colonial stolidobranch ascidians (e.g. *Botryllus*, *Botrylloides*), which starts from an extroflexion of the outer epithelia surrounding the gills and the integration of the mesendodermal cells and the blood surrounding the branchial basket.

- *Pyloric budding*, typical of the aplousobranch colonial ascidians (e.g. *Didemnum*, *Diplosoma*), starting from two opposite buds that develop simultaneously in the epicardial region at the base of an individual's endostyle or oesophagus. One bud differentiates into a new branchial sac and merges with the old abdomen, the other differentiates into a new abdomen that merges with the old branchial sac, thus forming two new individuals.

- *Budding by stolons*, practised by some aplousobranch and phlebobranch (e.g. *Perophora*) colonial ascidians, with buds regularly spaced along the stolon.

- *Strobilation*, a form of budding of some aplousobranch colonial ascidians (e.g. *Aplidium*), in which a portion detached from the posterior abdominal end will grow to form a new zooid.

- *Terminal budding*, which occurs in some phlebobranch colonial ascidians (e.g. *Perophora*), where a pelagic bud is released from the stolon's tip.

- *Vascular budding*, typical of the stolidobranch colonial ascidians (e.g. *Botryllus*, *Botrylloides*, *Symplegma*), which occurs by encapsulation by the vascular epithelium of cells derived from the blood, along the blood vessels that connect the zooids within the common tunic.

7.6.25 Vertebrates or Craniotes

In the recent zoological literature, the animals traditionally called vertebrates are preferentially called craniotes, based on the argument that one of their subgroups, the Myxinoidea, actually lack vertebrae. The term vertebrates should therefore be restricted to the craniotes other than myxinoids. Matters might be revised, however, in the light of evidence of the presence in myxinoids of homologues of the genes that control the formation of vertebral elements in the 'true vertebrates' (Ota 2018).

Vertebrates reproduce almost exclusively by sexual means, the only exception being the rare instances of polyembryony (Section 3.1.2.4). Sexes are mostly separate, but among the bony fish there are many hermaphrodites, as described in Chapter 3; one hermaphrodite species (the shark *Apristurus longicephalus*) is also known among the cartilaginous fishes (Iglésias *et al.* 2005). In gonochoric vertebrates, sex determination is more often chromosomal, but it is environmental (generally depending on temperature) in various bony fishes (Cichlidae, Atherinidae, Poeciliidae), some snakes, some lizards, several turtles and all the crocodiles in which the phenomenon has been studied up to now.

Both female and male heterogamety are known among the species of teleosts, amphibians and reptiles with genetic sex determination; in some snakes and lizards, the X0 system is also known. Birds have female, but mammals usually male heterogamety, although systems with multiple heterochromosomes are also known, among which the ones found in the monotremes stand out for their complexity (Section 6.1.1). For parthenogenesis and other metasexual forms of reproduction in vertebrates, see Chapter 3.

Fertilization is internal in chondroichthyes, in some osteichthyes (Poeciliidae, Anablepidae, Hemiramphidae, Goodeidae), in all reptiles including birds, and in all mammals. Within the Amphibia, fertilization occurs by means of spermatophores in most of the urodeles (except Sirenidae, Inobiidae and Cryptobranchidae, which practise external fertilization), is internal in caecilians and almost always external in the anurans. In the last group, internal fertilization is known only in *Ascaphus truei* (the 'frog with the tail', whose male has an eversible cloaca acting as a copulatory organ), in two kinds of African toads (*Nectophrynoides* and *Nimbaphrynoides*), in an extinct species that lived in Puerto Rico (*Eleuterodactylus jasperi*) and in the Sulawesi dicroglossid *Limnonectes larvaepartus*. Among the anurans with internal fertilization, only *L. larvaepartus* gives birth to tadpoles (Iskandar *et al.* 2014), while the few other species lay their eggs shortly after fertilization or, on the contrary, keep them in their body, where the offspring develop until reaching the final organization.

With a few exceptions (see Section 2.9), tetrapods are iteroparous, with a life cycle longer than one year. It is therefore worth noting the case of the chameleon *Furcifer labordi*, which is semelparous with an annual cycle. Males and females of this species live for about seven months as embryos and only 4–5 months as active animals, which makes their life cycle resemble that of many insects (Karsten *et al.* 2008).

Coda

We have come to the end of this excursion through the phenomena of reproduction. It was not a short trip, but the reader will certainly have realized that it could have been much longer. In every taxonomic group, reproduction involves a range of very different biological processes: from the creation of new genotypes to social behaviour, from resisting adverse environmental conditions to migration. Reproduction borders on these 'other aspects' of an organism's biology and often trespasses into them. So what is it that is common to all these reproductive phenomena? In other words, what exactly, in the end, is reproduction?

In the Introduction and in Chapter 1 we presented various arguments on the basis of which we contended that it is by no means an easy task to provide a rigorous definition of the concept of reproduction, or to mark its boundaries with respect to other biological processes. By adopting a pragmatic strategy, we then provided a number of operational definitions and clarifications on fundamental concepts of biology, with the aim of introducing a common language for all the different topics and taxa we would deal with. These initial choices guided us in the exploration of reproductive phenomena throughout the book, but we also promised that eventually, more informed, we would return to this topic.

What can we say now, after we have seen offspring produced solely for the purpose of nourishing their siblings, plants that entrust their pollen to bats, rotifers that have not known sex for tens of millions of years, and even apparently immortal trees? Or after we have seen reproductive processes merge into those of growth, regeneration and metamorphosis? Are we in a position to advance a new, better definition of reproduction than the one we sketched out a few hundred pages back? Then again, if we have done without such a definition all the way through the book, and yet have managed a reasonably coherent exploration of reproductive processes and their interactions with other biological processes, apparently without getting lost or wandering off into other topics, do we really need it? This doubt is similar to the one recently advanced by some authors regarding the need for a definition of the concept of 'development' (Pradeu *et al.* 2016). In truth, there are numerous concepts in biology that are fundamental work tools, but lack an unequivocal and shared definition, and are difficult to delineate rigorously. Among these are the concepts of 'species' and 'homology' – and even the concept of 'living'.

These problems, the discussion of which many biologists would willingly leave in the hands of philosophers, actually encroach on epistemology, and rivers of ink have been spilled over them. Without embarking on a discussion that would take us far from the theme of this book, as well as from the professional skills of its authors, we will just draw attention to something that represents not so much a solution to the problem, but rather an acknowledgment of the impossibility of solving it in the terms we would like.

There are some concepts and phenomena which by their very nature do not lend themselves to being circumscribed and defined.

In *Philosophical Investigations* (1953), the philosopher Ludwig Wittgenstein contemplates the use of a definition in relation to the apparently simple concept of a 'game'. Despite the familiarity of the concept, any attempt to find a set of necessary and sufficient conditions for an activity to be defined as a game seems doomed to failure. There is no function or feature shared by all games and only by games. A definition broad enough to include such disparate activities as table games, card games, ball games, role-playing games and sports games will also easily include many 'non-games'. On the other hand, a definition narrow enough to exclude all non-games will certainly exclude even many typical games. Instead of listing a set of necessary and sufficient conditions to separate games from non-games, Wittgenstein suggests we should think of a network of relationships that cross the multidimensional linguistic landscape occupied by the concept of 'game'. Some of these relationships may link multiple types, or instances, of games, while others link additional instances, and still others unite some (but not all) instances of different classes of games. In other words, the concept of game is identified by a relatively fluid set of characteristics, so that games are recognizable through a network of relationships that, by analogy, Wittgenstein called 'family resemblance'. In fact, members of a human family might resemble each other not because they all share a particular trait, but rather because each member shares some traits with some other members of the family. Some members may have similar eyes, others may have similar hair, or gesticulate or walk similarly, but none of these characteristics will be found in all family members, and in them exclusively. Some family members may not even resemble others at all, but be similar to them through a chain of similarities to other members.

Returning to reproduction and adopting Wittgenstein's point of view, it seems evident that our inability to arrive at a precise definition of 'reproduction' does not mean that we do not know what this phenomenon is, or that we lack imagination in trying to formulate a rigorous definition. The point is that there is no set of necessary and sufficient conditions to delimit all the different forms of reproduction that we have encountered. The concept of reproduction is in itself vague and nuanced.

This does not make the concept useless or rough, provided that it is treated for what it is. The only real risk is to want at all costs to make it too rigid, by attempting to apply an arbitrary and inflexible classification. This attempt would inevitably fail, and would alienate us from the true nature of the phenomenon. In the words of Heraclitus, 'nature loves to hide', and this is just one of many ways in which it does so.

We can thus relax, and think of reproduction as 'a family of related concepts', which revolve around the intuitive and familiar idea of reproduction and which, in one sense or another, involve the generation of the living.

Beyond classifications and theorizations, our excursion into the biology of reproduction brings us back in the end to the 'endless forms most beautiful' of the living celebrated by Darwin in *The Origin of Species* (1859), and in particular to the enormous variety of expedients through which, day after day, their continuity through time is accomplished, as we hope we have shown in the pages of this book.

Appendix: A Classification of Living Organisms

To help the reader place the numerous organisms mentioned in this book within the framework of an updated classification, we have generally indicated the name of a well-known higher taxon to which each genus or species mentioned belongs (except for the most obvious cases such as bee, *Drosophila*, mouse). We have specified, for example, that *Montacuta* is a bivalve mollusc, *Siboglinum* a polychaete, *Ginkgo* a gymnosperm. All these group names mentioned in the text are found in the following table, which represents a customized version of the hierarchical ordering of biological classification, according to current knowledge of phylogeny. When using this table, the reader should bear in mind that:

- Biological systematics is increasingly abandoning the traditional taxonomic ranks, according to which, for example, gastropods and mammals would be *classes*, lepidopterans and primates *orders*, etc. However, it is essential to pay attention to the nested relationships, indicated in the following table by increasing indentation of the name of a subordinate taxon with respect to the immediately higher-level taxon – so, for example, Ascomycota and Basidiomycota within the Dikarya.
- The following table includes all the groups mentioned in this book, but omits many other groups.
- The names given here in quotes ('Porifera', 'Hypermastigina', 'Gregarinia', 'Rotifera', 'Polychaeta', 'Oligochaeta', 'Crustacea', 'Cladocera', 'Homoptera', 'Gymnospermae') refer to very popular taxonomic groupings, which, however – in the light of current understanding of phylogeny – are no longer considered natural (monophyletic) groups. We alert the reader to this circumstance, but we believe that it is useful to keep these names temporarily in use. Moreover, to avoid excessive pedantry in the text, we have put these names in quotes only in the following table. Also note the inclusion of birds within reptiles.
- Where a taxon is shown in **bold**, a more detailed classification will be found later in the table.

Eubacteria
Archaea
Eukaryota
    Unikonta
        Amoebozoa
            Conosa
                Archamoebae
                Dictyostelea
                Myxogastrea
            Lobosa
    Opisthokonta
        Fungi
            Microsporidia
            Zygomycota
            Chytridiomycota
            Glomeromycota
            Dikarya
                Ascomycota, incl. Pucciniomycotina, Ustilaginomycotina
                Basidiomycota
        Choanozoa
        Animalia (= Metazoa) → p. 409
    Bikonta
        Euglenozoa
        Metamonada
            Trichomonadea
            Diplomonadida
            'Hypermastigina'
        Plantae s.s. → p. 413
    Chromalveolata
        Oomycetes
        Haptophyta, incl. Coccolithophyceae
        Heliozoa
        Cryptista, incl. Cryptomonadales
        Ochrophyta (= Heterokonta partim) → p. 414
        Opalinea
        Ciliata
        Dinoflagellata (= Dinophyceae)
        Apicomplexa (= Sporozoa)
            Haemosporidia
            Coccidea
            'Gregarinia'
        Radiolaria
        Foraminifera

Animalia (= Metazoa) (from p. 408)
  'Porifera'
  Placozoa
  Ctenophora
  Cnidaria
    Anthozoa
    Medusozoa
      Hydrozoa
      Cubozoa
      Scyphozoa
  Bilateria
    Acoelomorpha
      Acoela
      Nemertodermatida
    Protostomia → this page
    Deuterostomia → p. 412

Protostomia
  Lophotrochozoa
    Gastrotricha
    Micrognathozoa
    Gnathostomulida
    Syndermata
      'Rotifera'
        Seisonidea
        Monogononta
        Bdelloidea
      Acanthocephala
    Catenulida (= Platyhelminthes partim)
    Rhabditophora (= Platyhelminthes partim)
      Tricladida
      Trematoda (= Digenea)
      Monogenea
      Cestoda
    Cycliophora
    Ectoprocta (= Bryozoa)
    Entoprocta
    Orthonectida
    Rhombozoa, incl. Dicyemida
    Nemertea (= Nemertini)
    Phoronozoa
      Phoronida
      Brachiopoda

Lophotrochozoa (cont.)
    Mollusca
        Caudofoveata
        Solenogastres
        Polyplacophora
        Monoplacophora
        Bivalvia
        Scaphopoda
        Gastropoda
        Cephalopoda
    Annelida
        'Polychaeta', incl. Siboglinidae (= Pogonophora), Echiura, Sipuncula
        Clitellata
            'Oligochaeta'
            Hirudinea
Ecdysozoa
    Kinorhyncha
    Loricifera
    Priapulida
    Nematoda
    Nematomorpha
    Tardigrada
    Onychophora
    Arthropoda
        Chelicerata
            Pycnogonida
            Xiphosura
            Arachnida
                Acari
                Araneae
                Amblypygi
                Uropygi
                Schizomida
                Scorpiones
                Opiliones
                Pseudoscorpiones
                Solifuga
        Myriapoda
            Chilopoda
            Diplopoda
            Pauropoda
            Symphyla

Arthropoda (cont.)
    Pancrustacea
        'Crustacea'
            Remipeda
            Ostracoda
            Branchiopoda, incl. 'Cladocera'
            Copepoda
            Tantulocarida
            Cirripeda, incl. Rhizocephala
            Stomatopoda
            Isopoda
            Amphipoda
            Decapoda
        Hexapoda
            Collembola
            Protura
            Diplura
            Insecta
                Ephemeroptera
                Odonata
                Plecoptera
                Orthoptera
                Phasmatodea
                Dermaptera
                Dictyoptera
                    Blattodea
                    Mantodea
                    Isoptera
                Zoraptera
                Psocoptera
                Phthiraptera
                Hemiptera
                    'Homoptera'
                    Heteroptera
                Thysanoptera
                Coleoptera
                Strepsiptera
                Diptera
                Lepidoptera
                Hymenoptera

Deuterostomia (from p. 409)
    Echinodermata
        Crinoidea
        Asteroidea
        Ophiuroidea
        Holothuroidea
        Echinoidea
    Hemichordata
        Enteropneusta
        Pterobranchia
    Chordata
        Tunicata (= Urochordata)
            Ascidiacea
            Thaliacea
            Appendicularia
        Cephalochordata
        Vertebrata (= Craniota)
            Cyclostomata (= Agnatha)
            Chondrichthyes
            Actinopterygii, incl. Teleostei
            Sarcopterygii
                Coelacanthiformes
                Dipnoi
                Amphibia
                    Urodela
                    Apoda
                    Anura
                Reptilia
                    Lepidosauria
                    Sphenodontida
                    Squamata
                    Testudinata
                    Archosauria
                          Crocodylia
                          Aves
                Mammalia

Plantae s.s. (from p. 408)
    Glaucophyta
    Rhodophyceae
    Viridiplantae
        Chlorophyta
            Chlorodendrophyceae
            Chlorophyceae
                Chaetopeltidales
                Chlamydomonadales (= Volvocales)
                Oedogoniales
                Sphaeropleales
            Pedinophyceae
            Ulvophyceae
                Cladophorales
                Ulvales
            Pyramimonadophyceae, incl. Pseudoscourfieldiales
        Streptophyta
            Charophyta
                Charophyceae
                Chlorokybophyceae
                Conjugatophyceae (= Zygnematophyceae)
                    Desmidiales
                    Zygnematales
                Klebsormidiophyceae
                Mesostigmatophyceae
            Embryophyta
                Marchantiophyta (= Hepaticae)
                Anthocerotophyta
                Bryophyta s.s. (= Musci)
                Tracheophyta
                    Lycophytina
                    Equisetina
                    Marattiopsida (= Filicales partim)
                    Polipodiopsida (= Filicales partim)
                    Spermatopsida (= Spermatophyta)
                      ‘Gymnospermae’
                        Cicadales
                        Ginkgoales
                        Gnetales
                        Pinales (= Coniferae)
                  Magnoliopsida (= Angiospermae)

Ochrophyta (= Heterokonta partim) (from p. 408)
        Bacillariophyceae (= Diatomeae)
        Eustigmatophyceae
        Bolidophyceae
        Chrysophyceae
        Dictyochophyceae, incl. Pelagophyceae
        Phaeophyceae
        Phaeothamniophyceae, incl. Aurearenophyceae
        Picophageae, incl. Synchromophyceae
        Pinguiophyceae
        Raphidophyceae
        Schizocladiophyceae
        Xanthophyceae

References

Aanen, D., Beekman, M. & Kokko, H. (eds.) (2016). Weird sex: the underappreciated diversity of sexual reproduction. *Philosophical Transactions of the Royal Society of London. Series B, Biological Sciences*, 371 (1706).

Abbott, J. K., Norde, A. K. & Hansson, B. (2017). Sex chromosome evolution: historical insights and future perspectives. *Proceedings of the Royal Society of London. Series B, Biological Sciences*, 284: 20162806.

Ackermann, M., Stearns, S. & Jenal, U. (2003). Senescence in a bacterium with asymmetric division. *Science*, 300: 1920.

Adamson, M. & Ludwig, D. (1993). Oedipal mating as a factor in sex allocation in haplodiploids. *Philosophical Transactions of the Royal Society of London. Series B, Biological Sciences*, 341: 195–202.

Adamson, M. L. (1989). Evolutionary biology of the Oxyurida (Nematoda): biofacies of a haplo-diploid taxon. *Advances in Parasitology*, 28: 175–228.

Adler, L. & Jarms, G. (2009). New insights into reproductive traits of scyphozoans: special methods of propagation in *Sanderia malayensis* Goette, 1886 (Pelagiidae, Semaeostomeae) enable establishing a new classification of asexual reproduction in the class Scyphozoa. *Marine Biology*, 156: 1411–1420.

Adolfsson, S., Michalakis, Y., Paczesniak, D. *et al.* (2010). Evaluation of elevated ploidy and asexual reproduction as alternative explanations for geographic parthenogenesis in *Eucypris virens* ostracods. *Evolution*, 64: 986–997.

Agrawal, A. F. (2001). Sexual selection and the maintenance of sexual reproduction. *Nature*, 411: 692–695.

Agrawal, S. C. (2012). Factors controlling induction of reproduction in algae – review: the text. *Folia Microbiologica*, 57: 387–407.

Aisenberg, A. & Peretti A. V. (2011). Sexual dimorphism in immune response, fat reserve and muscle mass in a sex role reversed spider. *Zoology*, 114: 272–275.

Ajzenberg, D., Bañuls, A. L., Su, C. *et al.* (2004). Genetic diversity, clonality and sexuality in *Toxoplasma gondii*. *International Journal of Parasitology*, 34: 1185–1196.

Alberts, B., Johnson, A., Lewis, J. *et al.* (2015). *Molecular Biology of the Cell*, 6th edn. New York, NY: Garland Science.

Alby, K., Schaefer, D. & Bennett, R. J. (2009). Homothallic and heterothallic mating in the opportunistic pathogen *Candida albicans*. *Nature*, 460: 890–893.

Alcock, J. (2013). *Animal Behavior: an Evolutionary Approach*, 10th edn. Sunderland, MA: Sinauer Associates.

Andrade, M. C. B. (1996). Sexual selection for male sacrifice in the Australian redback spider. *Science*, 271: 70–72.

Archibald, J. M., Simpson, A. G. B. & Slamovits, C. H. (eds.) (2017). *Handbook of the Protists*. Cham: Springer International.

Arnold, A. P. (2012). The end of gonad-centric sex determination in mammals. *Trends in Genetics*, 28: 55–61.

Aron, S., de Menten, L., Van Bockstaele, D. R., Blank, S. M. & Roisin, Y. (2005). When hymenopteran males reinvented diploidy. *Current Biology*, 15: 824–827.

Arroyo, M. T. K. & Uslar, P. (1993). Breeding systems in a temperate Mediterranean-type climate montane sclerophyllous forest in central Chile. *Botanical Journal of the Linnean Society*, 111: 83–102.

Asplen, M. K., Whitefield, J. B., De Boer, J. G. & Heimpel, G. E. (2009). Ancestral state reconstruction analysis of hymenopteran sex determination mechanisms. *Journal of Evolutionary Biology*, 22: 1762–1769.

Avise, J. C. (2008). *Clonality: the Genetics, Ecology and Evolution of Sexual Abstinence in Vertebrate Animals*. New York, NY: Oxford University Press.

Avise, J. C. (2011). *Hermaphroditism: a Primer on the Biology, Ecology, and Evolution of Dual Sexuality*. New York, NY: Columbia University Press.

Bachtrog, D., Kirkpatrick, M., Mank, J. E. *et al.* (2011). Are all sex chromosomes created equal? *Trends in Genetics*, 27: 350–357.

Balon, E. K. (1975). Reproductive guilds in fishes: a proposal and definition. *Journal of the Fisheries Research Board of Canada*, 32: 821–864.

Balon, E. K. (1984). Patterns in the evolution of reproductive styles in fishes. In *Fish Reproduction: Strategies and Tactics*, ed. G. W. Potts & R. J. Wootton. London: Academic Press, pp. 35–53.

Barske, L. A. & Capel, B. (2010). Sex determination: an avian sexual revolution. *Nature*, 464: 171–172.

Bauer, R. T. (1986). Sex change and life history pattern in the shrimp *Thor manningi* (Decapoda: Caridea): a novel case of partial protandric hermaphroditism. *Biological Bulletin*, 170: 11–31.

Bauer, R. T. (2000). Simultaneous hermaphroditism in caridean shrimps: a unique and puzzling sexual system in the Decapoda. *Journal of Crustacean Biology*, 20 (Special Number 2): 116–128.

Bauer, R. T. & Newman, W. A. (2004). Protandric simultaneous hermaphroditism in the marine shrimp *Lysmata californica* (Caridea: Hippolytidae). *Journal of Crustacean Biology*, 24: 131–139.

Baurain, D., Brinkmann, H., Petersen, J. *et al.* (2010). Phylogenomic evidence for separate acquisition of plastids in cryptophytes, haptophytes, and stramenopiles. *Molecular Biology and Evolution*, 27: 1698–1709.

Bell, G. (1980). The costs of reproduction and their consequences. *American Naturalist*, 116: 45–76.

Bell, G. (1982). *The Masterpiece of Nature: the Evolution and Genetics of Sexuality*. London: Croom Helm.

Bell, G. (1988). *Sex and Death in Protozoa: the History of an Obsession*. Cambridge: Cambridge University Press.

Bell, G. & Praiss, M. (1986). Optimality and constraint in a self-fertilized alga. *Evolution*, 40: 194–198.

Benazzi, M. & Benazzi Lentati, G. (1992). Pseudogamy (gynogenesis) in planarians: annotations some forty years on. In *Sex: Origin and Evolution*, ed. R. Dallai. Selected Symposia and Monographs U.Z.I., 6. Modena: Mucchi, pp. 87–102.

Beukeboom, L. W. & Perrin, N. (2014). *The Evolution of Sex Determination*. Oxford: Oxford University Press.

Beukeboom, L. W. & van de Zande, L. (2010). Genetics of sex determination in the haplodiploid wasp *Nasonia vitripennis* (Hymenoptera: Chalcidoidea). *Journal of Genetics*, 89: 333–339.

Beukeboom, L. W. & Vrijenhoek, R. C. (1998). Evolutionary genetics and ecology of sperm-dependent parthenogenesis. *Journal of Evolutionary Biology*, 11: 755–782.

Beukeboom, L. W., Weinzierl, R. P., Reed, K. M. & Michiels N. K. (1996). Distribution and origin of chromosomal races in the freshwater planarian *Dugesia polychroa* (Turbellaria: Tricladida). *Hereditas*, 124: 7–15.

Bewley, J. D., Bradford, K. J., Hilhorst, H. W. M. & Nonogaki, H. (2013). *Seeds: Physiology of Development, Germination and Dormancy*. New York, NY: Springer.

Bickel, R. D., Cleveland, H. C., Barkas, J. *et al.* (2013). The pea aphid uses a version of the terminal system during oviparous, but not viviparous, development. *EvoDevo*, 4: 10.

Biddle, F. G., Eden, S. A., Rossler, J. S. & Eales, B. A. (1997). Sex and death in the mouse: genetically delayed reproduction and senescence. *Genome*, 40: 229–235.

Birkhead, T. R., Hosken D. J. & Pitnick S. S. (eds.) (2008). *Sperm Biology: an Evolutionary Perspective*. London: Academic Press.

Bode, S. N., Adolfsson, S., Lamatsch, D. K. *et al.* (2010). Exceptional cryptic diversity and multiple origins of parthenogenesis in a freshwater ostracod. *Molecular Phylogenetics and Evolution*, 54: 542–552.

Bogart, J., Bi, K., Fu, J., Noble, D. W. A. & Niedzwiecki, J. (2007). Unisexual salamanders (genus *Ambystoma*) present a new reproductive mode for eukaryotes. *Genome*, 50: 119–136.

Boldrin, F., Martinucci, G., Holland, L. Z., Miller, R. L. & Burighel, P. (2009). Internal fertilization in the salp *Thalia democratica*. *Canadian Journal of Zoology*, 87: 928–940.

Bonner, J. T. (2000). *First Signals*. Princeton, NJ: Princeton University Press.

Bonnet, X. (2011). The evolution of semelparity. In *Reproductive Biology and Phylogeny of Snakes*, ed. R. D. Aldridge, B. C. Jellen, D. S. Siegel & S. S. Wisniewski. Enfield, NH: Science Publishers, pp. 645–672.

Boschetti, C., Leasi, F. & Ricci, C. (2011). Developmental stages in diapausing eggs: an investigation across monogonont rotifer species. *Hydrobiologia*, 662: 149–155.

Bossinger, G., Spokevicius, A. V. (2011). Plant chimaeras and mosaics. In *Encyclopedia of Life Sciences (ELS)*. Chichester: Wiley. doi: 10.1002/9780470015902.a0002090.pub2.

Bouillon, J., Gravili, C., Pagès, F., Gili, J.-M. & Boero, F. (2006). *An Introduction to Hydrozoa*. Mémoires du Muséum national d'Histoire naturelle, 194. Paris: Muséum national d'Histoire naturelle.

Bourlière, F. (1964). *The Natural History of Mammals*, 3rd edn. New York, NY: Knopf.

Bowman, J. L., Sakakibara, K., Furumizu, C. & Dierschke, T. (2016). Evolution in the cycles of life. *Annual Review of Genetics*, 50: 133–154.

Boyden, A. (1950). Is parthenogenesis sexual or asexual reproduction? *Nature*, 166: 820.

Bradbury, J. W. & Vehrencamp, S. L. (2011). *Principles of Animal Communication*, 2nd edn. Sunderland, MA: Sinauer Associates.

Bradshaw, J. W. S., Baker, R., Lisk, J. C. (1983). Separate orientation and releaser components in a sex pheromone. *Nature*, 304: 265–267.

Brakefield, P. M. & Zwaan, B. J. (2011). Seasonal polyphenisms and environmentally induced plasticity in the Lepidoptera – the coordinated evolution of many traits on multiple levels. In *Mechanisms of Life History Evolution*, ed. T. Flatt & A. Heyland. Oxford: Oxford University Press, pp. 243–252.

Brien, P. (1973). Les démosponges: morphologie et reproduction. In *Traité de Zoologie, 3 (1)*, ed. P. P. Grassé. Paris: Masson, pp. 133–461.

Brown, F. D. & Swalla, B. J. (2012). Evolution and development of budding by stem cells: ascidian coloniality as a case study. *Developmental Biology*, 369: 151–162.

Bubley, W. J. & Pashuk, O. (2010). Life history of a simultaneously hermaphroditic fish, *Diplectrum formosum*. *Journal of Fish Biology*, 77: 676–691.

Buckley, D., Alcobendas, M., García-París, M. & Wake, M. H. (2007). Heterochrony, cannibalism, and the evolution of viviparity in *Salamandra salamandra*. *Evolution and Development*, 9: 105–115.

Bull, J. J. (1983). *Evolution of Sex Determining Mechanisms*. Menlo Park, CA: Benjamin/Cummings.

Bull, J. J. & Bulmer, M. G. (1981). The evolution of XY females in mammals. *Heredity*, 47: 347–365.

Burke, N. W. & Bonduriansky, R. (2017). Sexual conflict, facultative asexuality, and the true paradox of sex. *Trends in Ecology and Evolution*, 32: 646–652.

Buss, L. (1987). *The Evolution of Individuality*. Princeton, NJ: Princeton University Press.

Butterfield, N. J. (2000). *Bangiomorpha pubescens* n. gen. n. sp.: implications for the evolution of sex, multicellularity and the Mesoproterozoic/Neoproterozoic radiation of eukaryotes. *Paleobiology*, 26: 386–404.

Casanova, J. P. (1985) Description de l'appareil génital primitif du genre *Heterokrohnia* et nouvelle classification des Chaetognathes. *Comptes rendus de l'Académie des sciences Paris, Série III*, 301: 397–402.

Casselton, L. A. (2002). Mate recognition in fungi. *Heredity*, 88: 142–147.

Cavallin, M. (1971). La polyembryonie substitutive et la problème de l'origine de la lignée germinale chez le phasme *Carausius morosus* Br. *Comptes rendus de l'Académie des sciences Paris*, 272: 462–465.

Cavicchioli, R. (ed.) (2007). *Archaea: Molecular and Cellular Biology*. Washington, DC: ASM Press.

Cellerino, A., Valenzano, D. R. & Reichard, M. (2016). From the bush to the bench: the annual *Nothobranchius* fishes as a new model system in biology. *Biological Reviews*, 91: 511–533.

Chang, E. S., Orive, M. E. & Cartwright, P. (2018). Nonclonal coloniality: genetically chimeric colonies through fusion of sexually produced polyps in the hydrozoan *Ectopleura larynx*. *Evolution Letters*, 2–4: 442–455.

Chaparro, O. R., Schmidt, A. J., Pardo, L. M. *et al.* (2011). Reproductive strategy of the semelparous clam *Gaimardia bahamondei* (Bivalvia, Gaimardiidae). *Invertebrate Zoology*, 130: 49–59.

Chapman, H., Houliston, G. J., Robson, B. & Iline, I. (2003). A case of reversal: the evolution and maintenance of sexuals from parthenogenetic clones in *Hieracium pilosella*. *International Journal of Plant Sciences*, 164: 719–728.

Charlesworth, B. & Dempsey, N. D. (2001). A model of the evolution of the unusual sex chromosome system of *Microtus oregoni*. *Heredity*, 86: 387–394.

Charlesworth, D. (2002). Plant sex determination and sex chromosomes. *Heredity*, 88: 94–101.

Chemisquy, A. (2015). Peramorphic males and extreme sexual dimorphism in *Monodelphis dimidiata* (Didelphidae). *Zoomorphology*, 184: 587–599.

Chen, B. Y. & Henen, W. K. (1989). Evidence for spontaneous diploid androgenesis in *Brassica napus* L. *Sexual Plant Reproduction*, 2: 15–17.

Clark, J. R. (1983). Age-related changes in trees. *Journal of Arboriculture*, 9: 201–205.

Clark, W. R. (1996). *Sex and the Origin of Death*. New York, NY: Oxford University Press.

Clifton, K. E. (1997). Mass spawning by green algae on coral reefs. *Science*, 275: 1116–1118.

Cohan, F. M. (1999). Genetic structure of prokaryotic populations. In *Evolutionary Genetics: From Molecules to Morphology*, ed. R. S. Singh & C. B. Krimbas. Cambridge: Cambridge University Press, pp. 475–489.

Cook, J. M. (2002). Sex determination in invertebrates. In *Sex Ratios: Concepts and Research Methods*, ed. I. C. W. Hardy. Cambridge: Cambridge University Press, pp. 178–194.

Cordaux, R., Bouchon, D. & Grève, P. (2011). The impact of endosymbionts on the evolution of host sex-determination mechanisms. *Trends in Genetics*, 27: 332–341.

Craig, S. F., Slobodkin, L. B., Wray, G. A. & Biermann, C. H. (1997). The 'paradox' of polyembryony: a review of the cases and a hypothesis for its evolution. *Evolutionary Ecology*, 11: 127–143.

Crespi, B. J. (1992). Cannibalism and trophic eggs in subsocial and eusocial insects. In *Cannibalism: Ecology and Evolution Among Diverse Taxa*, ed. M. Elgar & B. J. Crespi. Oxford: Oxford University Press, pp. 176–213.

Cronberg, N., Natcheva, R. & Hedlund, K. (2006). Microarthropods mediate sperm transfer in mosses. *Science*, 313: 1255.

Dallai, R. (2014). Overview on spermatogenesis and sperm structure of Hexapoda. *Arthropod Structure and Development*, 43: 257–290.

Dallai, R., Gottardo, M., Mercati, D. *et al.* (2014). Giant spermatozoa and a huge spermatheca: a case of coevolution of male and female reproductive organs in the ground louse *Zorotypus impolitus* (Insecta, Zoraptera). *Arthropod Structure and Development*, 43: 135–151.

Danovaro, R., Dell'Anno, A., Pusceddu, A. *et al.* (2010). The first metazoa living in permanently anoxic conditions. *BMC Biology*, 8, 30.

Darwin, C. R. (1859). *On the Origin of Species by Means of Natural Selection or the Preservation of Favoured Races on the Struggle for Life*. London: Murray.

de Meeûs, T., Prugnolle, F. & Agnew, P. (2007). Asexual reproduction: genetics and evolutionary aspects. *Cellular and Molecular Life Sciences*, 64: 1355–1372.

de Reviers, B., Rousseau, F. & Silberfeld, T. (2015). Phaeophyceae. In *Syllabus of Plant Families*, 13th edn, ed. W. Frey. Berlin; Stuttgart: Borntraeger, Vol. 2/1, pp. 139–176.

Deakin, J. E., Hore, T. A., Koina, E. & Graves, J. A. M. (2008). The status of dosage compensation in the multiple X chromosomes of the platypus. *PLOS Genetics*, 4 (7): e1000140.

Debortoli, N., Li, X., Eyres, I., Fontaneto, D., Hespeels, B., Tang, C. Q. *et al.* (2016). Genetic exchange among bdelloid rotifers is more likely due to horizontal gene transfer than to meiotic sex. *Current Biology*, 26: 723–732.

den Bakker, H. C., VanKuren, N. W., Morton, J. B. & Pawlowska, T. E. (2010). Clonality and recombination in the life history of an asexual arbuscular mycorrhizal fungus. *Molecular Biology and Evolution*, 27: 2474–2486.

Devlin, R. H. & Nagahama, Y. (2002). Sex determination and sex differentiation in fish: an overview of genetic, physiological and environmental influences. *Aquaculture*, 208: 191–364.

Dournon, C., Houillon, C. & Pieau, C. (1990). Temperature sex-reversal in amphibians and reptiles. *International Journal of Developmental Biology*, 34: 81–92.

Drago, L., Fusco, G., Garollo, E. & Minelli, A. (2011). Structural aspects of leg-to-gonopod metamorphosis in male helminthomorph millipedes (Diplopoda). *Frontiers in Zoology*, 8: 19.

Dreyer, N, Høeg, J. T., Hess, M. *et al.* (2017). When dwarf males and hermaphrodites copulate: First record of mating behaviour in a dwarf male using the androdioecious barnacle *Scalpellum scalpellum* (Crustacea: Cirripedia: Thoracica). *Organisms Diversity and Evolution*, 18: 115–123.

D'Souza, T. G. & Michiels, N. K. (2010). The costs and benefits of occasional sex: theoretical predictions and a case study. *Journal of Heredity*, 101: S34–S41.

Dupré, J. (2010). The polygenomic organism. *The Sociological Review*, 58 (Supplement 1): 19–31.

Dutrillaux, A. M., Pluot-Sigwalt, D. & Dutrillaux, B. (2010). (Ovo-)viviparity in the darkling beetle, *Alegoria castelnaui* (Tenebrioninae: Ulomini), from Guadeloupe. *European Journal of Entomology*, 107: 481–485.

Eisen, J. A., Coyne, R. S., Wu, M. *et al.* (2006). Macronuclear genome sequence of the ciliate *Tetrahymena thermophila*, a model eukaryote. *PLOS Biology*, 4 (9): e286.

Ellegren, H. (2008). Sex chromosomes: platypus genome suggests a recent origin for the human X. *Current Biology*, 18: R557–R559.

Embley, T. M. (2006). Multiple secondary origins of the anaerobic lifestyle in eukaryotes. *Philos. Trans. R. Soc. B*, 361: 1055–1067.

Erickson, J. W. & Quintero, J. J. (2007). Indirect effects of ploidy suggest X chromosome dose, not the X:A ratio, signals sex in *Drosophila*. *PLOS Biology*, 5 (12): e332.

Evans, J. P., Kelley, J. L., Bisazza, A., Finazzo, E. & Pilastro, A. (2004). Sire attractiveness influences offspring performance in guppies. *Proceedings of the Royal Society of London. Series B, Biological Sciences*, 271: 2035–2042.

Evans, P. C., Lambert, N., Maloney, S. *et al.* (1999). Long-term fetal microchimerism in peripheral blood mononuclear cell subsets in healthy women and women with scleroderma. *Blood*, 93: 2033–2037.

Extavour, C. G. & Akam, M. (2003). Mechanisms of germ cell specification across the metazoans: epigenesis and preformation. *Development*, 130: 5869–5884.

Fahy, G. M. (2010). Precedents for the biological control of aging: experimental postponement, prevention, and reversal of aging processes. In *The Future of Aging. Pathways to Human Life Extension*, ed. G. M. Fahy. Dordrecht: Springer, pp. 127–225.

Farrar, D. R. (1990). Species and evolution in asexually reproducing independent fern gameto-phytes. *Systematic Botany*, 15: 98–111.

Fautin, D. G. (1992). Cnidaria. In *Reproductive Biology of Invertebrates, 5*, ed. K. G. Adiyodi & R. G. Adiyodi. Chichester: Wiley, pp. 31–52.

Fernando, D. D., Lazzaro, M. D. & Owens, J. N. (2005). Growth and development of conifer pollen tubes. *Sexual Plant Reproduction*, 18: 149–162.

Fields, C. & Levin, M. (2018). Are planaria individuals? What regenerative biology is telling us about the nature of multicellularity. *Evolutionary Biology*, 45: 237–247.

Finch, C. E. (1990). *Longevity, Senescence, and the Genome*. Chicago, IL: University of Chicago Press.

Fischer, E. (2009). Protracheophyta (Horneophytopsida), Tracheophyta p.p.: Rhyniophytina, Lycophytina, 'Trimerophytina', Moniliformopses ('Pteridophyta'), Radiatopses (Progymno-spermopsida). In *Syllabus of Plant Families*, 13th edn, ed. W. Frey. Berlin; Stuttgart: Borntraeger, Vol. 3, pp. 264–399.

Flatt, T. & Heyland, A. (eds.) (2011). *Mechanisms of Life History Evolution*. Oxford: Oxford University Press.

Flindt, R. (2003). *Amazing Numbers in Biology*. Berlin: Springer.

Flinn, K. M. (2006). Reproductive biology of three fern species may contribute to differential colonization success in post-agricultural forests. *American Journal of Botany*, 93: 1289–1294.

Flores-Renteria, L., Molina-Freaner, F., Whipple, A. V., Gehring, C. A. & Dominguez, C. A. (2013). Sexual stability in the nearly dioecious *Pinus johannis* (Pinaceae). *American Journal of Botany*, 100: 602–612.

Fransson, T., Jansson, L., Kolehmainen, T., Kroon, C. & Wenninger, T. (2017). EURING list of longevity records for European birds. https://euring.org/data-and-codes/longevity-list (accessed April 2019).

Fraser, C., Hanage, W. P. & Spratt, B. G. (2007). Recombination and the nature of bacterial speciation. *Science*, 315: 476–480.

Fraser, J. A. & Heitman, J. (2004). Evolution of fungal sex chromosomes. *Molecular Microbiology*, 51: 299–306.

Freeman, D. C., Harper, K. T. & Charnov, E. L. (1980). Sex change in plants: old and new observations and new hypotheses. *Oecologia (Berlin)*, 47: 222–232.

Freeman, D. C., Wachocki, B. A., Stender, M. J., Goldschlag, D. E. & Michaels, H. J. (1994). Seed size and sex ratio in spinach: application of the Trivers–Willard hypothesis to plants. *Ecoscience*, 1: 54–63.

Frey, W. & Stech, M. (2009). Marchantiophyta, Bryophyta, Anthocerotophyta. In *Syllabus of Plant Families*, 13th edn, ed. W. Frey. Berlin; Stuttgart: Borntraeger, Vol. 3, pp. 9–263.

Fritsch, F. E. (1935, 1945). *The Structure and Reproduction of the Algae, 1 (1935), 2 (1945)*. London; New York, NY: Cambridge University Press.

Fryer, G. (1961). The developmental history of *Mutela bourguignati* (Ancey) Bourguignat (Mollusca: Bivalvia). *Philosophical Transactions of the Royal Society of London. Series B, Biological Sciences*, 244: 259–298.

Funch, P. & Kristensen, R. M. (1995). Cycliophora is a new phylum with affinities to Entoprocta and Ectoprocta. *Nature*, 378: 711–714.

Fusco, G. & Minelli, A. (2010). Phenotypic plasticity in development and evolution: facts and concepts. *Philosophical Transactions of the Royal Society of London. Series B, Biological Sciences*, 365: 547–556.

Futuyma, D. J. & Kirkpatrick, M. (2018). *Evolution*, international 4th edn. New York, NY: Oxford University Press, Sinauer Associates.

Galaktionov, K. V. & Dobrovolskij, A. A. (2003). *The Biology and Evolution of Trematodes*. Dordrecht: Kluwer Academic Publishers.

Gardner, S. N. & Mangel, M. (1997). When can a clonal organism escape senescence? *American Naturalist*, 150: 462–490.

Gasparini, F., Manni, L., Cima, F. *et al.* (2015). Sexual and asexual reproduction in the colonial ascidian *Botryllus schlosseri*. *Genesis*, 53: 105–120.

Georgiades, P., Watkins, M., Burton, G. J. & Ferguson-Smith, A. C. (2001). Roles for genomic imprinting and the zygotic genome in placental development. *Proceedings of the National Academy of Sciences of the United States of America*, 98: 4522–4527.

Ghirardelli, E. (1953). L'accoppiamento in *Spadella cephaloptera* Busch. *Pubblicazioni della Stazione Zoologica di Napoli*, 24: 345–354.

Ghiselin, M. T. (1974a). *The Economy of Nature and the Evolution of Sex*. Berkeley, CA: University of California Press.

Ghiselin, M. T. (1974b). A radical solution to the species problem. *Systematic Zoology*, 23: 536–544.

Gianasi, B. L., Hamel, J.-F. & Mercier, A. (2018). Full allogeneic fusion of embryos in a holothuroid echinoderm. *Proceedings of the Royal Society of London. Series B, Biological Sciences*, 285: 20180339.

Gilbert, S. F. & Barresi, J. F. (2016). *Developmental Biology*, 11th edn. Sunderland, MA: Sinauer Associates.

Gilbert, S. F. & Epel, D. (2015). *Ecological Developmental Biology: the Environmental Regulation of Development, Health, and Evolution*, 2nd edn. Sunderland, MA: Sinauer Associates.

Gilbert, S. F., Sapp, J. & Tauber, A. I. (2012). A symbiotic view of life: we have never been individuals. *Quarterly Review of Biology*, 87: 325–341.

Gladyshev, E. A. & Arkhipova, I. R. (2010). Genome structure of bdelloid rotifers: shaped by asexuality or desiccation? *Journal of Heredity*, 101 (Supplement **1**): S85–S93.

Godfrey-Smith, P. (2009). *Darwinian Populations and Natural Selection*. New York, NY: Oxford University Press.

Gokhman, V. E. & Kuznetsova, V. G. (2018). Parthenogenesis in Hexapoda: Holometabolous insects. *Journal of Zoological Systematics and Evolutionary Research*, 56: 23–34.

Gordon, J. L., Armisen, D., Proux-Wera, E. *et al.* (2011). Evolutionary erosion of yeast sex chromosomes by mating-type switching accidents. *Proceedings of the National Academy of Sciences of the United States of America*, 108: 20024–20029.

Gorelik, R. (2012). Mitosis circumscribes individuals; sex creates new individuals. *Biology and Philosophy*, 27: 871–890.

Goto, T. & Yoshida, M. (1985). The mating sequence of the benthic arrowworm *Spadella schizoptera*. *Biological Bulletin*, 169: 328–333.

Grbic, M., Ode, P. J. & Strand, M. R. (1992). Sibling rivalry and brood sex ratios in polyembryonic wasps. *Nature*, 360: 254–256.

Green, R. F. & Noakes, D. L. G. (1995). Is a little bit of sex as good as a lot? *Journal of Theoretical Biology*, 174: 87–96.

Greene, D. F. & Johnson, E. A. (1994). Estimating the mean annual seed production of trees. *Ecology*, 75: 642–647.

Grosberg, R. K. & Strathmann, R. R. (1998). One cell, two cell, red cell, blue cell: the persistence of a unicellular stage in multicellular life histories. *Trends in Ecology and Evolution*, 13: 112–116.

Grosse-Veldmann, B. & Weigend, M. (2018). The geometry of gender: hyper-diversification of sexual systems in *Urtica* L. (Urticaceae). *Cladistics*, 34: 131–150.

Gunstream, S. E. & Chew, R. M. (1967). The ecology of *Psorophora confinnis* (Diptera: Culicidae) in southern California. II. Temperature and development. *Annals of the Entomological Society of America*, 60: 434–439.

Haig, D. (2016). Living together and living apart: the sexual lives of bryophytes. *Philosophical Transactions of the Royal Society of London. Series B, Biological Sciences*, 371: 20150535.

Halary, S., Malik, S. B., Lildhar, L. *et al.* (2011). Conserved meiotic machinery in *Glomus* spp., a putatively ancient asexual fungal lineage. *Genome Biology and Evolution*, 3: 950–958.

Hanelt, B., Bolek, M. G. & Schmidt-Rhaesa, A. (2012). Going solo: discovery of the first parthenogenetic gordiid (Nematomorpha: Gordiida). *PLOS One*, 7 (4): e34472.

Hansen, K. (1984). Discrimination and production of disparlure enantiomers by the gypsy moth and the nun moth. *Physiological Entomology*, 9: 9–18.

Harada, Y., Takagaki, M., Saito, T. *et al.* (2008). Mechanism of self-sterility in a hermaphroditic chordate. *Science*, 320: 548–550.

Harper, J. L. & White, J. (1974). The demography of plants. *Annual Review of Ecology and Systematics*, 5: 419–463.

Harrath, A. H., Sluys, R., Zghal, F. & Tekaya, S. (2009). First report of adelphophagy in flatworms during the embryonic development of the planarian *Schmidtea mediterranea* (Benazzi, Baguñà, Ballester, Puccinelli & Del Papa, 1975) (Platyhelminthes, Tricladida). *Invertebrate Reproduction and Development*, 53: 117–124.

Harrison, P. L. (2011). Sexual reproduction of scleractinian corals. In *Coral Reefs: an Ecosystem in Transition*, ed. Z. Dubinsky & N. Stambler. London; New York, NY: Springer, pp. 59–85.

Hechinger, R. F., Wood, A. C. & Kuris, A. M. (2011). Social organization in a flatworm: trematode parasites form soldier and reproductive castes. *Proceedings of the Royal Society of London. Series B, Biological Sciences*, 278: 656–665.

Hedtke, S. M., Stanger-Hall, K., Baker, R. J. & Hillis, D. M. (2008). All-male asexuality: origin and maintenance of androgenesis in the Asian clam *Corbicula*. *Evolution*, 62: 1119–1136.

Heiner, I. & Kristensen, R. M. (2008). *Urnaloricus gadi* nov. gen. et nov. sp. (Loricifera, Urnaloricidae nov. fam.), an aberrant Loricifera with a vivipaous pedogenetic life cycle. *Journal of Morphology*, 270: 129–153.

Heming, B. S. (2003). *Insect Development and Evolution*. Ithaca, NY: Comstock.

Henderson, K. A. & Gottschling, D. E. (2008). A mother's sacrifice: what is she keeping for herself? *Current Opinion in Cell Biology*, 20: 723–728.

Hinman, V. & Cary, G. (2017). Conserved processes of metazoan whole-body regeneration identified in sea star larvae. bioRxiv preprint. doi: 10.1101/118232.

Hojsgaard, D. H., Martínez, E. J. & Quarin, C. L. (2013). Competition between meiotic and apomictic pathways during ovule and seed development results in clonality. *New Phytologist*, 197: 336–347.

Hörandl, E. (2006). The complex causality of geographical parthenogenesis. *New Phytologist*, 171: 525–538.

Horne, D. J. & Martens, K. (1999). Geographical parthenogenesis in European non-marine ostracods: post-glacial invasion or Holocene stability? *Hydrobiologia*, 391: 1–7.

Hoving, H. J. T., Laptikhovsky, V. V. & Robison, B. H. (2015). Vampire squid reproductive strategy is unique among coleoid cephalopods. *Current Biology*, 25: R321–R323.

Hughes, P. W. (2017). Between semelparity and iteroparity: Empirical evidence for a continuum of modes of parity. *Ecology and Evolution*, 7: 8232–8261.

Hughes, R. N. (1989). *A Functional Biology of Clonal Animals*. New York, NY: Chapman & Hall.

Igea, J., Miller, E. F., Papadopulos, A. S. T. & Tanentzap, A. J. (2017). Seed size and its rate of evolution correlate with species diversification across angiosperms. *PLOS Biology*, 15 (7): e2002792.

Iglésias, S. P., Sellos, D. Y. & Nakaya, K. (2005). Discovery of a normal hermaphroditic chondrichthyan species: *Apristurus longicephalus*. *Journal of Fish Biology*, 66: 417–428.

Ignace, D. D., Dodson, S. I. & Kashian, D. R. (2011). Identification of the critical timing of sex determination in *Daphnia magna* (Crustacea, Branchiopoda) for use in toxicological studies. *Hydrobiologia*, 668: 117–123.

Iskandar, D. T., Evans, B. J. & McGuire, J. A. (2014). A novel reproductive mode in frogs: a new species of fanged frog with internal fertilization and birth of tadpoles. *PLOS One*, 9 (12): e115884.

Jablonka, E. & Lamb, M. J. (2005). *Evolution in Four Dimensions: Genetic, Epigenetic, Behavioral, and Symbolic Variation in the History of Life*. Cambridge, MA: MIT Press.

Jaeckle, W. B. (1994). Multiple modes of asexual reproduction by tropical and subtropical sea star larvae: an unusual adaptation for genet dispersal and survival. *Biological Bulletin*, 186: 62–71.

Jaklitsch, W., Baral, H.-O., Lücking, R. & Lumbsch, H. T. (2016). Ascomycota. In *Syllabus of Plant Families*, 13th edn, ed. W. Frey. Berlin; Stuttgart: Borntraeger, Vol. 1/2.

Janousek, B. & Mrackova, M. (2010). Sex chromosomes and sex determination pathway dynamics in plant and animal models. *Biological Journal of the Linnean Society*, 100: 737–752.

Jany, J. & Pawlowska, T. E. (2010). Multinucleate spores contribute to evolutionary longevity of asexual Glomeromycota. *American Naturalist*, 175: 424–435.

Janzen, D. H. (1977). What are dandelions and aphids? *American Naturalist*, 111: 586–589.

Jarne, P. & Auld, J. R. (2006). Animals mix it up too: the distribution of self-fertilization among hermaphroditic animals. *Evolution*, 60: 1816–1824.

Jaspers, C., Haraldsson, M., Bolte, S. *et al.* (2012). Ctenophore population recruits entirely through larval reproduction in the central Baltic Sea. *Biology Letters*, 8: 809–812.

Jetz, W., Sekercioglu, C. H. & Böhning-Gaese, K. (2008). The worldwide variation in avian clutch size across species and space. *PLOS Biology*, 6 (12): e303.

Johnson, G. D., Paxton, J. R., Sutton, T. T. *et al.* (2009). Deap-sea mystery solved: astonishing larval transformations and extreme sexual dimorphism unite three fish families. *Biology Letters*, 5: 235–239.

Jordal, B. H., Beaver, R. A., Normark, B. B. & Farrell, B. D. (2002). Extraordinary sex ratios and the evolution of male neoteny in sib-mating *Ozopemon* beetles. *Biological Journal of the Linnean Society*, 75: 353–360.

Jordal, B. H., Normark, B. B. & Farrell, B. D. (2000). Evolutionary radiation of an inbreeding haplodiploid beetle lineage (Curculionidae, Scolytinae). *Biological Journal of the Linnean Society*, 71: 483–499.

Juarez, C. & Banks, J. A. (1998). Sex determination in plants. *Current Opinion in Plant Biology*, 1: 68–72.

Kaiser, V. B. & Bachtrog, D. (2010). Evolution of sex chromosomes in insects. *Annual Review of Genetics*, 44: 91–112.

Kamiya, M., Lindstrom, S. C., Nakayama, T. *et al.* (2017). Photoautotrophic eukaryotic Algae. Rhodophyta. In *Syllabus of Plant Families*, 13th edn, ed. W. Frey. Berlin; Stuttgart: Borntraeger, Vol. 2/2.

Karsten, K. B., Andriamandimbiarisoa L. N., Fox S. F. & Raxworthy C. J. (2008). A unique life history among tetrapods: an annual chameleon living mostly as an egg. *Proceedings of the National Academy of Sciences of the United States of America*, 105: 8980–8984.

Kelly, D. R. (1996). When is a butterfly like an elephant? *Chemical Biology*, 3: 595–602.

Khosla, S., Mendiratta, G. & Brahmachari, V. (2006). Genomic imprinting in the mealybugs. *Cytogenetics and Genome Research*, 113: 41–52.

Kirk, D. L. (2001). Germ–soma differentiation in *Volvox*. *Developmental Biology*, 238: 213–223.

Kishore, K. (2014). Polyembryony. In *Reproductive Biology of Plants*, ed. K. G. Ramawat, J. M. Mérillon & K. R. Shivanna. Boca Raton, FL: CRC Press, pp. 355–370.

Knoflach, B. & van Harten, A. (2000). Palpal loss, single palp copulation and obligatory mate consumption in *Tidarren cuneolatum* (Tullgren, 1910) (Araneae, Theridiidae). *Journal of Natural History*, 34: 1639–1659.

Knoflach, B. & van Harten, A. (2001). *Tidarren argo* sp. nov. (Araneae: Theridiidae) and its exceptional copulatory behaviour: emasculation, male palpal organ as a mating plug and sexual cannibalism. *Journal of Zoology*, 254: 449–459.

Kočarek, P. (2009). A case of viviparity in a tropical non-parasitizing earwig (Dermaptera Spongiphoridae). *Tropical Zoology*, 22: 237–241.

Komma, D. J. & Endow, S. A. (1995). Haploidy and androgenesis in *Drosophila*. *Proceedings of the National Academy of Sciences of the United States of America*, 92: 11884–11888.

Kondrashov, A. S. (2018). Through sex, nature is telling us something important. *Trends in Genetics*, 34: 352–361.

Koopman, P. (2009). Sex determination: the power of *DMRT1*. *Trends in Genetics*, 25: 479–481.

Kopp, A. (2012). *Dmrt* genes in the development and evolution of sexual dimorphism. *Trends in Genetics*, 28: 175–184.

Kraak, S. B. M. & Pen, I. (2002). Sex-determining mechanisms in vertebrates. In *Sex Ratios: Concepts and Research Methods*, ed. I. C. W. Hardy. Cambridge: Cambridge University Press, pp. 158–177.

Krebs, J. E., Goldstein, E. S. & Kilpatrick, S. T. (2011). *Lewin's Genes X*. Sudbury, MA: Jones & Bartlett.

Kreulen, D. J. W. (1972). Spore output of moss capsules in relation to ontogeny of archesporial tissue. *Journal of Bryology*, 7: 61–74.

Kristensen, R. M. & Funch, P. (2000). Micrognathozoa: a new class with complicated jaws like those of Rotifera and Gnathostomulida. *Journal of Morphology*, 246: 1–49.

Kuo, L.-Y., Chen, C.-W., Shinohara, W. *et al.* (2017). Not only in the temperate zone: independent gametophytes of two vittarioid ferns (Pteridaceae, Polypodiales) in East Asian subtropics. *Journal of Plant Research*, 130: 255–262.

LaFave, M. C. & Sekelsky, J. (2009). Mitotic recombination: why? when? how? where? *PLOS Genetics*, 5 (3): e1000411.

Lanfear, R. (2018). Do plants have a segregated germline? *PLOS Biology*, 16 (5): e2005439.

Lapierre, P. & Gogarten, P. (2009). Estimating the size of the bacterial pan-genome. *Trends in Genetics*, 25: 107–110.

Larsen, K. (2005). *Deep-sea Tanaidacea (Peracarida) from the Gulf of Mexico*. Crustacean Monographs 5. Leiden: Brill.

Lee, R. E. (2008). *Phycology*. 4th edn. New York, NY: Cambridge University Press.

Leeb, M. & Wutz, A. (2010). Mechanistic concepts in X inactivation underlying dosage compensation in mammals. *Heredity*, 105: 64–70.

Lehtonen, J., Jennions, M. D. & Kokko, H. (2012). The many costs of sex. *Trends in Ecology and Evolution*, 27: 172–178.

Leliaert, F., Blindow, I. & Schudack, M. (2015a). Ulvophyceae (except Trentepohliales). In *Syllabus of Plant Families*, 13th edn, ed. W. Frey. Berlin; Stuttgart: Borntraeger, Vol. 2/1, pp. 267–280.

Leliaert, F., Lopez-Bautista, J. & De Clerk, O. (2015b). Charophyceae. In *Syllabus of Plant Families*, 13th edn, ed. W. Frey. Berlin; Stuttgart: Borntraeger, Vol. 2/1, pp. 294–300.

Levin, T. C. & King, N. (2013). Evidence for sex and recombination in the choanoflagellate *Salpingoeca rosetta*. *Current Biology*, 23: 2176–2180.

Li, S. I. & Purugganan, M. D. (2011). The cooperative amoeba: *Dictyostelium* as a model for social evolution. *Trends in Genetics*, 27: 48–54.

Lindås, A. C., Karlsson, E. A., Lindgren, M. T., Ettema, T. J. & Bernander, R. (2008). A unique cell division machinery in the Archaea. *Proceedings of the National Academy of Sciences of the United States of America*, 105: 18942–18946.

Lloyd, D. G. & Webb, C. J. (1977). Secondary sex characters in plants. *Botanical Review*, 43: 177–216.

Locey, K. J. & Lennon, J. T. (2016). Scaling laws predict global microbial diversity. *Proceedings of the National Academy of Sciences of the United States of America*, 113: 5970–5975.

Lundmark, M. & Saura, A. (2006). Asexuality alone does not explain the success of clonal forms in insects with geographical parthenogenesis. *Hereditas*, 143: 23–32.

Lushai, G. & Loxdale, H. D. (2002). The biological improbability of a clone. *Genetics Research*, 79: 1–9.

Lynch, M. (2010). Evolution of the mutation rate. *Trends in Genetics*, 26: 345–352.

Lynch, M., Koskella, B. & Schaack, S. (2006). Mutation pressure and the evolution of organelle genomic architecture. *Science*, 311: 1727–1730.

Makarova, K. S., Yutin, N., Bell, S. D. & Koonin, E. V. (2010). Evolution of diverse cell division and vesicle formation systems in Archaea. *Nature Reviews Microbiology*, 8: 731–741.

Malik, S. B., Pightling, A. W., Stefaniak, L. M., Schurko, A. M. & Logsdon, J. M. (2008). An expanded inventory of conserved meiotic genes provides evidence for sex in *Trichomonas vaginalis*. *PLOS One*, 3 (8): e2879.

Manabe, H., Ishimura, M., Shinomiya, A. & Sunobe, T. (2007). Field evidence for bidirectional sex change in the polygynous gobiid fish *Trimma okinawae*. *Journal of Fish Biology*, 70: 600–609.

Mann, T. (1984). *Spermatophores: Development, Structure, Biochemical Attributes, and Role in the Transfer of Spermatozoa*. Berlin: Springer.

Mantovani, B., Passamonti, M. & Scali, V. (1999). Genomic evolution in parental and hybrid taxa of the genus *Bacillus* (Insecta, Phasmatodea). *Italian Journal of Zoology*, 66: 265–272.

Mantovani, B. & Scali, V. (1992). Hybridogenesis and androgenesis in the stick-insect *Bacillus rossius-grandii benazzii* (Insecta, Phasmatodea). *Evolution*, 46: 783–796.

Marescalchi, O., Pijnacker, L. P. & Scali, V. (1993). Automictic parthenogenesis and its genetic consequence in *Bacillus atticus atticus* (Insecta Phasmatodea). *Invertebrate Reproduction and Development*, 24: 7–12.

Marin, I. & Baker, B. S. (1998). The evolutionary dynamics of sex determination. *Science*, 281: 1990–1994.

Marques, A. C., García, J. & Ames, C. L. (2015). Internal fertilization and sperm storage in cnidarians: a response to Orr and Brennan. *Trends in Ecology and Evolution*, 30: 435–436.

Martín-Durán, J. M. & Egger, B. (2012). Developmental diversity in free-living flatworms. *EvoDevo*, 3: 7.

Martinez, D. E. (1997). Mortality patterns suggest lack of senescence in hydra. *Experimental Gerontology*, 33: 217–225.

Maruyama, D., Hamamura, Y., Takeuchi, H. *et al.* (2013). Independent control by each female gamete prevents the attraction of multiple pollen tubes. *Developmental Cell*, 25: 317–323.

Matthes, D. (1988). *Tierische Parasiten: Biologie und Ökologie*. Braunschweig; Wiesbaden: Vieweg.

Matveevsky, S., Bakloushinskaya, I. & Kolomiets, O. (2016). Unique sex chromosome systems in *Ellobius*: how do male XX chromosomes recombine and undergo pachytene chromatin inactivation? *Scientific Reports*, 6: 29949.

Mauseth, J. D. (1988). *Plant Anatomy*. Menlo Park, CA: Benjamin/Cummings.

McCabe, J. & Dunn, A. M. (1997). Adaptive significance of environmental sex determination in an amphipod. *Journal of Evolutionary Biology*, 10: 515–527.

McCurdy, D. G., Painter, D. C., Kopec, M. T. *et al.* (2008). Reproductive behaviour of intersexes of an intertidal amphipod *Corophium volutator*. *Invertebrate Biology*, 127: 417–425.

McKone, M. J. & Halpern, S. L. (2003). The evolution of androgenesis. *American Naturalist*, 161: 641–656.

Michalik, P., Knoflach, B., Thaler, K. & Alberti, G. (2010). Live for the moment: adaptations in the male genital system of a sexually cannibalistic spider (Theridiidae, Araneae). *Tissue and Cell*, 42: 32–36.

Michalik, P. & Uhl, G. (2005). The male genital system of the cellar spider *Pholcus phalangioides* (Fuesslin, 1775) (Pholcidae, Araneae): development of spermatozoa and seminal secretion. *Frontiers in Zoology*, 2: 12.

Minelli, A. (2009). *Perspectives in Animal Phylogeny and Evolution*. Oxford: Oxford University Press.

Minelli, A. (2014). Developmental disparity. In *Towards a Theory of Development*, ed. A. Minelli & T. Pradeu. Oxford: Oxford University Press, pp. 227–245.

Minelli, A. (2018). *Plant Evolutionary Developmental Biology*. Cambridge: Cambridge University Press.

Minelli, A., Brena, C., Deflorian, G., Maruzzo, D. & Fusco, G. (2006). From embryo to adult: beyond the conventional periodization of arthropod development. *Development Genes and Evolution*, 216: 373–383.

Minelli, A. & Fusco, G. (2010). Developmental plasticity and the evolution of animal complex life cycles. *Philosophical Transactions of the Royal Society of London. Series B, Biological Sciences*, 365: 631–640.

Minelli, A. & Fusco, G. (2013). Arthropod post-embryonic development. In *Arthropod Biology and Evolution. Molecules, Development, Morphology*, ed. A. Minelli, G. Boxshall & G. Fusco. Heidelberg: Springer, pp. 91–122.

Ming, R., Bendahmane, A. & Renner, S. S. (2011). Sex chromosomes in land plants. *Annual Review of Plant Biology*, 62: 485–514.

Miya, M., Pietsch, T. W., Orr, J. W. *et al.* (2010). Evolutionary history of anglerfishes (Teleostei: Lophiiformes): a mitogenomic perspective. *BMC Evolutionary Biology*, 10: 58.

Mogie, M. (1992). *The Evolution of Asexual Reproduction in Plants*. London: Chapman & Hall.

Monaghan, P. & Haussmann, M. F. (2006). Do telomere dynamics link lifestyle and lifespan? *Trends in Ecology and Evolution*, 21: 47–53.

Moore, D. S. (2017). *The Developing Genome: an Introduction to Behavioral Epigenetics*. New York, NY: Oxford University Press.

Morison, I. M., Ramsay, J. P. & Spencer, H. G. (2005). A census of mammalian imprinting. *Trends in Genetics*, 21: 457–465.

Munday, P. L., Kuwamura, T. & Kroon, F. J. (2010). Bidirectional sex change in marine fishes. In *Reproduction and Sexuality in Marine Fishes: Patterns and Processes*, ed. K. S. Cole. Berkeley, CA: University of California Press, pp. 241–271.

Nath, P., Bouzayen, M., Mattoo, A. K. & Pech, J. C. (eds.) (2014). *Fruit Ripening: Physiology, Signalling and Genomics*. Wallingford: CABI.

Naurin, S., Hansson, B., Bensch, S. & Hasselquist, D. (2010). Why does dosage compensation differ between XY and ZW taxa? *Trends in Genetics*, 26: 15–20.

Neestupa, J. (2015). Chlorophyta, Streptophyta. In *Syllabus of Plant Families*, 13th edn, ed. W. Frey. Berlin; Stuttgart: Borntraeger, Vol. 2/1, pp. 191–247, 264–267, 282–294.

Neiman, M., Lively, C. M. & Meirmans, S. (2017). Why sex? A pluralist approach revisited. *Trends in Ecology and Evolution*, 32: 589–600.

Nguyen, K. B. & Smart, G. C. (1990). *Heterorhabditis* spp.: nematode parasites of insects. *Nematology Circular*, 173. Gainesville, FL: Florida Department of Agriculture & Consumer Service, Division of Plant Industry.

Nielsen, C. (2012). *Animal Evolution: Interrelationships of the Living Phyla*, 3rd edn. Oxford: Oxford University Press.

Nielsen, J., Hedeholm, R. B., Heinemeier, J. *et al.* (2016). Eye lens radiocarbon reveals centuries of longevity in the Greenland shark (*Somniosus microcephalus*). *Science*, 353: 702–704.

Nieuwenhuis, B. P. S. & James, T. Y. (2016). The frequency of sex in fungi. *Philosophical Transactions of the Royal Society of London. Series B, Biological Sciences*, 371: 20150540.

Noda, I. (1960). The emergence of winged viviparous female in aphids. VI. Difference in the rate of development between the winged and the unwinged forms. *Japanese Journal of Ecology*, 10: 97–102.

Nogler, G. A. (1984). Genetics of apospory in apomictic *Ranunculus auricomus*: 5. Conclusion. *Botanica Helvetica*, 94: 411–423.

Normark, B. B. (2003). The evolution of alternate genetic systems in insects. *Annual Review of Entomology*, 48: 397–423.

Noro, C., López-Greco, L. S. & Buckup, L. (2007). Gonad morphology and type of sexuality in *Parastacus defossus* Faxon 1898, a burrowing, intersexed crayfish from southern Brazil (Decapoda: Parastacidae). *Acta Zoologica*, 89: 59–67.

O'Gorman, C. M., Fuller, H. T. & Dyer, P. S. (2009). Discovery of a sexual cycle in the opportunistic fungal pathogen *Aspergillus fumigatus*. *Nature*, 457: 471–474.

Okasha, S. (2006). *Evolution and the Levels of Selection*. New York, NY: Oxford University Press.

Oliveira, R. F., Taborsky, M. & Brockmann, H. J. (eds.) (2008). *Alternative Reproductive Tactics: an Integrative Approach*. Cambridge: Cambridge University Press.

Oliver, B. (2007). Sex, dose, and equality. *PLOS Biology*, 5 (12): e340.

Omilian, A. R., Cristescu, M. E. A., Dudycha, J. L. & Lynch, M. (2006). Ameiotic recombination in asexual lineages of *Daphnia*. *Proceedings of the National Academy of Sciences of the United States of America*, 103: 18638–18643.

Ostrovsky, A. N., Lidgard, S., Gordon, D. P. *et al.* (2016). Matrotrophy and placentation in invertebrates: a new paradigm. *Biological Reviews*, 91: 673–711.

Ota, K. G. (2018). Recent advances in hagfish developmental biology in a historical context: implications for understanding the evolution of the vertebral elements. In *Reproductive and Developmental Strategies: Diversity and Commonality in Animals*, ed. K. Kobayashi, T. Kitano, Y. Iwao & M. Kondo. Tokyo: Springer, pp. 615–634.

Pacini, E. (2010). Relationship between tapetum, loculus, and pollen during development. *International Journal of Plant Sciences*, 171: 1–11.

Paemelaere, E. A. D., Guyer, C. & Dobson, F. S. (2011). A phylogenetic framework for the evolution of female polymorphism in anoles. *Biological Journal of the Linnean Society*, 104: 303–317.

Palevitz, B. A. & Tiezzi, A. (1992). Organization, composition, and function of the generative cell and sperm cytoskeleton. *International Review of Cytology*, 140: 149–185.

Pannell, J. R. (2017). Plant sex determination. *Current Biology*, 27: R191–R197.

Parker, G. A. (2014). The sexual cascade and the rise of pre-ejaculatory (Darwinian) sexual selection, sex roles, and sexual conflict. *Cold Spring Harbor Perspectives in Biology*, 6: a017509.

Parker, J. D. (2004). A major evolutionary transition to more than two sexes? *Trends in Ecology and Evolution*, 19: 83–86.

Pearcy, M., Hardy, O. & Aron, S. (2006). Thelytokous parthenogenesis and its consequences on inbreeding in an ant. *Heredity*, 96: 377–382.

Pearson, H. (2006). What is a gene? *Nature*, 441: 469–474.

Peccoud, J., Loiseau, V., Cordaux, R., Gilbert, C. (2017). Massive horizontal transfer of transposable elements in insects. *Proceedings of the National Academy of Sciences of the United States of America*, 114: 4721–4726.

Penman, D. J. & Piferrer, F. (2008). Fish gonadogenesis. Part I: genetic and environmental mechanisms of sex determination. *Reviews in Fisheries Science*, 16 (Supplement 1): 14–32.

Pennell, M. W., Mank, J. E. & Peichel, C. L. (2018). Transitions in sex determination and sex chromosomes across vertebrate species. *Molecular Ecology*, 27: 3950–3963.

Perrin, N. (2016). Random sex determination: when developmental noise tips the sex balance. *BioEssays*, 38: 1218–1226.

Perry, L. E., Pannell, J. R. & Dorken, M. E. (2012). Two's company, three's a crowd: experimental evaluation of the evolutionary maintenance of trioecy in *Mercurialis annua* (Euphorbiaceae). *PLOS One*, 7 (4): e35597.

Phadke, S. S. & Zufall, R. A. (2009). Rapid diversification of mating systems in ciliates. *Biological Journal of the Linnean Society*, 98: 187–197.

Pichot, C., El Mataoui, M., Raddi, S. & Raddi, P. (2001). Surrogate mother for endangered *Cupressus*. *Nature*, 412: 39.

Pietsch, T. W. & Orr, J. W. (2007). Phylogenetic relationships of deep-sea anglerfishes of the suborder Ceratioidei (Teleostei: Lophiiformes) based on morphology. *Copeia*, 2007: 1–34.

Piraino, S., Boero, F., Aeschbach, B. & Schmid, V. (1996). Reversing the life cycle: medusae transforming into polyps and cell transdifferentiation in *Turritopsis nutricula* (Cnidaria, Hydrozoa). *Biological Bulletin*, 190: 302–312.

Poethig, R. S. (2003). Phase change and the regulation of developmental timing in plants. *Science*, 301: 334–336.

Pommerville, J. C. (2011). *Alcamo's Fundamentals of Microbiology*, 9th edn. Sudbury, MA: Jones & Bartlett Learning.

Pradeu, T. (2010). What is an organism? An immunological answer. *History and Philosophy of the Life Sciences*, 32: 247–267.

Pradeu, T. (2012). *The Limits of the Self: Immunology and Biological Identity*. Oxford: Oxford University Press.

Pradeu, T. (2016). Organisms or biological individuals? Combining physiological and evolutionary individuality. *Biology and Philosophy*, 31: 797–817.

Pradeu, T., Laplane, L., Prévot, K. *et al.* (2016). Defining 'development'. *Current Topics in Developmental Biology*, 117: 171–183.

Premoli, M. C. & Sella, G. (1995). Sex economy in benthic polychaetes. *Ethology Ecology Evolution*, 7: 27–48.

Prévot, V., Jordaens, K., Sonet, G. & Backeljau, T. (2013). Exploring species level taxonomy and species delimitation methods in the facultatively self-fertilizing land snail genus *Rumina* (Gastropoda: Pulmonata). *PLOS One*, 8 (4): e60736.

Proctor, H. C. (1998). Indirect sperm transfer in arthropods. *Annual Review of Entomology*, 43: 153–174.

Queller, D. (2005). Males from Mars. *Nature*, 435: 1167–1168.

Radder, R. S., Pike, D. A., Quinn, A. E. & Shine, R. (2009). Offspring sex in a lizard depends on egg size. *Current Biology*, 19: 1102–1105.

Raigner, A. & van Bovan, J. (1955). Etude taxonomique, biologique et biométrique des *Dorylus* du sou-genre *Anomma* (Hymenoptera, Formicidae). *Annales du Musée Royal du Congo Belge. Nouvelle Série. Sciences Zoologiques*, 2: 1–359.

Raikov, I. B. (1994). The diversity of forms of mitosis in protozoa: a comparative review. *European Journal of Protistology*, 30: 253–269.

Ram, Y. & Hadany, L. (2016). Condition-dependent sex: who does it, when and why? *Philosophical Transactions of the Royal Society of London. Series B, Biological Sciences*, 371: 20150539.

Ramesh, M. A., Malik, S. B. & Logsdon, J. M. (2005). A phylogenomic inventory of meiotic genes: evidence for sex in *Giardia* and an early eukaryotic origin of meiosis. *Current Biology*, 15: 185–191.

Ramm, S. J., Poirier, M. & Scharer, L. (2015). Hypodermic self-insemination as a reproductive insurance strategy. *Proceedings of the Royal Society of London. Series B, Biological Sciences*, 282, 20150660.

Redecker, D. (2012). Glomeromycota. In *Syllabus of Plant Families*, 13th edn, ed. W. Frey. Berlin; Stuttgart: Borntraeger, Vol. 1/1, pp. 163–170.

Reinhard, F., Herle, M., Bastiansen, F. & Streit, B. (2003). *Economic Impact of the Spread of Alien Species in Germany*. Berlin: Federal Environmental Agency.

Reisinger, E., Cichocki, I., Erlach, T. & Szyskowitz, T. (1974). Ontogenetische Studien an Turbellarien: ein Beitrag zur Evolution der Dotterverarbeitung im ektolecithalen Ei. II. *Zeitschrift für Zoologische Systematik und Evolutionsforschung*, 12: 241–278.

Renner, S. S. (2014). The relative and absolute frequencies of angiosperm sexual systems: dioecy, monoecy, gynodioecy, and an updated online database. *American Journal of Botany* 101: 1588–1596.

Ricci, C. (2001). Dormancy patterns in rotifers. *Hydrobiologia*, 446/447: 1–11.

Richards, A. J. (1997). *Plant Breeding Systems*, 2nd edn. London: Chapman & Hall.

Rieger, R., Michaelis, A. & Green, M. M. (1976). *Glossary of Genetics and Cytogenetics*. Berlin: Springer.

Roark, E., Guilderson, T. P., Dunbar, R. B., Fallon, S. J. & Mucciarone, D. A. (2009). Extreme longevity in proteinaceous deep-sea corals. *Proceedings of the National Academy of Sciences of the United States of America*, 106: 5204–5208.

Rocha, F., Guerra, A. & González, A. F. (2001). A review of reproductive strategies in cephalopods. *Biological Reviews*, 76: 291–304.

Roff, D. A. (2002). *Life History Evolution*. Sunderland, MA: Sinauer Associates.

Rosenstiel, T. N., Shortlidge, E. E., Melnychenko, A. N., Pankow, J. F. & Eppley, S. M. (2012). Sex-specific volatile compounds influence microarthropod-mediated fertilization of moss. *Nature*, 489: 431–433.

Ross, C. N., French, J. A. & Ortí, G. (2007). Germ-line chimerism and paternal care in marmosets (*Callithrix kuhlii*). *Proceedings of the National Academy of Sciences of the United States of America*, 104: 6278–6282.

Rouse, G. W, Goffredi, S. K. & Vrijenhoek, R. C. (2004). *Osedax*: bone-eating marine worms with dwarf males. *Science*, 305: 668–671.

Rouse, G. W. & Pleijel, F. (2001). *Polychaetes*. Oxford: Oxford University Press.

Rouse, G. W., Wilson, N. G., Carvajal, J. I. & Vrijenhoek, R. C. (2016). New deep-sea species of *Xenoturbella* and the position of Xenacoelomorpha. *Nature*, 530: 94–97.

Russell, P. J. (2010). *iGenetics*, 3rd edn. San Francisco, CA: Benjamin/Cummings.

Ryland, J. S. (2005). Bryozoa: an introductory overview. *Denisia*, 16: 9–20.

Salje, J., Gayathri, P. & Löwe, J. (2010). The ParMRC system: molecular mechanisms of plasmid segregation by actin-like filaments. *Nature Reviews Microbiology*, 8: 683–692.

Sallon, S., Solowey, E., Cohen, Y. *et al.* (2008). Germination, genetics, and growth of an ancient date seed. *Science*, 320: 1464.

Salomon, M., Aflalo, E. D., Coll, M. & Lubin, Y. (2015). Dramatic histological changes preceding suicidal maternal care in the subsocial spider *Stegodyphus lineatus* (Araneae: Eresidae). *Journal of Arachnology*, 43: 77–85.

Sánchez, L. (2008). Sex-determining mechanisms in insects. *International Journal of Developmental Biology*, 52: 837–856.

Sano, N., Obata, M., Ooie, Y. & Komaru, A. (2011). Mitochondrial DNA copy number is maintained during spermatogenesis and in the development of male larvae to sustain the doubly uniparental inheritance of mitochondrial DNA system in the blue mussel *Mytilus galloprovincialis*. *Development Growth and Differentiation*, 53: 816–821.

Santelices, B. (1999). How many kinds of individual are there? *Trends in Ecology and Evolution*, 14: 152–155.

Santelices, B., Correa, J. A., Meneses, I., Aedo, D. & Varela, D. (1996). Sporeling coalescence and intra-clonal variation in *Gracilaria chilensis* (Gracilariales: Rhodophyta). *Journal of Phycology*, 32: 313–322.

Sato, S., Beakes, G., Idel, M., Nagumo, T. & Mann, D. G. (2011). Novel sex cells and evidence for sex pheromones in diatoms. *PLOS One*, 6 (10): e26923.

Scali, V., Passamonti, M., Marescalchi, O. & Mantovani, B. (2003). Linkage between sexual and asexual lineages: genome evolution in *Bacillus* stick insects. *Biological Journal of the Linnean Society*, 79: 137–150.

Schön, I. & Martens, K. (2017). Paradox of sex. In *Oxford Bibliographies in Evolutionary Biology*, ed. J. Losos. Oxford: Oxford University Press. www.oxfordbibliographies.com/view/document/obo-9780199941728/obo-9780199941728-0035.xml (accessed April 2019).

Schön, I., Martens, K. & van Dijk, P. (eds.) (2009). *Lost Sex: the Evolutionary Biology of Parthenogenesis*. Berlin: Springer.

Schrader, F. (1923). The origin of the mycetocytes in *Pseudococcus*. *Biological Bulletin*, 40: 259–270.

Schroeder, P. C. & Hermans, C. O. (1975). Annelida: Polychaeta. In *Reproduction of Marine Invertebrates, 3. Annelids and Echiurans*, ed. A. C. Giese & J. S. Pearse. New York, NY: Academic Press, pp. 1–213.

Schurko, A. M. & Logsdon, J. M. Jr. (2008). Using a meiosis detection toolkit to investigate ancient asexual 'scandals' and the evolution of sex. *BioEssays*, 30: 579–589.

Schurko, A. M., Neiman, M. & Logsdon, J. M. Jr. (2009). Signs of sex: what we know and how we know it. *Trends in Ecology and Evolution*, 24: 208–217.

Schut, E., Hemmings, N. & Birkhead, T. R. (2008). Parthenogenesis in a passerine bird, the zebra finch *Taeniopygia guttata*. *Ibis*, 150: 197–199.

Schwander, T. & Oldroyd, B. P. (2016). Androgenesis: where males hijack eggs to clone themselves. *Philosophical Transactions of the Royal Society of London. Series B, Biological Sciences*, 371: 20150534.

Scott, R. J., Armstrong, S. J., Doughty, J. & Spielman, M. (2008). Double fertilization in *Arabidopsis thaliana* involves a polyspermy block on the egg but not the central cell. *Molecular Plant*, 1: 611–619.

Segoli, M., Larari, A. R., Rosenheim, J. A., Bouslika, A. & Keasar, T. (2010). The evolution of polyembryony in parasitoid wasps. *Journal of Evolutionary Biology*, 23: 1807–1819.

Sender, R., Fuchs, S. & Milo, R. (2016). Revised estimates for the number of human and bacteria cells in the body. *PLOS Biology*, 14 (8): e1002533.

Shefferson, R. P., Jones, O. R. & Salguero-Gómez, R. (2017). *The Evolution of Senescence in the Tree of Life*. Cambridge: Cambridge University Press.

Shen-Miller, J., Mudgett, M. B., Schopf, J. W., Clarke, S. & Berger R. (1995). Exceptional seed longevity and robust growth: ancient sacred lotus from China. *American Journal of Botany*, 82: 1367–1380.

Smith, R. J., Matzke-Karasz, R., Kamiya, T. & De Deckker, P. (2016). Sperm lengths of non-marine cypridoidean ostracods (Crustacea). *Acta Zoologica*, 97: 1–17.

Smith, S. E. & Read, D. J. (2008). *Mycorrhizal Symbiosis*, 3rd edn. London: Academic Press.

Snell, T. W., Kubanek, J., Carter, W. *et al.* (2006). A protein signal triggers sexual reproduction in *Brachionus plicatilis* (Rotifera). *Marine Biology*, 149: 763–773.

Sodmergen, Z. Q. (2010). Why does biparental plastid inheritance revive in angiosperms. *Journal of Plant Research*, 123: 201–206.

Speijer, D., Lukeš, J. & Eliáš, M. (2015). Sex is a ubiquitous, ancient, and inherent attribute of eukaryotic life. *Proceedings of the National Academy of Sciences of the United States of America*, 112: 8827–8834.

Spratt, B. G. (2004) Exploring the concept of clonality in bacteria. *Methods in Molecular Biology*, 266: 323–352.

Stamps, J. & Krishnan, V. V. (1997). Sexual bimaturation and sexual size dimorphism in animals with asymptotic growth after maturity. *Evolutionary Ecology*, 11: 21–39.

Stanley, J. S. L. (2005). The *Entamoeba histolytica* genome: something old, something new, something borrowed and sex too? *Trends in Parasitology*, 21: 451.

Stearns, S. C. (ed.) (1987). *The Evolution of Sex and its Consequences*. Basel: Birkhäuser.

Stearns, S. C. (1992). *The Evolution of Life Histories*. Oxford: Oxford University Press.

Stenberg, P. & Saura, A. (2009). Cytology of asexual animals. In *Lost Sex*, ed. I. Schön, K. Martens & P. van Dijk, Berlin: Springer, pp. 63–74.

Sterelny, K. & Griffiths, P. E. (1999). *Sex and Death: an Introduction to the Philosophy of Biology*. Chicago, IL: University of Chicago Press.

Stern, D. L. (1994). A phylogenetic analysis of soldier evolution in the aphid family Hormaphididae. *Proceedings of the Royal Society of London. Series B, Biological Sciences*, 256: 203–209.

Stewart, D. T., Saavedra, C., Stanwood, R. R., Ball, A. O. & Zouros, E. (1995). Male and female mitochondrial DNA lineages in the blue mussel (*Mytilus edulis*) species group. *Molecular Biology and Evolution*, 12: 735–747.

Stewart, E., Madden, R., Paul, G. & Taddei, F. (2005). Aging and death in an organism that reproduces by morphologically symmetric division. *PLOS Biology*, 3 (2): e45.

Strasburger, E., Noll, F., Schenck, H. & Schimper, A. F. W. (2002). *Lehrbuch der Botanik für Hochschulen*, 35th edn. Heidelberg: Spektrum-Akademischer Verlag.

Sweeney, B. W. & Vannote, R. L. (1982). Population synchrony in mayflies: a predator satiation hypothesis. *Evolution*, 36: 810–821.

Taiz, L. & Zeiger, E. (2010). *Plant Physiology*, 5th edn. Sunderland, MA: Sinauer Associates.

Tekaya, S., Sluys, R. & Zghal, F. (1997). Sperm transfer and fertilization in the marine planarian *Sabussowia dioica* (Platyhelminthes, Tricladida, Maricola). *Invertebrate Reproduction and Development*, 32: 143–147.

Tindale, N. B. (1932). Revision of the Australian ghost moths (Lepidoptera Homoneura, family Hepialidae) Part I. *Records of the South Australian Museum*, 4: 497–536.

Tinti, F. & Scali, V. (1995). Allozymic and cytological evidence for hemiclonal, all-paternal and mosaic offspring of the hybridogenetic stick insect *Bacillus rossius-grandii grandii*. *Journal of Experimental Zoology*, 273: 149–159.

Togashi, T. & Cox, P. A. (2001). Tidal-linked synchrony of gamete release in the marine green alga, *Monostroma angicava* Kjellman. *Journal of Experimental Marine Biology and Ecology*, 264: 117–131.

Tripp, E. A. & Lendemer, J. C. (2018). Twenty-seven modes of reproduction in the obligate lichen symbiosis. *Brittonia*, 70: 1–14.

Trumbo, S. T. (2012). Patterns of parental care in invertebrates. In *The Evolution of Parental Care*, ed. N. J. Royle, P. T. Smiseth & M. Kölliker. Oxford: Oxford University Press, pp. 81–100.

Turke, P. W. (2013). Making young from old: how is sex designed to help? *Evolutionary Biology*, 40: 471–479.

van Dijk, P. (2009). Apomixis: basic for non-botanists. In *Lost Sex*, ed. I. Schön, K. Martens & P. van Dijk. Berlin: Springer, pp. 47–62.

van Voorhies, W. A. (1992). Production of sperm reduces nematode lifespan. *Nature*, 360: 456–458.

Vandel, A. (1928). La parthénogenèse geographique: contribution à l'étude biologique et cytologique de la parthénogenèse naturelle. *I. Bulletin Biologique de la France et de la Belgique*, 62: 164–281.

Vandel, A. (1931). *La parthénogenèse*. Paris: G. Doin.

Vanthournout, B., Greve, M., Bruun, A. *et al.* (2016). Benefits of group living include increased feeding efficiency and lower mass loss during desiccation in the social and inbreeding spider *Stegodyphus dumicola*. *Frontiers in Physiology*, 7: 18.

Veller, C., Nowak, M. A., Davis, C. C. & Blasius, B. (2015). Extended flowering intervals of bamboos evolved by discrete multiplication. *Ecology Letters*, 18: 653–659.

Verne, S., Johnson, M., Bouchon, D. & Grandjean, F. (2012). Effects of parasitic sex ratio distorters on host genetic structure in the *Armadillidium vulgare-Wolbachia* association. *Journal of Evolutionary Biology*, 25: 264–276.

Vershinina, A. O. & Kuznetsova, V. G. (2016). Parthenogenesis in Hexapoda: Entognatha and non-holometabolous insects. *Journal of Zoological Systematics and Evolutionary Research*, 54: 257–268.

Vila-Farré, M. & Rink, J. C. (2018). The ecology of freshwater planarians. In *Planarian Regeneration: Methods and Protocols*, ed. J. C. Rink. Methods in Molecular Biology, 1774. New York, NY: Springer, pp. 173–205.

Vogt, G. (2016). Structural specialties, curiosities and record-breaking features of crustacean reproduction. *Journal of Morphology*, 277: 1399–1422.

Voigt, K. (2012). Chytridiomycota, Zygomycota. In *Syllabus of Plant Families*, 13th edn, ed. W. Frey. Berlin; Stuttgart: Borntraeger, Vol. 1/1, pp. 106–162.

Volff, J.-N. & Schartl, M. (2001). Variability of genetic sex determination in poeciliid fishes. *Genetica*, 111: 101–110.

Vortsepneva, E., Tzetlin, A., Purschke, G. *et al.* (2008). The parasitic polychaetes known as *Asetocalamyzas laonicola* (Calamyzidae) is in fact the dwarf male of the spionid *Scolelepis laonicola* (comb. nov.). *Invertebrate Biology*, 124: 403–416.

Wang, H., Matsushita, M., Tomaru, N. & Nakagawa, M. (2017). Sex change in the subdioecious shrub *Eurya japonica* (Pentaphylacaceae). *Ecology and Evolution*, 7: 2340–2345.

Warburg, M. R. (2011). Scorpion reproductive strategies, allocation and potential: a partial review. *European Journal of Entomology*, 108: 173–181.

Waters, E., Hohn, M. J., Ahel, I. *et al.* (2003). The genome of *Nanoarchaeum equitans*: insights into early archaeal evolution and derived parasitism. *Proceedings of the National Academy of Sciences of the United States of America*, 100: 12984–12988.

Weismann, A. (1892). *Das Keimplasma. Eine Theorie der Vererbung*. Jena: Fischer.

Werner, B. (1955). On the development and reproduction of the anthomedusan *Margelopsis haeckeli* Hartlaub. *Annals of the New York Academy of Sciences*, 62: 1–29.

Werner, B., (1963). Effect of some environmental factors on differentiation and determination in marine Hydrozoa, with a note on their evolutionary significance. *Annals of the New York Academy of Sciences*, 105: 461–488.

Westneat, D. & Foz, C. (eds.) (2010). *Evolutionary Behavioral Ecology*. Oxford: Oxford University Press.

White, M. J. D. (1973). *Animal Cytology and Evolution*. Cambridge: Cambridge University Press.

White, M. J. D. (1984). Chromosomal mechanisms in animal reproduction. *Bollettino di Zoologia*, 51: 1–23.

Whiting, P. W. (1943). Multiple alleles in complementary sex determination of *Habrobracon*. *Genetics*, 28: 365–382.

Whittle, C. A., & Extavour, C. G. (2017). Causes and evolutionary consequences of primordial germ-cell specification mode in metazoans. *Proceedings of the National Academy of Sciences of the United States of America*, 114: 5784–5791.

Wilkinson, M., Sherratt, E., Starace, F. & Gower, D. J. (2013). A new species of skin-feeding caecilian and the first report of reproductive mode in *Microcaecilia* (Amphibia: Gymnophiona: Siphonopidae). *PLOS One*, 8 (3): e57756.

Williams, C. G. (2009). *Conifer Reproductive Biology*. Dordrecht: Springer.

Williamson, D. J. (2006). Hybridization in the evolution of animal form and life-cycle. *Biological Journal of the Linnean Society*, 148: 585–602.

Wilson, J. (1999). *Biological Individuality: the Identity and Persistence of Living Entities*. Cambridge: Cambridge University Press.

Wittgenstein, L. (1953). *Philosophical Investigations*. Cambridge: Basil Blackwell.

Wolff, N. C., Gandre, S., Kalinin, A. & Gemmell, N. J. (2008). Delimiting the frequency of paternal leakage of mitochondrial DNA in chinook salmon. *Genetics*, 179: 1029–1032.

Wolpert, L. (2007). *Principles of Development*, 3rd edn. New York, NY: Oxford University Press.

Wourms, J. P. (1981). Viviparity: the maternal–fetal relatioship in fishes. *American Zoologist*, 21: 473–515.

Wyatt, T. D. (2014). *Pheromones and Animal Behavior: Chemical Signals and Signatures*, 2nd edn. Cambridge: Cambridge University Press.

Yashina, S., Gubin, S., Maksimovich, S. *et al.* (2012). Regeneration of whole fertile plants from 30,000-y-old fruit tissue buried in Siberian permafrost. *Proceedings of the National Academy of Sciences of the United States of America*, 109: 4008–4013.

Yoshizawa, K., Ferreira, R. L., Kamimura, Y. & Lienhard, C. (2014). Female penis, male vagina, and their correlated evolution in a cave insect. *Current Biology*, 24: 1006–1010.

Zattara, E. E. & Bely, A. E. (2016). Phylogenetic distribution of regeneration and asexual reproduction in Annelida: regeneration is ancestral and fission evolves in regenerative clades. *Invertebrate Biology*, 135: 400–414.

Zimmer, R. L. (1997). Phoronids, brachiopods, and bryozoans, the lophophorates. In *Embryology: Constructing the Organism*, ed. S. F. Gilbert & A. M. Raunio. Sunderland, MA: Sinauer Associates, pp. 279–305.

Taxonomic Index

Entries for suprageneric taxa are given here in Latin form, but may occur in the text in anglicized form, as (e.g.) amphipods rather than Amphipoda.

Subject Index

accessory glands, 137

acrosomal
  filament, 141, 166
  reaction, 141, 166

acrosome, 141

adelphophagy, 222, 223, 231

adulthood, 78

aecidium, 378

aeciospore, 93, 378

aedeagus, 138, 153

agamogenesis, 187

agamont, 383

agamospermy, 181, 187

agamospory
  Braithwaite type, 186
  Döpp–Manton type, 186

age
  at production of first flower, 209
  maximum, 79

ageing, 33, 35

albumen, 141

Aldrovandi Ulisse, 8

allogamy, 271

allosperm, 155

amitosis, 250, 251, 293

amphigony, 149, 150, 271
  diploid, 175
  polyploid, 177
  uniparental, 16

amphimixis, 271

amphitoky, 176

ancient asexual scandals, 108

androdioecy, 134, 370

androgamone, 166

androgenesis, 150, 193, 193, 284, 289, 291

ameiotic, 290
  clonal, 290

andromonoecy, 135, 370

androspore, 93

anemochory, 235

anholocycle, 65

anisogamete, 55, 111

anisogamety, 111

anisogamy, 111

anisospore, 55

anisospory, 55

anther, 147

antheridiogen, 328

antheridiophore, 144

antheridium, 114, 122, 144, 145, 360–361, 376

antherozoid, 112, 148, 361, 363

antifertilisin, 166

androgamone, 166

anti-Müllerian factor (AMF), 336

aplanospore, 91–92, 356

apogamy, 187

apomeiosis, 187

apomixis, 43, 175–176, 187, 284
  gametophytic, 187, 190, 286
  polyploid, 177
  pseudogamic, 191
  sporophytic, 187, 189, 191

apophallation, 76

apospory, 189, 286

Appert Nicolas, 10

archegoniophore, 144

archegonium, 122, 144, 145, 149, 361

architomy, 45, 96, 96, 391

aromatase, 328

arrhenotoky, 176

ascocarp, 376